T0340075

Logistics Operations and Management

Logistics Operations and Management

Concepts and Models

Reza Zanjirani Farahani
Informatics and Operations Management
Kingston Business School
Kingston University, Kingston Hill
Kingston Upon Thames, Surrey KT2 7LB

Shabnam Rezapour
Industrial Engineering Department,
Urmia University of Technology, Urmia, Iran

Laleh Kardar
Department of Industrial Engineering, University of Houston,
Houston, TX, USA

ELSEVIER
AMSTERDAM • BOSTON • HEIDELBERG • LONDON • NEW YORK • OXFORD
PARIS • SAN DIEGO • SAN FRANCISCO • SINGAPORE • SYDNEY • TOKYO

Elsevier
32 Jamestown Road London NW1 7BY
225 Wyman Street, Waltham, MA 02451, USA

First edition 2011

British Library Cataloguing-in-Publication Data
A catalogue record for this book is available from the British Library

Library of Congress Cataloging-in-Publication Data
A catalog record for this book is available from the Library of Congress

ISBN: 978-0-323-16520-4

For information on all Elsevier publications visit our website at www.elsevierdirect.com

This book has been manufactured using Print On Demand technology. Each copy is produced to order and is limited to black ink. The online version of this book will show color figures where appropriate.

Contents

List of Contributors

Maryam Abbasi
Department of Industrial Engineering,
Amirkabir University of Technology,
Tehran, Iran

Hamid Afshari
Iran Khodro Industrial Group (IKCO),
Iran

Shokoofeh Asadi
Industrial Engineering Department,
Amirkabir University of Technology,
Tehran, Iran

Mohammad Bakhshayeshi Baygi
Mechanical and Industrial
Engineering Department,
University of Concordia,
Montreal, Canada

Fatemeh Hajipouran Benam
Iran Khodro Industrial Group (IKCO),
Iran

Farzaneh Daneshzand
Department of Industrial Engineering,
Amirkabir University of Technology,
Tehran, Iran

Wout Dullaert
Institute of Transport and Maritime
Management Antwerp,
University of Antwerp, Belgium and
Antwerp Maritime Academy,
Antwerp, Belgium

Mohammad Hassan Ebrahimi
Terminal Management System
Department,
InfoTech International Company,
Tehran, Iran

Gholamreza Esmaeilian
Department of Mechanical and
Manufacturing Engineering, University
Putra Malaysia, Serdang, Selangor,
Malaysia and Department of Industrial
Engineering, Payam Noor Universiti, Iran

Behnam Fahimnia
School of Management, Division of
Business, University of South
Australia, Adelaide, Australia

Samira Fallah
Department of Industrial Engineering,
Amirkabir University of Technology,
Tehran, Iran

Reza Zanjirani Farahani
Department of Informatics and
Operations Management,
Kingston Business School,
Kingston University, Kingston Hill,
Kingston Upon Thames,
Surrey KT2 7LB

Maryam Hamedi
Department of Mechanical and
Manufacturing Engineering,
University Putra Malaysia,
Serdang, Selangor, Malaysia

Sara Hosseini
Petrochemical Industries
Development Management Co.,
Tehran, Iran

Masoomeh Jamshidi
Industrial Engineering Department,
Amirkabir University of Technology,
Tehran, Iran

Laleh Kardar
Department of Industrial Engineering,
University of Houston,
Houston, TX, USA

Zohreh Khooban
Department of Industrial Engineering,
Amirkabir University of Technology,
Tehran, Iran

Reza Molaei
Department of Technology
Development,
Iran Broadcasting Services (IRIB),
Tehran, Iran

Seyyed Mostafa Mousavi
Centre for Complexity Science,
University of Warwick,
Coventry, UK

Ehsan Nikbakhsh
Department of Industrial Engineering,
Faculty of Engineering, Tarbiat
Modares University,
Tehran, Iran

Mahsa Parvini
Faculty of Industrial Engineering,
Amirkabir University,
Tehran, Iran

Mohsen Rajabi
Department of Industrial Management,
Faculty of Management,
Tehran University, Tehran, Iran

Fatemeh Ranaiefar
Institute of Transportation Studies,
University of California,
Irvine, CA, USA

Amelia Regan
Computer Science and Institute of
Transportation Studies, University
of California, Irvine, CA, USA

Shabnam Rezapour
Industrial Engineering Department,
Urmia University of Technology,
Urmia, Iran

Zahra Rouhollahi
Department of Industrial Engineering,
Amirkabir University of Technology,
Tehran, Iran

Hannan Sadjady
Department of Industrial Engineering,
Amirkabir University of Technology,
Tehran, Iran

Seyed-Alireza Seyed-Alagheband
Department of Industrial Engineering,
Amirkabir University of Technology,
Tehran, Iran

Maryam SteadieSeifi
Department of Industrial Engineering,
Amirkabir University of Technology,
Tehran, Iran

Amir Zakery
Department of Industrial Engineering,
Amirkabir University of Technology,
Tehran, Iran

Part I

Introduction

1 Overview

Reza Zanjirani Farahani[1], *Shabnam Rezapour*[2] *and Laleh Kardar*[3]

[1]Department of Informatics and Operations Management, Kingston Business School, Kingston University, Kingston Hill, Kingston Upon Thames, Surrey KT2 7LB
[2]Industrial Engineering Department, Urmia University of Technology, Urmia, Iran
[3]Department of Industrial Engineering, University of Houston, Houston, TX, USA

1.1 History

Many people believe that *logistics* is a word, but from a semantics point of view its origin was from ancient Greek and meant the "science of computation." In fact, it is originally from combat environments and not from business or academia. It seems the ancient Greeks referred the word *logistikos* to military officers who were expert in calculating the military needs for expeditions in war. As a science, it seems the first book written on logistics was by Antoine-Henri Jomini (1779–1869), a general in the French army and later in the Russian service, titled *Summary of the Art of War* (1838). The book was on the Napoleonic art of war [1,2].

1.2 Definition of Logistics

Jomini defined logistics as "the practical art of moving armies" and included a vast range of functions involved in moving and sustaining military forces: planning, administration, supply, billeting and encampments, bridge and road building, and even reconnaissance and intelligence insofar as they were related to maneuvers off the battlefield [1].

What is logistics? This section is an adoption of the first chapter in Farahani et al. (2009b) [3]. Many different definitions for logistics can be found. The most well known are the following: (a) "Logistics is ... the management of all activities which facilitate movement and the co-ordination of supply and demand in the creation of time and place utility" [8]. (b) "Logistics management is ... the planning, implementation and control of the efficient, effective forward and reverse flow and storage of goods, services and related information between the point of origin and the point of consumption in order to meet customer requirements" (CSCMP 2006) [7]

Logistics Operations and Management. DOI: 10.1016/B978-0-12-385202-1.00001-3

(c) "Logistics is... the *positioning* of resources at the right time, in the right place, at the right cost, at the right quality" (Chartered Institute of Logistics and Transport, UK, 2005) [7]. (d) "In civil organizations, logistics' issues are encountered in firms producing and distributing physical goods" [4]. (e) "Logistics is that part of the supply chain process that plans, implements, and controls the efficient, effective forward and reverse flow and storage of goods, services, and related information between the point of origin and the point of consumption in order to meet customers' requirements" (Council of Logistics Management 2003) [7].

1.2.1 Why Is Logistics Important?

In each country, a huge amount of money is spent annually in logistical activities. For instance, in 2003 US logistical activity costs were 8.5% of the country's GDP. Given that the US GDP in 2003 was approximately $12,400 billion, the logistical activity cost was approximately $1054 billion (Seventeenth Annual State of Logistics Report of USA 2006)! [9].

1.3 Evolution of Logistics Over Time

Logistics has an ancient history. A quick look back can be enlightening. Its history dates to the wars of the Greek and Roman empires in which the military officials called *logistiks* were responsible for supplying and distributing needed resources and services. Providing them had an important and essential role in the outcomes of these wars. These logistiks also worked to damage the stores of their enemies while defending their own. This gradually guided the development of current logistics systems.

Logistics systems developed extensively during World War II (1939–1945). Throughout this war, the United States and its allies' armies were more efficient than Germany's. German army stores were damaged extensively, but Germany could not impose the same destruction on its enemies' stores. The US army could supply whatever was needed by its forces at the right time, at the right place, and in the most economical way. From that time, several new and advanced military logistic techniques started to take off. Gradually, logistics started to evolve as an art and science.

Today, experts in logistics perform their duties based on their skills, experiences, and knowledge. In modern industries, the task of logistics managers is to provide appropriate and efficient logistics systems. They guarantee that the right goods will be delivered to the right customers, at the right time, at the right place, and in the most economical way.

Although logistics is a dilemma for many companies, logistical science can bring some relief to them. In today's business environment, logistics is a competitive strategy for the companies that can help them meet the expectations of their customers. Logistics helps members of supply chains integrate in an efficient way.

Logistics does not consist of one single component but involves a group of various activities and disciplines such as purchasing, planning, coordinating, warehousing, distributing, and customer service [5].

1.4 Other Logistical Books

As noted earlier, logistics was traditionally used in the military environments. Therefore, it is rational that the first books explicitly or implicitly relevant to logistics were combat oriented. The oldest one was Jomini's *Summary of the Art of War* (1838). Another and more recent example was the book coauthored by Lieutenant General William Gus Pagonis, the director of logistics during the 1991 Gulf War, and Jeffrey Cruikshank, *Moving Mountains* [6].

Nowadays logistics is being used in business environments as widely as in wars, and we can find different books recently written by researchers in academia. However, although many books talk about logistical processes individually—such as transportation, warehousing, distribution, vehicle routing problems (VRPs), and packaging—few comprehensive books encompass all of the logistical processes. Two examples of complete books that are basically applicable to private organizations are those by Riopel et al. (2005) and Ghiani et al. (2004) [4].

Sometimes, we can see cooperation between logistical areas among several private organizations, governmental organizations, and also militants. For example, in case of a natural disaster such as an earthquake, tsunami, or hurricane or typhoon, all of these organizations will be involved. Integration and coordination of materials, information, and financial flows between two or more private organizations can promote a traditional logistical system to become an advanced supply chain. To see a book in this area, interested readers can refer to [3].

1.5 The Focus of This Book

The question that might arise is, what is specifically different about this book? We explicitly highlight the following issues as the main point of this book.

- We have worked to include updated sources such as journal papers, conference proceedings, books, and Internet sources, so you may see references from recent years.
- Many of the references highlight some of the logistical processes such as the VRP and transportation. We have tried to equally cover all of the main logistics processes and thus have allocated separate chapters to each.
- There are two main classifications in a book such as this: (1) qualitative concepts and (2) quantitative models. We have tried to view both equally. Of course, we believe different topics need different degrees of focus. For instance, when talking about information and communication technology in logistical systems, most texts look at these areas from qualitative angles whereas the VRP is viewed mainly in quantitative terms. However, aside from the nature of a chapter, by default we have considered the importance of quantitative models and qualitative concepts at the same time.

- We are covering some topics in logistics that are not predominant in most large and private enterprises, for instance, disaster logistics and retail logistics. Moreover, some approaches and modeling concepts such as robustness and risk are highlighted in separate chapters.
- Then last but not the least, some chapters such as those covering logistical parties, logistical philosophies, and logistical future trends will interest readers and are not found in other sources.

1.6 Organization

This book is organized in 4 parts and 21 chapters such that the reader can study each chapter not only independently but also as part of a whole. If someone wants to study the book more deeply, our suggestion is observing the strategy in Figure 1.1.

Part I, Introduction, has two chapters. Chapter 1 (Overview) and Chapter 2, Physical Flows, which looks at the physical entities of a logistical system, including fixed and static components and moving entities. To do this, the author focuses on transportation modes, including land, air, water, and pipeline as well as warehousing systems. This chapter also summarizes intermodal, multimodal, and material-handling equipment.

Part II, Strategic Issues, includes four chapters. Chapter 3, Logistics Strategic Decisions, covers the strategic decisions that should be made in a logistical system such as network design, outsourcing, and integration. It also includes the objectives of making a strategic decision and informs interested readers on how to do that. Chapter 4, Logistical Philosophies, introduces different approaches to logistics and

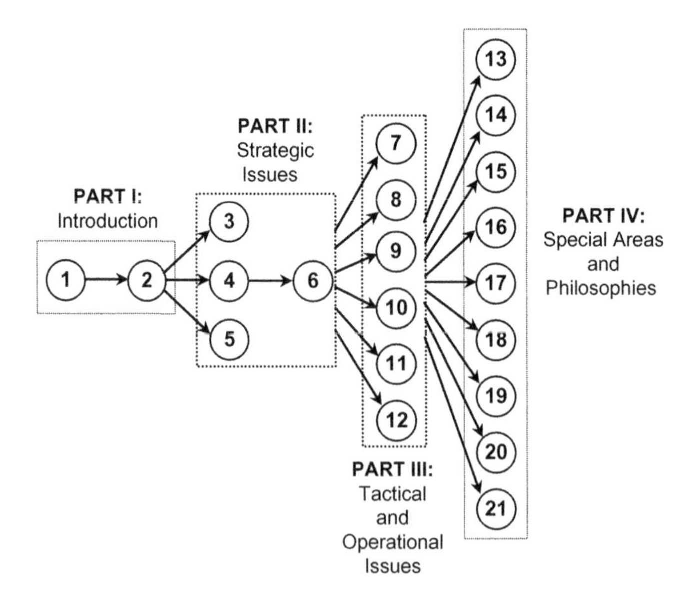

Figure 1.1 Sequencing the chapters dependently.

their advantages and disadvantages. These philosophies are mainly lean logistics, cross docking, just-in-time, agile logistical quick response, efficient consumer response, vendor-managed inventory (VMI). Chapter 5, Logistical Parties, examines definitions, activities, advantages, disadvantages, and types of first-, second-, third-, fourth-, and fifth-party logistical providers. Finally, Chapter 6, Logistics Future Trends, introduces the main future trends of logistics and considers emerging technologies, trends, new strategies in industries, and recent technical reports and surveys, and it predicts logistical future focusing, especially on globalization, information technology (IT) and e-commerce, and new technologies.

Part III, Tactical and Operational Issues, includes five chapters. Chapter 7, Transportation, discusses how transportation systems move materials between facilities using different vehicles and equipment. In this chapter, we talk about the basic aspects of these systems and the classification of transportation problems. Chapter 8, Vehicle Routing Problems, includes different methods of product distribution between customers. The chapter is dedicated to introducing brief information of the most studied kinds of VRPs. Chapter 9, Packaging and Material Handling, discusses the movement, storage, control, and protection of materials, goods, and products throughout the process of manufacturing, distribution, consumption, and disposal. The focus is on methods, mechanical equipment, systems, and related controls used to achieve these functions. Chapter 10, Storage, Warehousing, and Inventory Management, examines the process of coordinating incoming goods, the subsequent storage and tracking of these goods, and, finally, the distribution of the goods to their proper destinations.

Chapter 11, Customer Service, is about order management, customer service, and the reasons for their importance. Then the elements of customer service are introduced with an emphasis on order cycle time. Given the importance of determining proper service levels in logical cost, the steps for developing proper service levels based on current frameworks are the next to be introduced.

Part IV, Special Areas and Philosophies, includes 10 chapters. In Chapter 12, Logistics System: Information and Communication Technology, the role of information in logistical systems is reviewed briefly and the effects of IT on a company's logistical operations are discussed. In Chapter 13, Reverse Logistics, we propose a comprehensive investigation into reverse logistics and related subjects. After introducing the subject and providing a literature review, we try to answer the following questions: Why and how are things returned? What kind of returns take place? Chapter 14, Retail Logistics, explains the essential concepts of retailing and then demonstrates different types of retailing and some of the more common and applicable techniques. In Chapter 15, Humanitarian Logistics Planning in Disaster Relief Operations, classifications of different types of disasters and their effects on human lives are given. After introducing the concept of the disaster-management system cycle, humanitarian logistics and their characteristics and main stages are discussed. Then mathematical modeling of the required relief logistical decisions and their optimization solution techniques are discussed. Next, concepts of coordination and performance measurement in the context of humanitarian logistics are talked about. In Chapter 16, Freight Transportation Externalities, we begin by investigating different types of freight transportation externalities with a focus on road transportation. The results of related studies in the United States and Europe are presented and compared, although consistent comparison is challenging because of the differences in the times and locations of various studies as well as variable currency exchange rates and basic assumptions such as the statistical value of life. Later we present practical policies that have been introduced to reduce these externalities in the United States and other parts of the world. Chapter 17, Robust Optimization of Uncertain Logistics Networks, discusses the first literature review of this topic and then an optimization method under uncertainty and robust optimization of logistical networks are investigated. Chapter 18, Integration in Logistics Planning and Optimization, identifies the key issues in this area and then formulates a complex integrated logistical planning model. Following a discussion on the available tools and techniques for optimizing complex large-scale logistics planning problems, genetic algorithms are chosen to optimize the proposed integrated model. A medium-size case study is finally presented to demonstrate the capability of the developed optimization model in achieving the global optimal solution.

Chapter 19, Optimization in Natural Gas Network Planning, presents a survey on the role of optimization methods and operation research techniques in different fields of natural gas network planning. These fields have received more attention from researchers because of their enormous effects on reducing costs. To make a good comparison between what has been done and what should be improved on in the future, model characteristics and solution methods are discussed, and the application of mathematical models to the most important problems of this field has been highlighted. To present the efficiency of developed models in the real world, two case studies are presented. In Chapter 20, Risk Management in Gas Networks, after presenting an introduction to gas networks, we explain the vulnerability and resulted risks in the gas industry. Then we present a six-step process to manage and control the risks in gas networks. Chapter 21, Modeling the Energy Freight Transportation Network, presents the importance of energy around the world and the main energy freight transportation planning and management issues, describing the associated literature such as levels of planning and briefly reviewing the components such as networks. Modes of transporting energy are then discussed. Afterward, the chapter continues with a detailed explanation of the components of energy transportation network, the ones on which the models are based. The components are followed by the energy freight transportation models, which attempt to analyze real cases in order to solve real problems.

1.7 Audiences

The target audience of this book is composed of professionals and researchers working in various fields such as management, industrial engineering, applied operations search, and business at all levels, particularly undergraduates in their final year of study and graduate students. Specific courses for which our book is written can be logistics, logistics planning, and logistics systems; supply chain management; inventory management; business management; operations management, and information technology.

The book can also be used by professionals and practitioners of different organizations. Some topics—such as transportation, VRP, packaging and material handling, storage, warehousing and inventory management, order management, and customer services—are applicable to most of the enterprises, whereas others—such as reverse logistics, retail logistics, and logistics planning in case of disasters—are applicable to certain organizations or particular circumstances.

Acknowledgments

We would like to thank our friend Dr. Wout Dullaert (Associate Professor) from the Institute of Transport and Maritime Management Antwerp (ITMMA), University of Antwerp, Belgium, and also Dr. Dong-Wook Song (Reader) from the Logistics Research Centre, Heriot-Watt University, Edinburgh, in the United Kingdom, for their valuable comments on the organization of this book.

We would also like to express our appreciation to Dr. Anita Koch at Elsevier for her invaluable assistance in connecting us to the right place. In this right place, our contact persons were Ms. Lisa Tickner (our publisher at Serials and Elsevier Insights) and Ms. Joanne Tracy (Vice President, Editorial Director, Science). Managing typesetting was done by Paul Prasad Chandramohan, who is the Development Editor and Sujatha Thirugnana Sambandam, who is the Senior Project Manager of our book and are based in the Chennai office, India. When Lisa was busy, Ms. Zoe Kruze (Associate Acquisitions Editor) on Serials at Elsevier was helpful in following up when necessary. Then last but not least, Este Johnson who edited some of the chapters.

References

[1] Available from: http://www.britannica.com/EBchecked/topic/346423/logistics.
[2] Available from: http://en.wikipedia.org/wiki/Antoine-Henri_Jomini.
[3] R.Z. Farahani, N. Asgari, H. Davarzani (Eds.), Supply Chain and Logistics in National, International and Governmental Environmental, Physica-Verlag, Heidelberg, 2009.
[4] G. Ghiani, G. Laporte, R. Musmanno, Introduction to Logistic Systems Planning and Control, Wiley, Chichester, 2004.
[5] Available from: http://www.bestlogisticsguide.com/logistics-history.html.
[6] W.G. Pagonis, J.L. Cruikshank, Moving Mountains: Lessons in Leadership and Logistics from the Gulf War, Harvard Business School Press, Boston, 1992.
[7] D. Riopel, A. Langevin, J.F. Campbell, The network of logistics decisions, chapter published in: A. Langevin, D. Riopel (Eds.), Logistics Systems: Definition and Optimization, Springer, New York, 2005.
[8] J.L. Heskett, A.N. Glaskowsky, R.M. Ivie, Business Logistics−Physical Distribution and Materials Management, Ronald Press, New York, 1973.
[9] Available from: http://www.inboundlogistics.com/articles/trends/trends0706.shtml.

2 Physical Flows

Hannan Sadjady

Department of Industrial Engineering, Amirkabir University of Technology, Tehran, Iran

The objective of logistics process is to get the right quantity and quality of materials (or services) to the right place at the right time, for the right client, and at the right price. As customers, many people tend to neglect the direct or indirect effects of logistics on almost every sphere of their lives until one of these "rights" goes wrong. The logistics concept was introduced as a response to the increasing necessity of an integrated system, which plans and coordinates the materials flow from the source of supply to the point of consumption instead of managing theses flows as series of independent tasks. The Council of Supply Chain Management Professionals (CSCMP) defines the logistics management as follows:

> Logistics management is that part of supply chain management that plans, implements, and controls the efficient, effective forward and reverses flow and storage of goods, services and related information between the point of origin and the point of consumption in order to meet customers' requirements. [1]

The entire process of logistics, which deals with the moving of materials into, through, and out of a firm, can be divided into three parts: (1) *inbound logistics*, which represents the movement and storage of materials received from suppliers; (2) *materials management*, which covers the storage and flows of materials within a firm; and (3) *outbound logistics* or *physical distribution*, which describes the movement and storage of products from the final production point to the customer [2]. These terms as well as some of the other associated logistics terminologies are indicated in Figure 2.1.

As Figure 2.1 illustrates, logistics is concerned with two types of flow: *physical flow* and *information flow*. It is common to consider physical flow as the forward flow throughout the logistics network, the main direction of which is from the point of origin to the point of consumption. Also, the information flow is considered to be backward, so its main direction is from downstream to upstream elements. However, in practical terms, the directions of physical and information flows are not one way. Materials and information flow from both upstream and downstream. In regard to physical flow, the backward flow of product is referred to as *reverse*

Logistics Operations and Management. DOI: 10.1016/B978-0-12-385202-1.00002-5

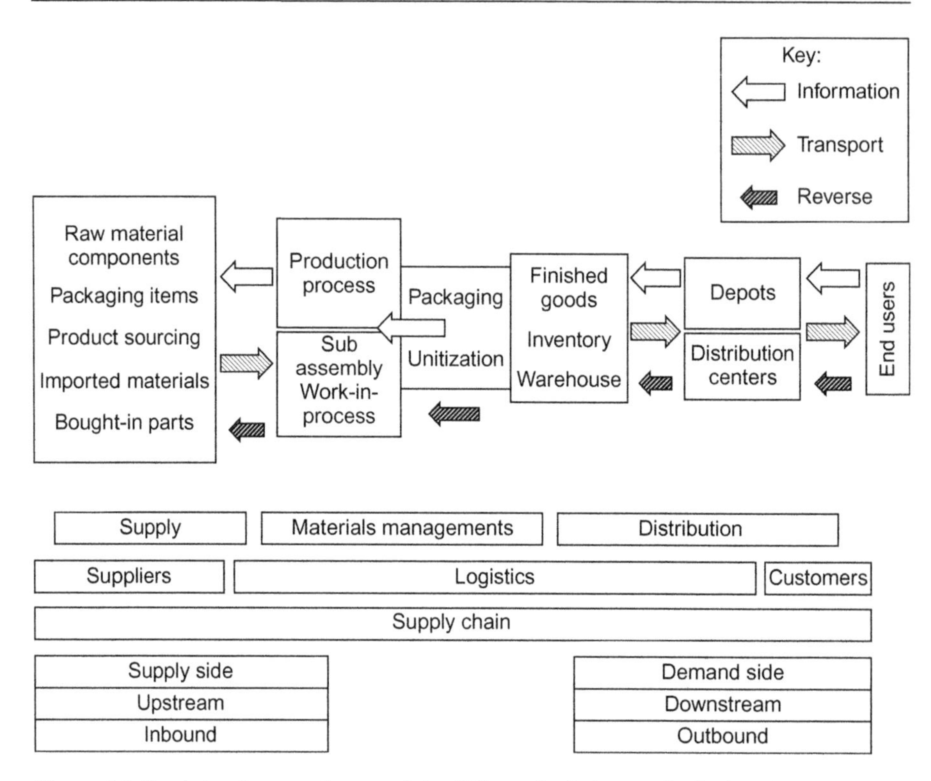

Figure 2.1 Logistics flows and some of the different logistics terminologies [3].

logistics. It is the flow of returned goods and used products as well as salvage, scrap disposal, and returnable packaging back through the system.

In this chapter, the emphasis is on the physical flows (also known as *material* or *inventory flows*). Information flows are discussed in Chapter 12.

Physical flows involve the entire process and activities of logistics systems; however, to explore the concept of physical flows systematically, the major components of logistics systems can be categorized into five functional areas, based on Ailawadi and Singh [4]:

- Network design
- Information
- Transportation
- Inventory
- Warehousing, material handling, and packaging[1]

Considering these functional areas, physical flow is more involved with the transportation and warehousing, material handling, and packaging. These two functional areas are discussed in Sections 2.1 and 2.4, respectively. Also, the physical

[1] For further information about the logistics functional areas, see reference [4], pp. 11–16.

nature of the product is investigated in Section 2.2, followed by some explanations *about distribution channels* in Section 2.3.

2.1 The Transportation System

Transportation accounts for between one-third and two-thirds of total logistics costs; for most firms, it is the most important single element of logistics costs [5]. Firms and their products' markets are often separated geographically. Transportation increases the time and place utility of products by delivering them at the right time and to the right place where they are needed. By doing so, the customers' level of satisfaction increases, which is a key factor for successful marketing.

A comprehensive discussion of transportation is beyond the scope of this text, so we focus here on essential issues of transportation systems, which are more related to the physical flows of materials.

2.1.1 Transport Modes and Their Characteristics

Various options for moving products from one place to another are called *transportation modes*. Road, rail, air, water, and pipelines are considered the five basic modes of transportation by most sources (see, e.g., [2,4–8]). In addition, digital or electronic transport is referred to as the sixth mode of transportation in some texts (see, e.g., [9]). Any one or more of these six distinct modes could be selected to deliver products to customers (Figure 2.2). However, all transport modes may not be applicable or feasible options for all markets and products.

Road

Road transport—also known as *highway*, *truck*, and *motor carriage*—steadily increased its share of transportation. Throughout the 1960s, road transport became the dominant form of freight transport in the United States, replacing rail carriage [10], and it now accounts for 39.8% of total cargo ton-miles, which is more than 68% of actual tonnage [11].

The key advantages of road transport over other transportation modes are its *flexibility* and *versatility*. Trucks are flexible because they offer door-to-door services without any loading or unloading between origin and destination. Trucks' versatility

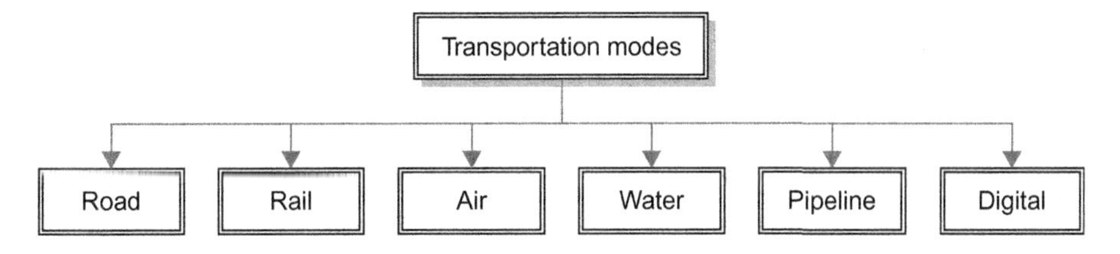

Figure 2.2 Basic modes of transportation.

is made possible by having the widest range of vehicle types, enabling them to transport products of almost any size and weight over any distance [10].

Road transport also offers *reliable* and *fast* service to the customers. The loss and damage ratios for road transport are slightly higher than for the air shipment, but are too far lower than for the rail carriage. Road transport generally offers faster service than railroads, especially for small shipments (less than truckload, or LTL).[2] For large shipments (truckload, or TL), they compete directly with each other on journeys longer than 500 miles. However, for shipments larger than 100,000 pounds, rail is the dominant mode. Also, as motor carriers are more efficient in terminal, pickup, and delivery operations, they compete with air carriers, for both TL and LTL shipments that are transported 500 miles or less [7].

In regard to economic aspects, road transport has relatively small fixed cost, because it operates on publicly maintained networks of high-speed and often toll-free roads. However, the variable cost per kilometer is high because of fuel, tires, maintenance, and, especially, labor costs (a separate driver and cleaner are required for each vehicle) [4]. Road transport is best suited for small shipments and high-value products, moving short distances. Legislative control and driver fatigue are some problems of motor carriers' long journeys [6].

Rail

Rail carriage accounts for 37.1% of total freight ton-miles (more than 14% of actual tonnage) in the United States [11], which places railroads after motor carriers as the second dominant mode of transportation. However, in some countries such as the People's Republic of China, the countries of the former Yugoslavia, and Austria, rail remains the dominant transportation mode [10].

Although rail service is available in almost every major city around the world, the railroad network is not as extensive as the road networks in most countries. Thus, rail system lacks the flexibility and versatility of the road transport. Indeed, rail carriers offer terminal-to-terminal service rather than the door-to-door service provided by motor carriers. Therefore, railroads, like water, pipelines, and air transport, need to be integrated with trucks to provide door-to-door services. Also, railroads offer less-frequent services compared to motor carriers.

Rail transportation is relatively slow and quite unreliable, as the loss and damage ratios of rail transport for many shipments are higher than other modes. As a result, the railroad is a slow mover of both raw materials (e.g., coal, lumber, and chemicals) and low-value finished goods (e.g., tinned food, paper, and wood products) [8].

Railroads have high fixed costs and relatively low variable costs. Expensive equipment, multishipment trains, multiproduct switching yards and terminals, and right-of-way maintenance result in high fixed costs [4,5]. However, the variable costs are low, especially for long hauls, so rail carriage generally costs less than motor and air transport on a weight basis. It would be explained later in this chapter,

[2] Less than truckload: Any quantity of freight weighing less than the amount required for the application of a truckload rate.

that is how we might combine the economy of rail or water movement with truck flexibility, thus using trailer-on-flatcar (TOFC) or container-on-flatcar (COFC) services (see Section 2.1.2).

Air

Air carriers transport only around 0.1% of ton-mile traffic in the United States [11]. Although airfreight offers the shortest time in transit (especially over long distances) of any transport mode, most shippers consider air transport as a premium emergency service because of its higher costs. However, the high cost of air transport may be traded off with inventory and warehousing reductions or justified in some situations: (1) for high-value products, (2) for perishables, (3) in limited marketing periods, and (4) in an emergency [4].

The portion of total product costs dedicated to transportation is an important issue for most shippers. The high price of airfright consumes a greater portion of low-valued products' total costs, so it is not economically justifiable for these items. This could be why air carriers usually handle high-value items.

Total transit time (from pickup at the vendor to delivery to the customer) is important to shippers and the customers. From this point of view, well-managed surface carriers can compete favorably with air carriers, especially on short and medium hauls. Even though air carriers provide rapid time in transit from terminal to terminal, they may spend too much time on the ground (e.g., for pickup, delivery, delays and congestions, and waiting for scheduled aircraft departures) [10].

Loss and damage ratios resulting from transportation by air are considered lower than the other modes. The classic study by Lewis et al. [12] shows that the ratio of claim costs to freight revenue was only about 60% of those for road and rail.

Airline companies generally own neither airways nor airports. Air spaces and air terminals are usually developed and maintained with public funds, so fixed airfreight costs (including aircraft purchases, specialized handling systems, and cargo containers) are lower than rail, water, and pipeline. Air-transport variable expenses are extremely high because of fuel, maintenance, and the labor intensity of both inflight and ground crew [4]. Variable costs are reduced by the length of journey because takeoffs and landings are the most inefficient phases of aircraft operation. Moreover, increasing shipment sizes reduces the variable operating cost per ton-mile. Hence, variable costs are influenced by both distance and shipment size [5].

Water

Water carriage—as the oldest mode of transportation—accounts for 5% of total freight ton-miles (around 3.3% of actual tonnage) in the United States [11]. Sampson et al. [13] describe the nature and characteristics of water carriage as follows:

> Water carriage by nature is particularly suited for movements of heavy, bulky, low-value-per-unit commodities that can be loaded and unloaded efficiently by mechanical means in situations where speed is not of primary importance, where

the commodities shipped are not particularly susceptible to shipping damage or theft, and where accompanying land movements are unnecessary.

As already mentioned, the majority of commodities transported by water are semiprocessed and raw materials; thus, water transportation competes primarily with rail and pipeline. Water carriage can be broken into the following distinct categories [10]:

1. Inland waterways (such as rivers and canals)
2. Lakes
3. Coastal and intercoastal oceans
4. International deep sea

Water transportation service is limited in scope, mainly for two reasons: its limited range of operation and speed. Water service is confined to waterway systems; thus, unless the origin and the destination of movement are located on waterways, it needs to be supplemented by another transportation mode (rail or motor carrier). In addition, the average speed of water carriage is less than rail transport, and the availability and dependability of its service are greatly influenced by weather [4,5].

Containers[3] are used for many domestic and most international water shipments. Moving freight in containers on containerized ships affects the intermodal transfer by reducing handling time and shortening total transit time. It also reduces staffing needs and allows shippers to take advantage of volume shipping rates. Finally, containers reduce loss and damage [5,7]. For all these reasons, high-value commodities (especially those in foreign shipments) are shipped in containers and containerized ships.

Loss and damage costs for water carriage are lower in comparison with other transportation modes because damage is not much of a concern with low-valued bulk commodities. Also, because large inventories are often maintained by buyers, losses from delays are not serious. For high-valued products, claims are much higher: approximately 4% of ocean-ship revenues. Most damages are caused by rough handling during loading and unloading operations, so substantial packaging is needed to protect goods [5].

Regardless of the limitations inherent in water transportation, water is the least expensive mode for transporting high-bulk, low-value freights. The fixed cost of water carriage is mainly found in terminal facilities and transport equipment. Although water carriers have to develop and operate their own terminals, rights-of-way and harbors are developed and maintained publicly. This moderates water-transport fixed costs, putting the mode between rail and motor carriages. Water-transport variable costs, including waterway charges and transport equipment operation costs, are very low. Because of the high fixed cost and low line-haul costs of water carriage, its

[3] Containers are standardized boxes that are typically 8 feet high, 8 feet wide, and of various lengths (usually 10, 20, and 40 feet). The freight is handled as a unit in containers, which are easily transferred as units to other transportation modes [5].

costs per ton-mile decrease significantly as the distance and shipment size increase [4,5].

Pipeline

Pipeline systems were mainly developed for transporting large volumes of products, often over long distances. Pipelines tend to be product specific, which means they are used for only one particular type of product throughout their design life [6]. A limited number of products can be transported by pipelines, including natural gas, crude oil, refined petroleum products, chemicals, water, and slurry products.[4]

Although product movement through pipelines is very slow (only 3 to 4 miles per hour), their effective speed is much greater than the other modes because they operate 24 hours a day, 7 days a week. For transit time, pipeline service is the most dependable of all modes because of the following factors: [10]

- Pumping equipment is highly reliable, so losses and damage because of pipeline leaks or breaks are extremely rare.
- Climatic conditions have minimal effects on products moving in pipelines, so weather is not a significant factor.
- Pipelines are not labor intensive, so strikes or employee absences have little effect on their operations.
- Computers are used to monitor and control the flows of products within the pipelines.

Losses and damage costs from transporting by pipeline systems are low because (1) liquid and gases are not subject to damage to the same degree as manufactured products, and (2) there are fewer types of danger throughout a pipeline operation [5].

Pipelines have the highest fixed cost and the lowest variable cost among transportation modes. High fixed costs result from right-of-way, construction, and requirements for control station and pumping capacity. To spread these high capital costs, and to be competitive with other modes, pipelines must operate at high volumes. The variable costs are extremely low and mainly include the power for moving products, because, as noted, pipelines are not labor intensive [4].

Digital

Digital or electronic transport is the fastest mode of transportation. Besides its high speed, digital transport is cost efficient and benefits from its high accessibility and flexibility. However, only a limited range of products can be shipped by this mode, including electric energy, data, and products such as texts, pictures, music, movies, and software, all of which are composed of data [9].

Most logistics references do not cite digital transport as a transportation mode because of its limited product options. However, someday, technology may allow

[4] "Slurry systems involve grinding the solid material to certain particle size, mixing it with water to form a fluid, muddy substance, pumping that substance trough a pipeline, and then decanting the water and removing it, leaving the solid material." [2]

us to convert matter to energy, transport it to desired destination, and convert it back to matter again.

Any one or more of the six above-mentioned transportation modes can be a viable option for a company or individual who wants to move products from one point to another. Shippers take several factors into account in selecting the proper transportation modes. The company and its customers' needs, the characteristics of the transportation modes, and the nature of traffic are the main factors that should be considered in the modal choice. Table 2.1 summarizes the general and service characteristics of the six transportation modes, based on references [7,14].

In addition to the six basic modes of transportation, several intermodal combinations are available to shippers. Such combinations can lead to transportation services with cost and service characteristics that rank between those of the single modes. In fact, intermodalism combines the cost and service advantages of two or more transportation modes. Deveci et al. [15] quoted the definition of intermodal transport from reference [16] as follows: "The movement of goods in one and the same loading unit or vehicle that uses successively several modes of transport without handling of the goods themselves in changing modes."

If we exclude digital or electronic transport, which has a very low intermodal capability, we have 10 possible intermodal service combinations: (1) rail–road, (2) rail–water, (3) rail–air, (4) rail–pipeline, (5) road–air, (6) road–water, (7) road–pipeline, (8) water–air, (9) water–pipeline, and (10) air–pipeline. These are combinations in theory, but in practice only a few of them turn out to be convenient. The most frequent combined intermodal services are rail–road ("piggyback"), road–water ("fishyback"), and road–air ("birdyback"). Road–water combinations are gradually gaining acceptance, especially for international shipments of high-valued products. However, only rail–road combinations have seen widespread use throughout the world [5,8]. The more popular combinations that we have explored in this section are:

1. Trailer on flatcar (TOFC)
2. Container on flatcar (COFC)
3. Roadrailers

Piggyback (TOFC/COFC)

Transporting a motor carrier trailer on a rail flatcar is referred to as TOFC service. It is also possible to transport only the container on a flatcar to omit the deadweight of understructures and wheels. Such combination is referred to as COFC service. Although these two services are technically different, they are both referred to as *piggyback* service by most logistics executives [10]. In piggyback service, first terminal-to-terminal transportation is achieved by placing truck trailers or containers on railroad flatcars and transporting them over longer distances than trucks normally haul. Temporary axles can be employed under the containers so they can be distributed via trucks or tractors. Finally, to achieve point-to-point distribution, the pickup and delivery functions are performed by motor carriers at the terminal facilities.

Table 2.1 Characteristics of Transportation Modes

	Road	Rail	Air	Water	Pipeline	Digital
General characteristics						
Product options	Very broad	Broad	Narrow	Broad	Very narrow	Very narrow
Predominant traffic	All types	Low-moderate value, moderate-high density	High value, low-moderate density	Low value, high density	Low value, high density	All types of data
Market coverage	Point to point	Terminal to terminal	Terminal to terminal	Terminal to terminal	Terminal to terminal	Point to point (computer to computer)
Average length of haul	Short to long	Medium to long	Medium to long	Medium to long	Medium to long	Short to long
Capacity	Low	Moderate	Low	Very high	Very high	Moderate
Service characteristics						
Cost	Moderate	Low	High	Low	Low	Very low
Speed (time in transit)	Moderate	Slow	Fast	Very slow	Very slow	Very fast
Availability	High	Moderate	Moderate	Low	Low	Very high
Delivery time consistency	High	Moderate	High	Low-moderate	High	High
Loss and damage	Low	Moderate-high	Low	Low-moderate	Low	Very low
Flexibility	High	Moderate	Low-moderate	Low	Low	High
Intermodal capability	Very high	Very high	Moderate	Very high	Very low	Very low

Piggyback service combines the convenience and flexibility of short-haul trucking and the long-haul economy of rail transportation. The cost of this combination is less than for trucking alone and has permitted truck movement to expand its economical range. Likewise, rail carriage has been allowed through this combination to share in some traffic that normally would move by truck alone. Moreover, this combination brings door-to-door service convenience to shippers over long distances at reasonable rates. The above-mentioned features can interpret why piggyback service is the most popular intermodal combination [5].

Stock and Lambert [7] mentioned the partnership between the Burlington Northern Santa Fe (BNSF) Railroad and J. B. Hunt Transportation Services as an interesting example of intermodalism. This partnership, which began in late 1989, combined a large railroad company with a national TL motor carrier. As a result, door-to-door intermodal services between California and the Midwest are now available to shippers.

Roadrailers

Roadrailer, also called *trailertrain*, is an innovative intermodal concept that was first introduced in the late 1970s. Although roadrailers appear similar to conventional truck trailers, they have both rubber truck tires and steel rail wheels, thus providing a combination of rail and motor transport in a single piece of equipment (Figure 2.3). The trailers are shipped in the normal way via tractor over highways. By changing wheels for rail movement, the trailer rides directly on the railroad instead of being placed on a flatcar.

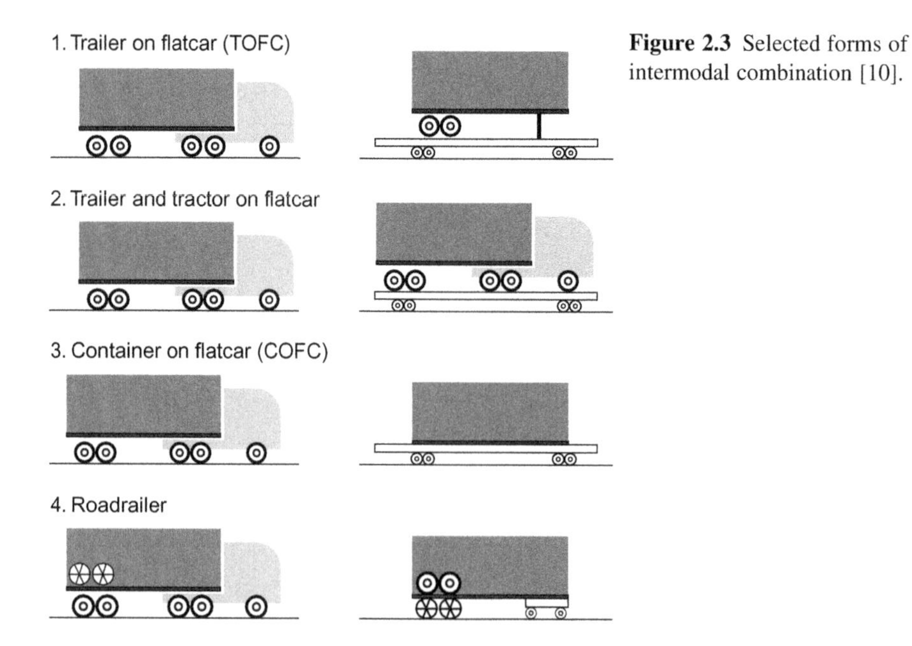

Figure 2.3 Selected forms of intermodal combination [10].

In comparison with piggyback service, the main advantage of roadrailers is that rail flatcars are not required. Moreover, the required time for switching between highway and rail wheels is less than loading or unloading the trailer from the flatcar. The major disadvantage of this intermodal form is the additional weight of rail wheels, which reduces fuel efficiency and leads to higher costs for the highway portion of the shipment. As a direct result of high operation and equipment costs, the use of roadrailers is limited [10].

2.1.2 Other Transport Options

In addition to the options previously explained, there exist other important entities in transportation systems. These entities, whether unimodal or multimodal in scope, include nonoperating third parties that provide various services to shippers. The major alternatives are:

1. Freight forwarders
2. Shippers' associations
3. Intermodal marketing companies
4. Brokers
5. Small package carriers
6. Third-party logistics service providers

Freight Forwarders

Freight forwarders or forwarding agents are agencies that organize the freight shipments of other companies or individuals. They often do not own transport equipment except for pickup and delivery operations. Freight forwarders purchase long-distance transport services from truck, rail, air, and water carriers. Then they consolidate numerous small shipments of different shippers into large shipments. After transporting the bulk load through one or more of the basic modes to a destination, they split the load into the original smaller quantities. The transportation cost per pound of small shipments is higher than that of the large shipments. The difference between the large and small shipments' rates offsets the operating costs of these companies. This is why forwarding agents offer lower rates to the shippers than they can obtain directly from the carriers. Moreover, these companies can also provide more complete and faster services to the shippers.

Freight forwarders can be classified as surface or air forwarders, based on the transportation modes they use. Also, a forwarding agent can be considered as an international forwarder if it is specialized in shipments to other countries or as a domestic forwarder if it specializes in shipments within the country [5,7].

Shippers' Association or Cooperative

A shippers' association is a nonprofit transportation membership cooperative that organizes the domestic or international shipments for member companies. These associations consolidate the small shipments of their members into vehicle-load freight so that small and medium shippers can also benefit from the economies of scale. They contract with motor, rail, air, and water carriers to physically move their members' cargo, benefiting both shippers and carriers. Shippers take advantage of the lower rates, and the carriers benefit from better equipment utilization, as well as the economies of large and often long-distance shipments. Shippers' associations are not classified as common carriers, and in the United States the Interstate Commerce Commission (ICC) has never had jurisdiction over them [10].

Intermodal Marketing Companies

Shippers' agents or intermodal marketing companies (IMCs) are important intermodal links between shippers and carriers. These agencies are much like shippers' associations in their operations, but they offer specialized TOFC or COFC services to shippers. They purchase large quantities of piggyback services at discount rates and then resell the available services in smaller quantities to the shippers. Similar to shippers' associations, these companies are not licensed by the ICC, and their importance is increasing as the use of intermodal transportation is growing in today's world [7].

Brokers

Brokers are the intermediaries that organize the transportation of products for shippers, consignees, and carriers and charge a fee to do so. Besides providing timely information about rates, routes, and capabilities to bring shippers and carriers together, brokers also provide other services such as rate negotiation, billing, and tracking. These agents are subject to the same regulations that apply to carriers, and they are all licensed by the ICC.

Brokers help shippers, especially those with no traffic department or minimal traffic support, to negotiate rates, supervise their shipments, and perform what they may not be able to carry out because of resource constraints. Brokers can also help carriers find business or obtain back hauls and return loads that increase their efficiency as they transport "full" equipments rather than "empty" ones [5,10].

Small Package Carriers

Small-shipment delivery services can be important transportation options for many shippers. Electronics firms and cosmetic companies, as well as book distributors and catalog merchandisers, are examples of these shippers. Well-known small package carriers include the US Postal Service's parcel post, United Parcel Service (UPS), and air-express companies.

Parcel post is a delivery service provided to companies that ship small packages. Low cost and wide geographical coverage are the competitive advantages of parcel post because it offers both surface and air services, domestically and internationally. Size and weight limitations, transit time variations, and relatively high loss and damage ratios are the main disadvantages of this service. Another disadvantage of this service is its inconvenience to shippers, because packages must be paid for in advance and deposited at a postal facility [7].

UPS is a private package-delivery company. It transports small packages, so it competes directly with parcel post for shipping small parcels, especially in the United States. The primary business of UPS is the time-definite delivery of documents and packages internationally. UPS has extended its services in three main segments: domestic package services in the United States, international package services, and supply-chain and freight services. UPS's advantages include its low cost and low time-in-transit variability, as well as its wide geographical coverage. The disadvantages of UPS include specific size and weight limitations and inconvenience because small shippers must deposit their parcels at a UPS facility. However, UPS provides pickup for larger shippers [10].

Since its inception in 1973, the air-express industry has expanded significantly, mainly because of its high levels of customer service. Because these companies can offer overnight or second-day delivery services nationally or internationally, they are valuable shipping options for those shippers who need to transport their products quickly. Federal Express (FedEx), UPS, Airborne, and Emery are some of the most well-known examples of the air-express industry. Substantial revenues and considerable profits of these companies illustrate the importance of rapid-transit services with high consistency to the shippers [10].

Third-Party Logistics Service Providers

Nowadays, more companies are outsourcing their logistics functions to third-party logistics service providers, as the emphasis on supply-chain management has increased. Third-party logistics providers, commonly referred to as 3PLs, provide their clients with several logistics services, such as freight forwarding, packaging, transportation, and inventory management, as well as warehousing and cross docking. Because these services are bundled together by 3PLs, most companies consider these service providers as one-stop outsourcing solutions that can do the jobs more efficiently, allowing the companies to focus on their core business.

Third-party options can lead to cost reductions and customer-service improvements, especially for those small and mid-sized companies that cannot afford to develop their own distribution networks. Instead, they outsource their product distribution to 3PL providers so they can compete in today's global market. The cost savings gained through this channel is mainly because of reduced transportation charges. In addition to cost saving in transportation, fixed capital investments and labor and operating costs are reduced through this option. Moreover, these companies may benefit from the available cash previously tied up in inventory [17]. It should be mentioned that the ultimate objective of outsourcing logistics functions must be enhancing customer satisfaction through the improvement of delivery systems. However, often too much attention is paid to cost reduction and in making logistics alliances, rather than in improving delivery performance and customer satisfaction [18].

Freight forwarders, shippers' associations, shippers' agents, brokers, small-package carriers, and 3PL companies are all viable transport alternatives for a shipper in the same way as the six basic transportation modes and the intermodal combinations. The optimal combination of shipping options should be determined by a company's logistics executive. This decision depends on several issues, including the master production schedule, customer-service objectives, the existing physical facility network, and standards and regulations [19]. Another substantial issue in determining the right modal choice or combination of transport alternatives for a company is the product characteristics or the *physical nature of the product*, which is explored in the following section.

2.2 Physical Nature of the Product

A product's physical nature substantially affects almost every aspect of logistics and distribution systems, including packaging, material handling, storage, and transportation. In fact, both the structure and the cost of a distribution system for a given product are directly affected by the product's particular characteristics. These characteristics can be classified into four main categories, based on [3]:

- Volume-to-weight ratio
- Value-to-weight ratio
- Substitutability
- Special characteristics

2.2.1 Volume-to-Weight Ratio

Both the volume and weight characteristics of a product significantly affect distribution costs. Products with low volume-to-weight ratios tend to fully utilize the weight-constrained capacities of road freight vehicles, handling equipment, and storage space. Therefore, distribution systems deal with these kinds of products, including dense products such as sheet steel and books, more efficiently. In contrast, high-ratio products, such as many food items, paper tissues, and feathers, use up a lot of space, which results in underutilized distribution components, raising both transportation and storage costs [20].

In general, storage rates are volume based and value based, but transportation rates are more dependent on the type of transportation mode. For example, water carriers normally charge the same price for 1 ton as for 1 cubic meter, but 1 ton costs the same as 6 cubic meters for airfreight. Hence, the transportation of heavy products by air is relatively more expensive. However, in most cases, overall distribution costs (including transportation and storage costs) tend to decrease as the volume-to-weight ratio decreases. To avoid abnormally low rates, carriers and warehouses often stipulate a minimum charge for the transportation and storage of very light or very heavy products, respectively [3,21].

2.2.2 Value-to-Weight Ratio

This ratio shows the value per unit weight of a given product. High-value, low-weight products, such as electronic equipments and jewelry, have greater potential for absorbing the distribution costs because the relative transport cost of these products to their overall value is not significant. Therefore, criteria other than price play a significant role in determining the proper distribution system for high-value products. In contrast, only inexpensive transport alternatives can be viable shipping options for products with low value-to-weight ratios, including ore, coal, and food. However, the storage and inventory holding costs for products with high value-to-weight ratios tend to be high in comparison with low ratio products because the capital tied up in the stock is higher, and more expensive and secure warehousing is required [3,20].

Substitutability

The degree to which a given product can be substituted by an alternative from another source is referred to as its *substitutability*. Highly substitutable products, such as soft drinks and junk food, are those that customers would readily substitute with another brand or type of products if the initially desired products are not available. The distribution system should ensure the availability of these products at all times, otherwise the sale would be lost. This could be achieved through maintaining high inventory levels to decrease the stock-out probability or by using efficient and reliable transportation modes for on-time replenishments. Both of the preceding options are high cost because they would raise the average inventory level and enforce a transport system with higher costs, respectively. However, for products with low substitutability degrees, less-expensive distribution systems with lower average inventory levels and slower transportation modes can be used [3,21].

Special Characteristics

Certain other characteristics of products imply a degree of risk in their distribution. These characteristics, including fragility, perishability, hazard and contamination potential, time constraints, and extreme value, pose some requirements and restrictions on a distribution system. Therefore, a special transport, storage, and handling system is required to minimize this risk or even satisfy the legal obligations, which means the company will incur extra charges, as it is the case with any form of specialization. Examples of these specifications could be packaging requirements of fragile products, necessary inventory controls, and refrigerated storage and transportation facilities for perishable products, as well as special packaging and stringent regulations (such as controlled temperature, restricted stacking height, and isolation from other products) for contaminant and hazardous products. Moreover, time-constrained products, such as foods, newspapers, and seasonal and fashion goods, have significant implications for distribution systems and often require fast and expensive transportation modes to meet their time deadlines. Finally, extremely valuable products, or small items that are vulnerable to theft, require special stock control and distribution systems with high security [3,20,21].

Many different product characteristics significantly affect almost every logistics function; some of them have been explained in this section. Because these logistics functions are interrelated, the requirements and restrictions imposed by these characteristics often lead to complicated alternatives. These alternatives vary in cost and service attributes and should be thoroughly evaluated by logistics executives for determining an appropriate distribution system.

2.3 Channels of Distribution

2.3.1 Distribution Channels and Their Types

Another crucial, and often challenging, decision that should be taken by logistics executives is determining a product's *distribution channels*—the alternative ways or path through which a product reaches its market. In contrast to some decisions, such as advertising and promotion programs, which can be readily changed by companies, distribution channel decisions tend to be hard to change because they usually involve long-term and often strong commitments to intermediaries, including brokers, wholesalers, and retailers. Therefore, in determining distribution channels, many different factors of today's and tomorrow's business environment should be taken into account by companies [18]. In general, there exist two types of distribution channels: physical and trading or transaction.

As the name suggests, a physical distribution channel deals with the physical aspects of a product distribution, including all the methods, means, and entities through which the product is distributed from the supplier's or manufacturer's outlet to the end user. In fact, products are physically transferred through these channels, reaching their desired destination, which could be a factory outlet, a retail store, or even a customer's house. However, trading or transaction channels are concerned with the nonphysical aspects of distributing products from their point of origin to their point of consumption. When a product transfers through distribution channels, the ownership of the product is transferred along with its physical movement. The sequence of negotiation and the exchange of product's ownership are the distribution aspects that the trading or transaction channel is concerned with [3].

Manufacturing firms face the same questions for both the physical and transaction channels: Do they transfer and sell their products directly to end users? Should intermediaries participate in the product distribution? Although intermediaries add a markup to the product cost, they provide several benefits to both producers and customers, three of which are specialized distribution functions, improved product assortment, and increased transactional efficiency [22].

Intermediaries benefit from their great expertise in distribution functions, so they can perform distribution activities more efficiently than producers, allowing them to concentrate on their core businesses. Also, intermediaries can provide distribution services more economically than individual manufacturers because they handle large-sized shipments and benefit from larger economies of scale.

Intermediaries provide a second benefit by converting the assortments of products made by manufacturers into the assortments demanded from consumers. Producers tend to generate narrow assortments of goods (similar types of products)

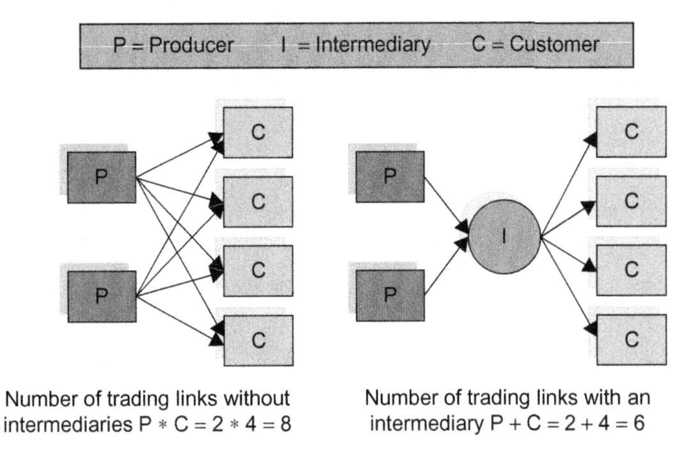

Figure 2.4 Distribution systems with and without intermediaries.

in large quantities, whereas consumers typically want broad assortments of products (different types of products) in small quantities. Therefore, intermediaries take the supply of many different producers in large quantities and then break them down into smaller quantities of wider assortments demanded by consumers [18].

Transactional efficiency is the third benefit provided by intermediaries. Figure 2.4 shows how using intermediaries can improve the efficiency of a distribution channel by reducing the number of trading links. As illustrated in the figure, when producers use direct marketing to reach their customers (in the absence of intermediaries), the number of contact lines equals the number of producers multiplied by the number of customers. The number of these trading links can be reduced by adding an intermediary (an agent or broker, a wholesaler, or a retailer) between producers and customers. In this case, the number of contact lines is calculated by adding the number of producers and customers. Hence, the presence of intermediaries can eliminate the duplicate efforts of both producers and customers and increase the efficiency of distribution systems [18,22].

2.3.2 Physical Distribution Channel

There exist several alternative distribution channels that can be used separately or in combination with each other to bring a product or group of products to the end user. Distribution channels contain different numbers of intermediary levels that are referred to as the *length* of those channels. Each member of a distribution channel that has an impact on transferring the product and its ownership to the ultimate user is considered to be a channel level [18]. Therefore, both the producer and the consumer are members of every distribution channel. Figure 2.5 shows the main alternative channels of distribution, based on references [3,8,23]. The physical transference of products between channel members is illustrated by the hand-shaped icons in the

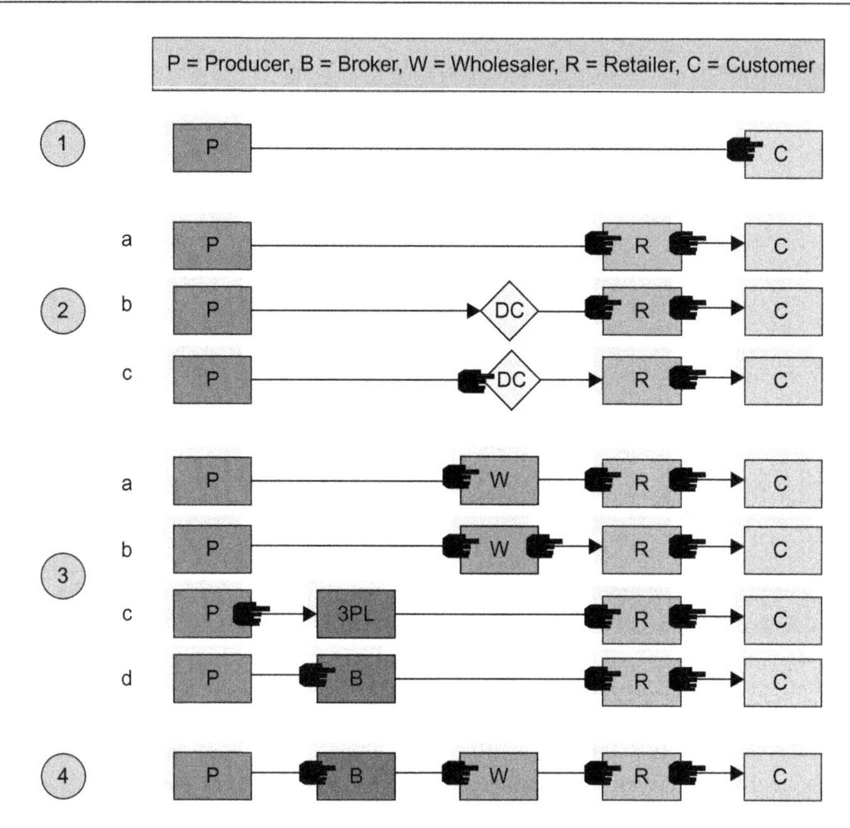

Figure 2.5 Alternative distribution channels.

figure. Although these channels are mainly for consumer products, industrial marketing channels are quite the same and will be explained later in this section.

Producer Direct to Consumer

The direct-marketing channel is the simplest and shortest distribution channel, and it has no intermediary levels (channel 1). This channel can be a part of direct-selling marketing strategy, which consists of a producer selling directly to the final consumer. Products that are customized for specific customers and those ordered through catalog or newspaper advertising are examples of goods commonly distributed through this channel. Also, customers who now shop from home, thanks to the Internet, are other users of this channel. Moreover, products composed of data, such as text, software, music, and films, can be directly distributed from computer to computer [3].

Producer to Retailer to Consumer

In contrast to channel 1, the remaining channels in Figure 2.5 contain one or more intermediary levels, which are referred to as *indirect-marketing channels*. In channel

2a, producers deliver their own goods directly to large retail stores, which then sell these products to end users. In general, this channel is suitable for manufacturer distributing their products in TL size. Channel 2b is quite similar to channel 2a, but producers deliver their own products in large shipments to distribution centers, which could be one central distribution center (CDC) or a number of regional distribution centers (RDCs). The products are then broken down into smaller orders that are transported to retailers on the manufacturers' own means. The only difference between channels 2b and 2c is that in the latter products from several suppliers are delivered to distribution centers run by retail organizations. The different types of products are consolidated in these centers and then delivered to retail stores in full truck loads, using the retailers' own vehicles or those of third-party providers.

Channel 3 has two intermediary levels. These intermediaries are retailers along with wholesalers in channels 3a and 3b, and third-party distribution service providers and brokers in channels 3c and 3d, respectively.

Producer to Wholesaler to Retailer to Consumer

The use of wholesalers as intermediaries is popular whenever the limited lines and financial resources of some small manufacturers would not allow them to distribute their products on their own; therefore, they rely on wholesalers' specialized distribution services, which supply a great number of retailers. Channel 3a is also suitable for those small retailers that cannot afford to buy large quantities of products. Wholesalers use their own delivery vehicles and distribution centers, and they benefit from the price advantage of buying bulk shipments from suppliers.

Channel 3b illustrates the concept of cash-and-carry organizations. In contrast to traditional wholesaling, this channel consists of several small retailers that are collecting their own demands from wholesalers. As the order quantities of small retailers and shops are very small, producers, suppliers, and even wholesalers do not deliver their demands directly to their stores; as a result, the use of cash-and-carry organizations tends to increase nowadays.

Producer via 3PL or Broker to Retailer to Consumer

Third-party distribution service providers and brokers are the other intermediaries supplying manufacturers' products to retail stores. The use of 3PL providers has increased as the distribution of products became more expensive and more complicated. The constantly changing legislation and restrictive rules and regulations on product distribution can justify the growing need for third-party distribution service providers, companies that are experts in distribution and warehousing, as well as other logistics functions. Although most of the 3PLs provide general distribution services, there are companies offering services for special types of products such as small parcels carriers that provide "specialist" services for products in the form of small parcels (see channel 3c).

Brokers or agents are independent intermediaries that bring buyers and sellers together. Brokers are not often fully concerned with physical distribution of products, and they never take title to goods. They are more concerned with products' marketing, and thus they may be considered as trading or transaction channels sometimes (see channel 3).

Producer via Broker to Wholesaler to Retailer to Consumer

Channel 4 represents a physical distribution channel with three intermediaries. This channel is similar to previous channels except that producers are represented by brokers who deliver their products through wholesalers. In general, a broker may represent either a manufacturer or a wholesaler by searching markets for its goods or by seeking supply source for its orders.

Distribution channels with more than three intermediary levels can be imagined, though they are not common channels. As the number of intermediary levels increases, the channel becomes more complex and the producer's control over the product flows decreases. The distribution channels illustrated in Figure 2.5 are the main alternatives of consumer marketing channels; however, distribution channels for industrial goods have quite the same structures as explained in the following section.

Business-to-Business Channels

The first industrial distribution channel a business marketer can use is the direct-marketing channel, which is quite similar to channel 1, except the final consumer is replaced with a business customer. Most industrial goods such as raw materials, equipment, and component parts are sold through this business channel. There is no need for wholesalers or other intermediaries in this channel because the goods are sold in large quantities. In the case of small accessories, producers sell their products to wholesalers or industrial distributors, which in turn sell them to business customers. Brokers and sales agents are also common intermediaries in industrial marketing channels. Often small producers are represented by independent intermediaries called *manufacturers' representatives* to market their products to large wholesalers or to final business customers [18,23].

Any one or more of the above-mentioned alternative distribution channels might be used by manufacturers to make their products available to final customers. An individual producer may choose different marketing channels with respect to its different types of products or customers.

2.4 Warehousing and Storage

Another important logistics functional area, which is strongly related to physical flow, is warehousing. In contrast to transportation, which primarily takes place on network arcs, warehousing and product storage mainly take place at nodal points. Warehousing, storage, and material-handling activities, which are often referred to as "transportation at zero miles per hour," take around 20% of total logistics distribution costs; therefore, they compel logistics executives to give them serious consideration [5].

Because demand for products cannot be predicted with certainty and they cannot be supplied immediately, storing inventories is inevitable. Companies store inventories to reduce their overall logistics costs and to reach higher levels of customer service through better coordination between supply and demand. Therefore, warehousing has become an important part of companies' logistics systems, which stores goods at and between the origin and destination points and provides the management with information about the status, disposition, and condition of inventories. These inventories may belong to different phases of the logistics process and can be categorized into three groups [7]:

1. Physical supply (raw materials, components, and parts)
2. Physical distribution (finished goods)
3. Goods in process (constitute small portion of total inventories)

This section is intended to provide a concise introduction to some of the basic *warehousing functions*, which will be explored broadly later in Chapter 10. It continues with a brief discussion about *packaging and unit loads*, as well as the *handling systems*. These issues are also investigated in more detail in Chapter 9.

2.4.1 Warehousing Functions

Warehousing plays a critical role in logistics systems, providing the desired customer-service levels in combination with other logistics activities. A wide variety of operations and tasks are performed in warehousing; these can be categorized under three basic functions [10]:

1. Movement (material handling)
2. Storage (inventory holding)
3. Information transfer

Traditionally, the storage function was considered as the primary role of warehouses because they were perceived as places for long-term storage of products. However, today's organizations try to improve their inventory turns and move orders more quickly through supply-chain networks; therefore, nowadays, long-term storage role of warehouses has diminished, and their movement function has received more attention.

Movement

The movement or material-handling function is represented by four primary activities.

- *Receiving and put away*: This activity includes unloading goods from the transportation equipment as well as verifying their count and specifications against order records, inspecting them for damage, and updating warehouse inventory records. Receiving also includes sorting and classification of products and prepackaging bulk shipments into smaller ones before moving them to their warehouse storage location. Finally, the physical movements of products to storage areas, locations for specialized services (such as consolidation areas), and outbound shipment places are referred to as *pass-away activities* [5,24].

- *Order filling or order picking*: This is a fundamental movement activity in warehousing and involves identifying and retrieving products from storage areas according to customer orders. Order filling also includes accumulating, regrouping, and packaging the products into customers' desired assortments. Moreover, generating packing slips or delivery lists may also take place at this point [7,24]. Order-picking activities are time consuming and labor intensive. A study in the United Kingdom revealed that around 63% of warehouse operating costs are the result of order picking [25].
- *Cross docking*: In this process, receiving products from one source are occasionally consolidated with products from other sources with the same destination and immediately sent to customers, without moving to long-term storage. A pure cross-docking operation only organizes the transfer of materials from inbound receiving dock to the outbound dock, eliminating nonvalue-adding activities such as put away, storage, and order filling. In practice, however, there might be some delay, and the items may remain in the facility between 1 and 3 days [26,27].
- *Shipping*: This activity involves physically moving and loading assembled orders onto transportation carriers, checking the content and sequence of orders, and updating inventory records. It may also include sorting and packaging the products for specific customers or bracing and packing the items to prevent them from damage.

Storage

The storage function of warehouses is simply about the inventory accumulation over a period of time. The storage of inventory may take place in different locations and for different lengths of time in warehouses, depending on the storage purpose. In general, four primary storage functions have significant impacts on the storage facilities' design and structure: *holding*, *consolidation*, *break-bulk*, and *mixing*.[5] Warehouses may be designed to satisfy one or more of these functions, and their layout and structure will vary based on their emphasis on performing these storage functions.

The storage of inventory in warehouses can be categorized into two main groups, according to the length of storage time: *temporary* or short-term storage and *semipermanent* or long-term storage. In temporary storage, only products required for basic inventory replenishment are stored. The amount of temporary inventory required to be stored in warehouses is determined based on the extent of variability in lead time and demand. Also, the design of logistics systems may affect the inventory extent. The emphasis of temporary storage is on the movement function of warehousing, and pure cross docking tends to use only this kind of storage. However, semipermanent or long-term storage includes the storage of products in excess of that necessary for basic replenishment. Semipermanent storage is justified in some common situations, including [7]:

1. Seasonal or erratic demand
2. Conditioning of products (e.g., fruits and meats)

[5] These functions are broadly explored in reference [5], pp. 472–477.

3. Special deals (e.g., quantity discounts)
4. Speculation or forward buying

Information Transfer

Precise and timely information is a must for managers to administer the warehousing operation; therefore, they attach great importance to the information-transfer function. This function takes place concurrently with the other warehousing functions—movement and storage—and provides the warehouse manager with information on the inventory and throughput levels,[6] locations where products stored, as well as inbound and outbound shipments. These types of information along with the data on space utilization, customer and personnel information, and other pertinent information are essential for ensuring a successful warehousing operation. Recognizing the crucial importance of these types of information, companies are continually improving the speed and accuracy of their information-transfer function by using computerized and modern processes such as bar coding their products, and using the Internet or electronic data interchange (EDI) systems for transferring their information [7].

2.4.2 Packaging and Unit Loads

Almost all the products flowing through logistics networks are packaged, mainly to promote or protect the product. The former goal is achieved by one type of packaging referred to as *interior* or *consumer packaging*. This packaging is brightly colored and contains marketing and promotional materials. Although the *exterior* or *industrial packaging* is the plain box or pallet that includes basic information about the item for organizations, it is designed to protect the product and make its handling easier. In general, the main reasons for packaging goods can be summarized as follows [3,27]:

- To protect or preserve items
- To identify the product and provide basic information
- To facilitate the handling and storage of products
- To improve the product appearance, and assist in promoting, marketing, and advertising it

Products may also be packed at different levels. *Primary* or *elementary packaging* creates the smallest handling unit of any system, enclosing the product directly and keeping it unchanged throughout the logistics network. *Secondary* or *compounded packaging* is created by bundling a number of primary packages together. Finally, similar to secondary packaging, *outer packaging* takes place to make the handling of products easier. These packages disappear after the products are unpacked at destination points [28]. Customers may order the products at any of these levels, and the logistics and distribution systems must satisfy their demand cost-effectively. Therefore, the concept of load unitization—storing and handling

[6] The amount of material moving through a warehouse.

goods in standard modules—has become a fundamental issue in today's supply-chain networks. Moving standard unit loads is much easier than moving a variety of products with different sizes and shapes. Thus, smaller logistics units are collected and bundled together to form standard unit loads. Determining the optimal type and size of unit loads decreases both the products movement rates and their loading and unloading times. It also brings the chance of using standard handling and storage equipment to the company, so that it can be set up to work efficiently and be optimally utilized [3,27]. Different types of unit loads are designed for application in three basic areas of supply networks: manufacturing, storage, and distribution [29]. *Small containers* (such as tote bins), *intermediate bulk containers* (IBCs), *dollies*, *roll-cages*, and *cage and box pallets* are examples of the most frequently used storage-unit loads (see references [3,24] for more details). However, the most commonly used storage-unit load is probably the *wooden pallet*. Wooden pallets are intended to be made to standard sizes; however, the existence of different standards (e.g., in the United States, the United Kingdom, and continental Europe) caused international movements to encounter some problems. Moreover, these pallets may also be made of metal or plastic, and they can be two- or four-way, open- or close-boarded, and single- or double-sided. Because these pallets are the most significant unit loads in warehousing, specific storage and handling systems are designed for them; these are explored later in this section. Systems for nonpalletized loads are examined in the concluding section of this chapter.

2.4.3 Storage and Handling Systems

Storage and handling systems fall into two main categories: palletized and nonpalletized. Some of the most common examples of the various types of storage and handling equipment available for these two systems are introduced in this section.

Palletized Storage and Handling Systems

As mentioned earlier, the most frequently used unit load in warehouses is the wooden pallet. Wooden pallets are popular mainly because they allow the use of standard storage and handling equipment, regardless of the size and characteristics of the goods on the pallet. Products in these types of systems either arrive on pallets or are palletized at the receiving areas, so they can benefit from the convenient-size load for their movement and storage.

Pallet Movement

There exists a wide variety of equipment for horizontal movement of pallets in warehouses. The most regular types include the following:

- *Pallet trucks* generally fall into two types: *hand pallet trucks* and *powered pallet trucks*. Hand pallet trucks are manually operated trucks with two forks and a steering unit. The forks enter into the pallet slots and then can be raised slightly by a hand-operated hydraulic pump or by using a mechanical system of levers until the load is lifted off the floor, ready for transportation. Powered pallet trucks are similar to hand pallet trucks, except they are electrically powered. They can be pedestrian- or rider-controlled, depending on their design. These types of trucks are faster than hand pallet trucks, and they are more suitable for moving greater loads over longer distances and more frequently [30,31].

- *Tow and platform trucks* are suited for horizontal movements of loads over long distances, and they can be manually operated or battery powered. Using a tug towing several trailers can reduce the number of required journeys in warehouses. In situations where fumes or oil spillage may not be a hazard to operators or cause contamination of products, diesel tow trucks can also be used [30].
- *Conveyors* are devices suited for continuously transporting material, especially where unit loads are uniform and the path and rate of their movements tend to remain unchanged. Thus, conveyors are used mainly in cases where the frequent movement of material between specific points is required and the flow volume would justify the conveyors' fixed costs. Conveyors may be driven using some source of power (traction conveyors) or without power—for example, as with gravity roller conveyors known as *tractionless conveyors*. The latter is suitable for short distances, while the former is suited for longer and more controlled movement of material [30].
- *Automated guided vehicles* (AGVs) are driverless battery-powered vehicles controlled by computer. They are mainly used for material movement between determined points and in cooperation with other handling systems such as conveyors. They are programmable devices, and a variety of means may be used for their guidance such as underfloor wires or magnets, optical guidance through painted lines or strips, or, more recently, by laser-guided systems. AGVs benefit from obstacle detectors, so when there is an obstacle (e.g., a person or a truck) in their way, they stop [3,31].

These types of equipment are mainly used to move material horizontally. However, as today's warehousing usually involves with stacking, some lifting mechanism should be applied to place the pallets into their storage positions. We describe stacking equipment for palletized loads in the following section, but note that many of these lifting trucks are also commonly used for horizontal movement around the warehouse.

Pallet Stacking

To utilize the warehouse space more efficiently nowadays, pallet stacking methods are employed in most warehouses. Pallets are placed on top of each other or, more commonly, they are placed in storage racks by means of equipment capable of lifting pallets or loads. Again, a wide range of stacking equipment and lifting trucks exist that may also be used to horizontally move goods around the warehouse. Some of the more common types are as follows:

- *Counterbalanced forklift trucks (CB truks)* carry their loads outside the chassis area and forward of the front wheels so that they can accommodate a great range of loads. Because a lack of counterbalance would produce a turning moment that would tend to tip the truck forward, all the truck's heavy components such as the engine and battery are located at the rear of the machine. A counterbalance weight also is built there to balance the overturning movement. Counterbalanced trucks are one of the most common devices used in warehousing because they are general purpose machines that can handle many different types of loads with the aid of various auxiliary attachments. CB trucks are robust and flexible, and they are available off the shelf from many different suppliers and in a large range of capacities, ranging from 1 ton to about 45 tons. However, because they are large and heavy trucks, wide aisles of about 3.5 meters or more are required for their operation. Therefore, although counterbalanced trucks are great "yard trucks" for loading and unloading the vehicles, they are less suitable for indoor warehousing activities [3,30].

- *Pallet stackers* are quite similar to pallet trucks, except they have a greater range of lifting motion and can do more than just lift a pallet high enough to move it. A wide range of pallet stackers are available from simple pedestrian or so-called walkie stackers to stand-on and ride-on types. To store or retrieve pallets from racks, the truck legs are straddled around the bottom pallet or driven into the space under it. Unlike counterbalanced trucks, pallet stacker trucks are fairly lightweight, and they can operate even in 90-degree turning aisles of 2 meters or less. The maximum capacity of these trucks is usually up to around 2 tons, and their lift height limitation is about 6 meters [3,24].
- *Reach trucks* are one of the most popular types of material-handling devices, lighter and smaller than the counterbalanced trucks. They carry their load within their wheelbase, so they can maneuver and operate in relatively smaller areas. In fact, typical models of reach trucks can handle standard pallets in aisles of between 2.7 and 3 meters. To place loads into or to retrieve them from storage racks, the truck turns 90 degrees facing the load location; a mast with a pantograph or scissors mechanism is then used to reach the storage rack to place or retrieve the load. Then the mast is retracted into the area enclosed by the wheels; hence, when the truck travels, the load rests on the outrigger arms. Therefore, unlike CB trucks, no counterbalance weight is required and the truck length can be reduced. However, to support the load when the mast is extended beyond the outrigger arms, the counterbalance of the truck is used. Modern reach trucks are available with lift heights of up to 11.8 meters, and their capacities typically range from 1 ton up to 3.5 tons [30,32].
- *Turret trucks* provide greater stacking height (up to 12 meters or so), but they also require greater investment cost. Turret trucks are suitable for narrow aisles and confined spaces. In fact, they can operate in aisles of between 1.5 and 2 meters. The rotating forks and mast allow the pallets to be picked up and retrieved from either side of the storage aisle. The truck itself does not rotate during the storage or retrieval, so their bodies should be longer to increase their counterbalance capabilities.
- *Specialist pallet stacking equipment* exists in a wide range of equipment for specific types of storage and handling systems. These include narrow-aisle trucks, which have minimum aisle width requirement; double-reach trucks with telescopic forks that allow two pallets to be handled from the same side of an aisle; and stacker cranes, which are commonly used in automated storage and retrieval systems (AS/RSs).[7]

Nonpalletized Storage and Handling Systems

There are many types of products that cannot be palletized because of their special characteristics. Being too large, too long, or even too small or having some handling limitations such as being lifted from the top, many types of products are not suitable for palletization. Some electronic items, nut and bolts, steel bars, drums, paper reels, hanging garments, and carpets are examples. This section introduces some of the equipment and devices that are commonly applied in various storage and handling systems of these items.

[7] Further details about these devices, along with the appropriate storage systems are described in the chapter 15 of reference [3].

Truck Attachments

A wide range of attachments are available for fitting to forklift trucks so that they can handle goods that cannot be touched by forks. These truck attachments enable additional degrees of movement for handling unit loads. However, the extra weight of these attachments should be considered in calculating the payload capacity of trucks and in determining the weight that can safely be carried by the truck. Some of the most common types follow:

- *Clamp* attachments can be used instead of forks to handle loads such as cardboard boxes, cartons and bales, drums, kegs, paper reels, and similar materials. They consist of shaped or flat side arms that are powered by a truck's hydraulic system. The side-arm pressure is adjustable, but the clamps often exert severe pressure on loads, so they must have enough strength to resist such pressure [3,30].
- *Rotating head* attachments are suitable for situations in which the orientation of loads needs to be changed. A common example is newsprint reels, which are stored with their axes vertical, even though they are presented horizontally to printing machines [3].
- *Load push–pull* devices are alternatives to forks, by which materials are positioned on slip sheets. Slip sheets take up little space in warehouses or shipping containers, and they avoid the cost of one-way pallets. Also, such devices enable nonpalletized loads to be pulled on to a platen, so they can be lifted, moved, and placed effectively as if they were a pallet load. The major drawback of this type of material handling is its longer operation time [30].
- *Drum-handling equipment* are devices for handling drums or barrels. There are several types of forklift truck attachments, including cradle attachments, horizontal carry attachments, vertical carry attachments, and drum tines. These attachments are used to lift a number of horizontally or vertically oriented drums at one time.
- *Booms* are available in various types. They replace the truck forks for handling rolled items, such as carpet rolls, steel coils, and horizontal reels.

Cranes

The typical application of cranes in warehousing is for handling very heavy loads, such as metal bars, or for relatively lighter loads that are just too heavy to be handled manually. Cranes can provide movement of loads only in limited and predetermined areas. Some of the more commonly used types are as follows.

- *Bridge cranes* or *overhead traveling cranes* consist of a hoist mounted on a bridge made up of one or more horizontal girders that are supported at each end by carriages that travel along a pair of parallel runways that are installed at right angles to the bridge. The hoist moves along the bridge, while the bridge itself travels along the runway, so that the working area is fully covered [31].
- *Gantry cranes* consist of hoists fitted in a traverse trolley that can move horizontally along the bridge section and supporting columns or legs at each end of the bridge. These legs may run on ground rails, or they may be supported on wheels that allow the whole crane to traverse. Therefore, mobile gantry cranes are also suitable for outdoor applications.
- *Jib cranes* are lifting devices that travel along a horizontal boom—a pivoted arm called the *jib*—that is mounted on a pillar or on a wall. These types of crane are relatively inexpensive and can handle loads within the arc defined by the radial jib [3,31].

- *Stacker cranes* have a forklift-type mechanism and are usually used in automated storage and retrieval systems (AS/RSs). The crane traverses on a track in the warehouse aisles, and the fork can then be lowered or raised to any levels of storage racks on either side of the aisle. The fork can then be extended into the rack to store or retrieve products.

Typical types of crane are electrically powered, and they may be used with a wide range of attachments for specific purposes. Different types of hooks, magnets, and mechanical clamps are common examples of these attachments.

Conveyors

As mentioned previously, conveyors are widely used in warehousing to move both palletized and nonpalletized loads over fixed routes. They may also be used to sort or accumulate products (short-term buffering) or as an integral part of packaging and order-picking operations. In general, two types of conveyer exist: *nonpowered* or *gravity conveyers* and *powered conveyors*.

In comparison to powered conveyors, gravity conveyors are more basic and less costly. They are normally used to move loads weighing up to several tons over short distances. The major types of gravity conveyors include the following [31,32].

- *Spiral chute conveyors* are normally used to convey goods between floors or to link two handling devices. They may have double or triple spiral runways for sorting and transferring items to different levels.
- *Wheel conveyors* are constructed of a series of skate wheels mounted on a shaft or common axle. The spacing of the wheels and the slope for gravity movement depend on the type and weight of the loads being transported. Wheel conveyors are commonly used for vehicle loading and unloading, and they are more economical than roller conveyors for light-duty applications.
- *Gravity roller conveyors* are an alternative to skate-wheel conveyors for heavy-duty applications. Roller conveyors can handle items with a rigid riding surface, and their slope depends on the load weight.

Powered conveyors are normally used for moving heavier loads over longer distances. Some of the more frequently used types include the following:

- *Live roller conveyors* comprise a series of rollers, and they are suitable for moving heavy goods or loads with irregular shapes and sharp corners. To provide accumulation, rollers can be disengaged using force-sensitive transmission features. Powered roller conveyors can move loads horizontally and up 5- to 7-degree slopes [31,32].
- *Belt conveyors* comprise a continuous belt running on supporting rollers that provides complete support under the loads being transported. They are generally used for moving light and medium weight loads, and they provide considerable control over their orientation and placement. Belt conveyors are suitable for paths with inclines or declines, and they can move loads with unusual shapes and configurations; however, no accumulation is provided [3,32].
- *Slat conveyors* consist of separately spaced slats, and they are generally used for heavy loads with abrasive surfaces that may damage the belt. They can provide control over the orientation and placement of the load [31,32].

- *Chain conveyors* carry loads directly on one or more endless chains. They are primarily used to move heavy loads or to transfer loads between sections of roller conveyors [3,31].
- *Trolleys conveyors* consist of chains or cable suspended from equally spaced trolleys running in a closed loop path. Overhead trolley conveyors can handle loads up to several tons, and they are commonly used in systems with fixed path and paced flow [31].

Automated Guided Vehicles

Automated guided vehicles are introduced briefly under the section "Palletized storage and handling systems". In addition to palletized loads, AGVs may be used for moving nonpalletized loads, especially large and heavy loads, including paper reels and automobile bodies. There exists a wide range of AGV types, and they can be guided using physical guide path (such as wire, tape, paint), or by nonphysical guide path (software). Changing vehicle path in latter method is easier, because the path is not physically fixed; however, absolute position estimates (e.g., from lasers) are required in those methods [3,32].

References

[1] Council of Supply Chain Management Professionals. Supply chain management definitions [online]. http://cscmp.org/aboutcscmp/definitions.asp, 2000 (cited 14.11.10).

[2] J.C. Johnson, D.F. Wood, D.L. Wardlow, P.R. Murphy, Contemporary Logistics, seventh ed., Prentice Hall, Upper Saddle River, NJ, 1999, pp. 1–21.

[3] A. Rushton, P. Crouche, P. Baker, The Handbook of Logistics and Distribution Management, third ed., Kogan Page, London, 2006.

[4] S.C. Ailawadi, R. Singh, Logistics Management, Prentice Hall of India, New Delhi, 2005.

[5] R.H. Ballou, Business Logistics/Supply Chain Management: Planning, Organizing, and Controlling the Supply Chain, fifth ed., Pearson-Prentice Hall, Upper Saddle River, NJ, 2004.

[6] A. West, Managing Distribution and Change: The Total Distribution Concept, John Wiley & Sons, New York, 1989, pp. 181–195.

[7] J.R. Stock, D.M. Lambert, Strategic Logistics Management, fourth ed., Irwin McGraw-Hill, New York, 2001.

[8] G. Ghiani, G. Laporte, R. Musmanno, Introduction to Logistics Systems Planning and Control, John Wiley & Sons, NJ, 2004, pp. 6–20.

[9] M. Hugos, Essentials of Supply Chain Management, John Wiley & Sons, Hoboken, NJ, 2003, pp. 1–15.

[10] D.M. Lambert, J.R. Stock, Strategic Logistics Management, third ed., Irwin, Homewood, IL, 1993.

[11] U.S. Department of Transportation, Research and Innovative Technology Administration, Bureau of Transportation Statistics, and U.S. Census Bureau. 2007 Commodity Flow Survey; Transportation Commodity Flow Survey, preliminary [online]. http://www.census.gov/compendia/statab/2010/tables/10s1036.pdf, 2008 (cited 14.11.10).

[12] H.T. Lewis, J.W. Culliton, J.D. Steel, The Role of Air Freight in Physical Distribution, Division of Research, Graduate School of Business Administration, Harvard University, Boston, MA, 1956, p. 82.
[13] R.J. Sampson, M.T. Farris, D.L. Shrock, Domestic Transportation: Practice, Theory, and Policy, fifth ed., Houghton Mifflin, Boston, MA, 1985, pp. 68–85.
[14] KnowThis LLC. Principles of marketing tutorials [online]. http://www.knowthis.com/principles-of-marketing-tutorials/managing-product-movement/modes-of-transportation-comparison, 1998 (cited 14.11.10).
[15] D.A. Deveci, A.G. Cerit, O. Tuna, Determinants of intermodal transport and Turkey's transport infrastructure, in: METU International Conference in Economics VI. Ankara, Turkey, September 11–14, 2002.
[16] Institute of Logistics, Understanding European Intermodal Transport—A User's Guide, Guideline No. 4, Institute of Logistics, Corby, UK, 1994.
[17] M. Yan, H. Yu, Logistics solutions for company A's Latin American market. Research project, Simon Fraser University, Faculty of Business Administration, 2004, pp. 62–73.
[18] P. Kotler, G. Armstrong, J. Saunders, V. Wong, Principles of Marketing, second European ed., Prentice Hall Europe, London, 1999, pp. 894–907.
[19] D. Riopel, A. Langevin, J.F. Campbell, The network of logistics decisions, in: A. Langevin, D. Riopel (Eds.), Logistics Systems: Design and Optimization, Springer, New York, 2005, pp. 12–17.
[20] M. Browne, J. Allen, Logistics of food transport, in: R. Heap, M. Kierstan, G. Ford (Eds.), Food Transportation, Blackie Academic & Professional, London, 1998, pp. 22–25.
[21] R.G. Kasilingam, Logistics and Transportation: Design and Planning, first ed., Springer, London, 1999, pp. 24–26.
[22] A.C. McKinnon, Physical Distribution Systems, first ed., Routledge, London and New York, 1989, pp. 26–30.
[23] L.E. Boone, D.E. Kurtz, H.F. MacKenzie, K. Snow, Contemporary Marketing, second Canadian ed., Nelson Education Ltd, Toronto, ON, Canada, 2010, pp. 390–395.
[24] J.A. Tompkins, J.A. White, Y.A. Bozer, J.M.A. Tanchoco, Facilities Planning, fourth ed., John Wiley & Sons, Hoboken, NJ, 2010.
[25] J. Drury, Towards More Efficient Order Picking, IMM Monograph No. 1, The Institute of Materials Management, Cranfield, 1988.
[26] J.K. Higginson, J.H. Bookbinder, Distribution centers in supply chain operations, in: A. Langevin, D. Riopel (Eds.), Logistics Systems: Design and Optimization, Springer, New York, 2005, pp. 71–78.
[27] D. Waters, Logistics: An Introduction to Supply Chain Management, Palgrave Macmillan, New York, 2003, pp. 283–290.
[28] T. Gudehus, H. Kotzab, Comprehensive Logistics, first ed., Springer-Verlag, Berlin, 2009, pp. 342–344.
[29] J.M.A. Tanchoco, Material Flow Systems in Manufacturing, first ed., Chapman & Hall publication, London, 1994, pp. 107–109.
[30] K.C. Arora, V. Shinde, Aspects of Materials Handling, first ed., Laxmi Publications, New Delhi, 2007, pp. 146–170.
[31] K. Bagadia, Definitions and classifications, in: R.A. Kulwiec (Ed.), Materials Handling Handbook, second ed., John Wiley & Sons, New York, 1985, pp. 103–110.
[32] C. Hiregoudar, B.R. Reddy, Facility Planning and Layout Design, first ed., Technical Publications Pune, Pune, Maharashtra, 2007, pp. 31–56.

Part II

Strategic Issues

3 Logistics Strategic Decisions

Maryam SteadieSeifi

Department of Industrial Engineering, Amirkabir University of Technology, Tehran, Iran

3.1 Strategy

Strategy originates from the Greek word *strategos* ("general" of the army), but its contemporary definition refers to a plan for achieving chosen objectives. Therefore, strategy as planning and positioning is the traditional definition of strategy [1].

In other words, strategies are the directional, focused efforts that establish a consistent and planned approach for a business organization. In fact, strategies are the human management decisions that are made with partial information—assumptions about conditions, interactions, attitudes, behaviors, and actions and reactions in the environment—by a company in advance.

Strategies are commonly agreed to be complex but not completely deliberate, to involve different thought processes (conceptual and analytical), to address both content and process, to concern both the organization and its environment, and to affect the overall welfare of the organization. Strategies as explained by Hines exist on four different levels [1].

1. *Corporate strategy*: By establishing this strategy, a firm creates the structure of its organization and its business interests, as well as the share of its portfolio in those businesses. The company's goals define how to develop such infrastructure.
2. *Business strategy*: This level of strategy determines the type of products or services that the organization wants to offer. Also, business strategy determines what each business unit needs to do and where to do it in order to reach the business objectives at the corporate level.
3. *Operational strategy*: This strategy is about how to achieve business objectives set at the corporate and business levels by extensive planning of available resources, processes, products, technologies used, and so on. This strategy also consists of setting certain market policies in order to achieve the organization's long-term competitive strategy.
4. *Competitive strategy*: These strategies tell a company how to meet its customers' desirable demands as defined by cost, quality, reliability, and so on. In a competitive, customer-driven market, an organization cannot ignore necessary competitive strategies.

Figure 3.1 shows the hierarchies of strategies. As the figure shows, developing strategies must consider the other strategies.

Logistics Operations and Management. DOI: 10.1016/B978-0-12-385202-1.00003-7

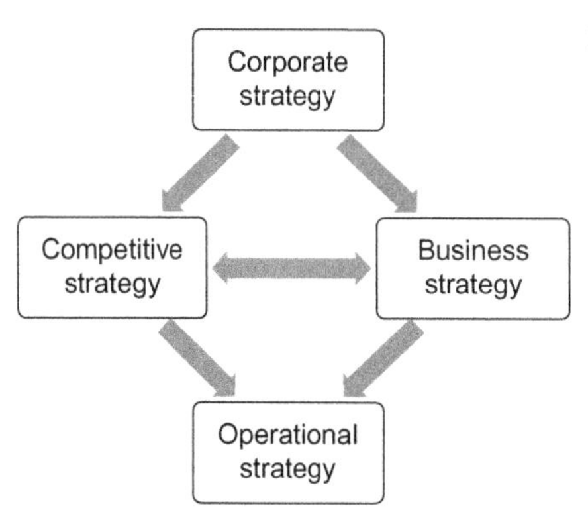

Figure 3.1 Hierarchies of strategies.

While planning the business process, corporate strategy and objectives are identified, and a specific competitive strategy is determined [2].

A strategic study includes two phases: (1) reviewing external environment and (2) analyzing internal factors. External environment is the study of economic climate, political climate and regulations, technological developments, and evaluation of primary competitors, especially service and logistics industries. It is done usually by a political, economic, sociocultural, and technological (PEST) analysis. As for analyzing internal factors, a logistics business should typically perform a strengths, weaknesses, opportunities, and threats (SWOT) analysis to review its position within the relevant market, its potential demand, and its offered services in comparison to its competitors and to determine its overall corporate or business strategy. "Logistics is a part of a firm's corporate strategy" [3].

Competitive strategy is also an important part of a company's plans. It affects the whole configuration of the company's logistics strategy: the extent of globalization, the type of adopted competitive positioning, and the degree to which the supply chain is integrated [2].

3.2 Strategic Planning

As noted in Section 3.1, strategies are used to determine how to get to chosen objectives. Therefore, strategic planning is the process of identifying those long-term goals (objectives) and the necessary steps to achieve those goals over a long-term horizon. Strategic planning incorporates the concerns and future expectations of major stockholders [4].

In the previous section, we defined the levels of strategies. Strategic planning is to set these strategies for the next 3–5 years and to create a plan to reach the objectives defined in those strategies. Different analyses such as SWOT and PEST help managers identify future paths and directions.

But what are objectives? They are planned outcomes that are as specific as possible, measurable, achievable, relevant to a company's mission, and timed [1]. Setting objectives is a systematic process that involves human judgment and decision making.

What is the difference between strategic planning and strategic management? [1] Planning is important, but if the managers want to move the organization to its desired destination, then managing the plan is equally important and required.

3.3 Logistics

We will not define logistics again in this section, but let us review some of the definitions. In its modern form, the concept of logistics dates back to the second half of the twentieth century [2].

Logistics is the entire process of planning, implementing, and controlling the efficient flow and storage of materials and products, services, information, energy, people, and other resources that move into, through, and out of a firm (in both the public and private sectors) from the point of origin to the point of consumption and with the purpose of meeting customer requirements [3].

According to Rushton et al., "Logistics is ... the positioning of resource at the right time, in the right place, at the right cost, at the right quality, while optimizing a given performance measure and satisfying a given set of constraints" [2].

Logistics profoundly affects living standards. A late food delivery to a store, an article of clothing in limited sizes and colors, and an expensive piece of furniture are tangible examples of logistics problems.

A logistics system based on its definition and nature includes the following [2]:

1. Storage, warehousing, and material handling
2. Packaging and unitization
3. Inventory
4. Transport
5. Information and control

The logistics planning department of a firm includes highly professional people. The department's management has a highly complex and challenging position in planning and controlling the system.

These definitions look at logistics as a part, section, or unit in a business, but keep in mind that there are logistics organization in most supply chains. These organizations do not include logistics as a part of their business, because logistics becomes their business, therefore all their strategies are about logistics, unlike the former where logistics strategy is a part of their competitive, business or operational strategies. You should distinguish the concept of logistics from supply chain, although many recent publications have used them as one concept. The following section tries to find the difference between them.

3.3.1 Logistics Differences to Supply Chain

We defined logistics and mentioned its importance. It is now generally agreed that for better planning and to realize the real benefits of logistics, its logic should be extended upstream to suppliers and downstream to final customers [5].

In managing a supply chain, factors such as partnership and the degree of linkage and coordination between chain entities are considered. Rushton et al. [2] mention four differences between classic logistics and supply-chain management:

1. From systematic point of view, the supply chain is viewed as a whole rather than as a series of distinguished elements such as procurement, manufacturing, and distribution. Moreover, both suppliers and end users are included in the planning process.
2. Supply-chain management is a highly strategic planning process, based on strategic decisions rather than operational ones.
3. Supply-chain management has another view of inventory. Instead of bulking large inventories in a traditional way as a safety stock for each entity in a chain, supply-chain management uses inventory as a last resort to balance the integrated flow of product through the chain.
4. In a supply chain, it is crucial to construct an integrated information system in which all entities have access to information on demand and stock levels. If a supply chain was going to be the sum of entities, not their integration, this flow of information would not have existed, while it is a necessity for the success of the chain.

Despite efforts to define the difference between the concepts of logistics and supply chain, most businesses now try to move their logistics into a supply chain—or, perhaps a better term, a *demand chain*. Therefore, most of the literature of logistics is changing.

3.4 Logistics Decisions

Logistics is a part of a firm's corporate strategy, but planning a logistics system has its own definitions, components, rules, and so on.

According to the planning horizon, logistics decisions are traditionally classified as strategic, tactical, and operational [6]. Logistics decisions are generally made hierarchically, in an iterative manner from the strategic to the tactical and the operational (Figure 3.2). But because this chapter is about logistics strategic decision making and planning, we describe these three logistics decisions in reverse order.

3.4.1 Operational Decisions

Operational decisions are made in real time on a daily or weekly basis, so their scope is narrow. Decisions such as vehicle loading or dispatching, shipment, and warehouse routines are among the many types of operational decisions. These kinds of decisions are based on lots of detailed data and usually made by supervisors.

3.4.2 Tactical Decisions

Tactical decisions are made on a longer-term basis, whether monthly, quarterly, or even annually. Production planning, transportation planning, and resource planning

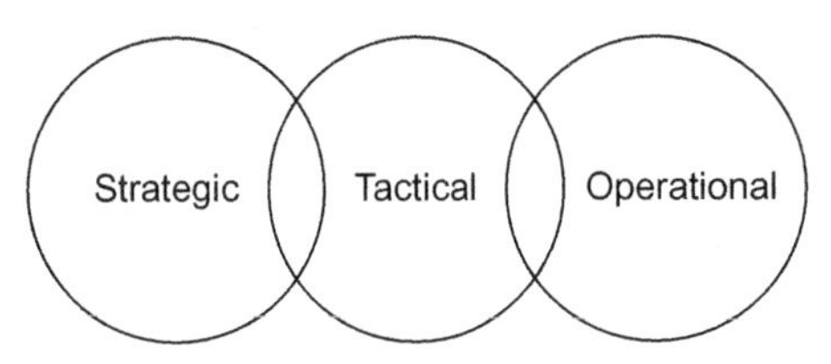

Figure 3.2 How logistics decisions are interrelated.

are the best known types of logistics tactical decisions. These decisions are often made by middle managers or logistics engineers and often with disaggregated data.

3.4.3 Strategic Decisions

As mentioned earlier in this chapter, strategic decisions are business objectives and mission statements, as well as marketing and customer-service strategies. Therefore, they are long-term kinds of decisions made over one or more years. These decisions are made by executive administrators, top managers, and stockholders. The data at hand for such decisions are often imprecise, incomplete, and need forecasts.

Strategic decisions are made to optimize three main objectives [6]:

1. Capital reduction (the level of investment, which depends on owned equipment and inventories)
2. Cost reduction (the total cost of transportation and storage)
3. Service-level improvement (customer satisfaction and order cycle time)

According to Stock and Lambert, "Strategic plans provide direction and control for tactical plans and daily operations" [4].

3.5 Logistics Planning

Logistics planning for any business is based on the three levels of decisions described in the previous section. Logistics planning starts from strategic decision making and hierarchically covers tactical and operational decisions. Remember that the scope and structure of logistics planning can change from one business to another based on its nature and size, and the strategies it uses. Factors such as the time frame, resources required, and level of managerial responsibilities affect this difference. For example, distribution planning may be a part of the strategic decision making of one firm but be a tactical plan in another firm. What is clear is that these decisions overlap and are interrelated.

An appropriate strategic logistics plan consists of an overview of logistics strategy in general terms and its relationship to the other functions, the relationship between logistics objectives and cost and services for both products and customers, a description of necessary individual unit strategies such as warehousing strategy, forecasting of all required resources such as labor, and the role of logistics strategy in corporate profits and customer-service performance [4].

As Stock and Lambert [4] note, the key inputs in developing an effective logistics plan are the following.

1. *Marketing inputs*: knowledge of products, pricing programs, sales programs and forecasts, and customer-service policies
2. *Manufacturing inputs*: manufacturing capabilities and locations
3. *Purchasing inputs*: new sources, materials, services, and technologies
4. *Financial inputs*: the source of the cost data and the availability of capital
5. *Logistics inputs*: location of current logistics facilities

Finally, as discussed in Ballou [7], strategic decisions determine what our distribution system should be, tactical decisions are how the distribution system can be utilized, and operational decisions implement action—"Let's get the goods out." In the managerial hierarchy of a logistics system, strategic decisions are made by top managers, tactical decisions are made by middle managers, and operational ones are made by supervisory personnel.

A logistics plan is about implementing the logistics strategy, so this strategy should be translatable to both tactical aims and operational actions. An organization will not achieve its goals with a badly designed strategy or an inappropriate execution plan.

3.6 Logistics Strategic Decisions

As discussed in the previous section, depending on the type of business, logistics decision groupings are diverse. However, all of these decision categorizations comprise the following three basic types of strategic decisions:

1. Customer service
2. Logistics network design
3. Outsourcing versus vertical integration

The following sections describe each of these logistics strategic decisions.

3.6.1 Customer Service

Customer service is the first and foremost class of logistics strategic decisions. As defined earlier, logistics involves delivering the right product to the right customer at the right place, at the right time, and with the right cost and quality. Therefore, customer service is the output of logistics.

Traditionally, businesses have determined their customer service based on what their customers want rather than what they really need. Identifying the customer's need is the primary step in establishing a logistics system.

Two basic factors require a trade-off: cost and level of service. It is almost impossible to provide a customer-service plan that has an optimal cost–service balance [2]. Some businesses prefer a cost-minimization strategy, and some prefer a service-maximization strategy. In the former strategy, the company delivers the same product but at a cheaper cost, whereas in the later strategy the company provides products or services that no other competitors can give. Deciding on which to choose depends on the type of business, the products it offers, and the market it competes. These two approaches are usually discussed and compared as *lean* and *agile* strategies.

Competitive strategies help an organization understand its competitors and the market in which it competes, so it can audit them and find existing gaps and opportunities available to close gaps in customer requirements.

In general, a customer-service plan is divided into three phases [2]: pretransaction, transaction, and posttransaction.

The purpose of defining customer-service strategies is to provide customers with the services they need. This is usually done by defining the *perfect order* concept. Christopher [8] has defined three elements for the perfect order: on time, in full, and error free.

Also, it is important to determine the necessary customer-service standard so that the performance can be measured and compared to the desired defined strategies.

3.6.2 Logistics Network Design

To achieve corporate strategies, an organization must ensure that its structure and flow of materials and information are appropriate. Therefore, logistics network (or, in some references, logistics channel) design is divided into two groups: the physical facility (PF) network and the communication and information (C&I) network [9]. These decisions are critical because the largest part of invested capital belongs to them. Rushton et al. [2] have also included process and organizational designs in their logistics strategy, but in this section we only take a look at physical and information network strategies.

Physical Facility Network

Keep in mind that the physical facility location of logistics is not similar to the one in supply-chain management. Physical facility location is about determining the number, size, location, and necessary equipment of new facilities together with alteration of existed ones. In many facility location decisions, allocation decisions are now also included. The usual objective of such decision problems is minimizing total system costs, but some businesses, particularly those in public sectors, consider maximizing the service level or even balancing both objectives.

Facility location decisions are clearly made at the start of a business, but Ghiani et al. [6] say that these decisions should not be made only at the start but also with a long-term view in mind; system changes should also be reconsidered as business proceeds. The reasons are obvious: When facilities are located, products are allocated to retailers but when the demand trends change, so should the system.

Location problems are classified based on the following criteria [6]:

1. Time horizon (single period vs. multiperiod)
2. Facility typology (single type vs. multitype)

3. Hierarchy (single level vs. multilevel)
4. Material flow (single commodity vs. multicommodity)
5. Interaction among facilities (with interactions vs. without interactions)
6. Dominant material flow (single echelon vs. multi-echelon)
7. Demand divisibility (divisible vs. indivisible)
8. Influence of transportation on location decisions (location problems vs. location-routing problems)
9. Retail location (with competition vs. without competition)

The complexity of physical facility location problems has attracted lots of research, and many mathematical models have been developed. Single-facility location problem is one of the most basic problems of this kind. Some other famous mathematical models are median location problems, center location problems, covering problems, and hub location problems [10].

Communication and Information Network

The importance of an integrated communication and information system to success makes planning for such system a part of strategic decisions. These decisions are about the establishment and maintenance of an effective communication system and planning for information sharing throughout the system. Centralizing or distributing the information, the technology used for such system, integration of the information flow (such as the use of enterprise resource planning, or ERP, systems), standardization of hardware, software, development environment, vendors, and the role of e-commerce are some of most important decisions made in this category [8].

Designing an integrated information and communication system means building an information-sharing system as well as programming an interorganizational collaborative planning procedure [11].

3.6.3 Outsourcing versus Vertical Integration

Decisions related to outsourcing bring greater flexibility, lower investment risk, improved cash flow, and lower potential labor costs. Instead, the business might lose control over its process, might have long lead times or shortages, or choose the wrong supplier. Outsourcing decisions determine which functions should be outsourced, as well as the nature and extent of outsourcing agreements [12].

In contrast, decisions on vertical integration (also known as *insourcing*) have higher control over inputs, higher visibility over the process, and so on. However, more integration requires higher volume and higher investment, and there is less flexibility in using equipment. Vertical integration decisions include the nature of the integration, its direction (downward toward customers or upward toward suppliers), and its extent (which activities, parts, or components should be included).

When a firm is unable to build an item (especially a routine one) or is uncertain about the volume required and suppliers offer favorable costs or have specialized research on the job, then outsourcing the job to a third party seems the best choice.

Insourcing is preferred when the firm wants to integrate plant operations, needs to have direct control over production and quality, desires some secrecy, lacks reliable suppliers, or has items or production technology that is strategic to the firm [12].

Since 1980, many firms have realized that they cannot do it all, so they have made use of outsourcing so that they can avoid parts or activities that do not have any prime competency and value for their business.

3.7 Tools of Strategic Decision Making

There are lots of tools for strategic decision making. The most applicable ones can be classified into the following categories [6]:

1. *Benchmarking*: In management science, benchmarking is about comparing the performance of a logistics system to a *best-practice* standard (e.g., a successful logistics firm). Another use of benchmarking, as mentioned previously, is auditing the performance of the competitors in the market and finding their gaps in serving customers.
2. *Optimization programming*: Like many decision-making problems, most logistics strategic problems can be cast as mathematical problems. Unfortunately, these optimization problems are among nondeterministic polynomial-time hard (NP-hard) problems, which has resulted in the development of fast heuristic algorithms. These algorithms are intended to search the solution space for good but not necessarily the best solutions. Tabu search, genetic algorithm, and simulated annealing are some of the most famous algorithms.
3. *Continuous approximation*: This method can be used whenever customers are so numerous that demand can be seen as a continuous spatial function. Approximation often yields closed-form solutions, and it can be used as a simple heuristic.
4. *Simulation*: Simulation evaluates the behavior of the system or a particular configuration under different alternative conditions. With each simulation run, these conditions are set one by one, and the results show the probable reaction of the system to these scenarios. This tool comes really handy for strategic decision making because managers can evaluate their strategies before spending capital, building facilities, and establishing their logistics system.
5. *Forecasting*: Forecasting is an attempt to determine in advance the most likely outcome of an uncertain variable. Logistics requirements to be predicted include customer demands, raw material prices, labor costs, and lead times. There are long-term, medium-term, and short-term forecasts. Medium- and long-term demand forecasts are hardly ever left to the logistician. More frequently, this forecasting task is assigned to marketing managers who try to influence the demand. However, the logistician will often produce short-term demand forecasts. Because in most cases customers are geographically dispersed, it is worth estimating not only when but also where demand volume will occur. Forecasting approaches can be classified into two main categories: *qualitative* and *quantitative* methods. Qualitative methods are most likely surveys, market research, the Delphi method, and sales force assessment. Quantitative methods are casual methods (regression, econometric models, input–output models, life-cycle analysis, computer simulation models, and neural networks) based on past and current data and time-series extrapolation (elementary techniques, moving averages, exponential smoothing techniques, decomposition

approaches, and the Box–Jenkins method). The proper choice of method is based on the type of information available.

These are not the only tools available. As discussed earlier in this chapter, tools such as SWOT and PEST analyses are two other applicable tools that help managers in their strategic decision-making process. Each category and its tools are taught as academic courses in universities. Students of industrial engineering, management engineering, systems engineering, operations research, and management programs learn to use these tools for various problems and projects. Many books also describe and comprehensively explain these tools.

3.8 Logistics Strategic Flexibility

Logistics strategic decisions are usually made over 3–5 years, but this does not mean that strategic decision making is a one-time job.

In the current uncertain and rapidly changing business climate, managers must monitor these changes and prepare their organization to respond and even take advantage of them. If managers do not act proactively, then they must constantly react to these changes and may not be able to push their organizations forward [13]. We call this proactive ability *strategic flexibility*.

Abrahamsson et al. [14] call such a logistics system a *logistics platform* to show its systematic integrity. Logistics platforms have a dynamic ability to reoptimize a system faster and more efficiently than those of competitors.

One of the most important aspects of strategic flexibility is logistics flexibility toward the market. Logistics managers should have a complete knowledge of the life cycle of a product and the role of market changes in the life cycle so that they can respond to changes and decide when to abandon or improve the product. Some organizations have the strategy of not limiting the businesses to one market, so changes in one market do not affect their whole profit package.

Strategic flexibility needs to be evaluated and measured in terms of three major dimensions: speed of change, cost of change, and amount of change.

One approach toward a flexible logistics system is to develop differently configured logistics structures. It is now commonly accepted that one particular strategy is limited and cannot cover all possible scenarios.

In conclusion, logistics platforms nowadays are not optimized for cost minimization like other types of logistics systems. Instead, they are optimized for high strategic flexibility [12].

3.9 Summary

In this chapter, we glimpsed strategy and strategic planning in a business and described different types of strategies and the place of logistics among business strategies. Then we defined three levels of logistics strategic decisions; because the concept of the chapter was the strategic types of these decisions, we introduced three primary strategic decision categories. Many references have illustrated various categorizations, but we believe that all strategic decisions can be classified into these three: customer care, logistics network design, and outsourcing versus vertical integration. Next, we introduced important tools of strategic decision making for logistics systems in brief. Finally, at the end of the chapter, we brought up logistics strategic flexibility and compared it to typical strategic goals of a logistics system.

To briefly summarize, first steps of a logistics system management, like every other type of businesses, are logistics strategic decisions. From here we find out about our customers and their needs, determine our logistics hard and soft infrastructures, and, last but not least, specify whether or not to outsource our logistics functions.

References

[1] T. Hines, Supply Chain Strategies: Customer-Driven and Customer-Focused, Elsevier, Oxford, 2006.

[2] A. Rushton, P. Croucher, P. Baker, The Handbook of Logistics and Distribution Management, Kogan Page, London, 2006.

[3] J.C. Johnson, D.F. Wood, D.L. Wardlow, P.R. Murphy, Contemporary Logistics, Prentice Hall, New Jersey, 1999.

[4] J.R. Stock, D.M. Lambert, Strategic Logistics Management, McGraw Hill, New York, 2001.

[5] D. Waters, Global Logistics and Distribution Planning—Strategies for Management, Kogan Page Limited, London, 2003.

[6] G. Ghiani, G. Laporte, R. Musmanno, Introduction to Logistics Systems Planning and Control, John Wiley & Sons Ltd., West Sussex, 2004.

[7] R.H. Ballou, Basic Business Logistics, Prentice Hall Int. Inc., NJ, 1987.

[8] M. Christopher, Logistics and Customer Value, in Book: Logistics and Supply Chain Management—Creating Value Adding Networks, Prentice Hall, London, 2005, pp. 43–80.

[9] A. Langevin, D. Riopel, Logistics Systems: Design and Optimization, Springer, New York, 2005.

[10] R.Z. Farahani, M. Hekmatfar, Facility Location: Concepts, Models, Algorithms and Case Studies, Physica-Verlag, Berlin, 2009.

[11] S.C. Kulp, E. Ofek, J. Whitaker, Supply chain coordination: how companies leverage information flows to generate value, in: T.P. Harrison, H.L. Lee, J.J. Neal (Eds.), The Practice of Supply Chain Management, Springer, New York, 2004, pp. 91–108.

[12] R. Monczka, R.J. Trent, R.B. Handfield, Purchasing and Supply Chain Management, International Thomson Publishing, OH, 1998.

[13] D.M. Lambert, J.R. Stock, L.M. Ellram, Fundamentals of Logistics Management, McGraw Hill, New York, 1998.

[14] M. Abrahamsson, N. Aldin, F. Stahre, Logistics platforms for improved strategic flexibility, Int. J. Logist. Res. Appl. 6(3) (2003) 85–106.

4 Logistics Philosophies

Zahra Rouhollahi

Department of Industrial Engineering, Amirkabir University of Technology, Tehran, Iran

4.1 Lean Logistics

Waste is the basic concept of lean philosophy. Lean philosophy uses a number of simple concepts and tools to eliminate waste in all supply-chain activities. *Lean manufacturing* was introduced and implemented in the manufacturing area in the book *The Machine That Changed the World* by Womack et al. [1] in 1990. In this book, the models the Japanese were using in their car industry was conceptualized by a group of MIT researchers. Since then, lean concepts and tools have been adjusted and used in other areas such as transportation and warehousing.

True lean achievement requires major changes and improvement in a company. In manufacturing processes, *changes* cover all activities from design to actual manufacturing. Generally, in logistics, both the inbound and outbound sides require changes [2].

Before we discuss lean philosophy, let us look at other philosophies that preceded and led to the foundation of lean philosophy.

4.1.1 Japanese Philosophy

One of the most important philosophies that was accepted for many years is the *just-in-case* philosophy. For many years, suppliers held extra inventory as safety stock in case products were needed. The Japanese believed that holding inventory kept management away from seeing manufacturing process problems such as bottlenecks and quality problems. They argued that keeping extra inventory is like the water of a deep lake. With deep water, a captain is not worried about hazardous rocks below the surface of the lake (Figure 4.1)[3]. To confront these rocks (problems) and find a solution, the first step is inventory reduction. With this belief, the Japanese developed the *kanban* concept (which is also known as the Toyota production system, or TPS) that started in assembly-type operations. The philosophy of kanban is that parts and materials should be supplied right at the moment they are needed in the production process [4].

In fact, kanban is a *pull system*, which means that parts needed for production must only be pulled through the chain in response to demand from the end customer.

Logistics Operations and Management. DOI: 10.1016/B978-0-12-385202-1.00004-9

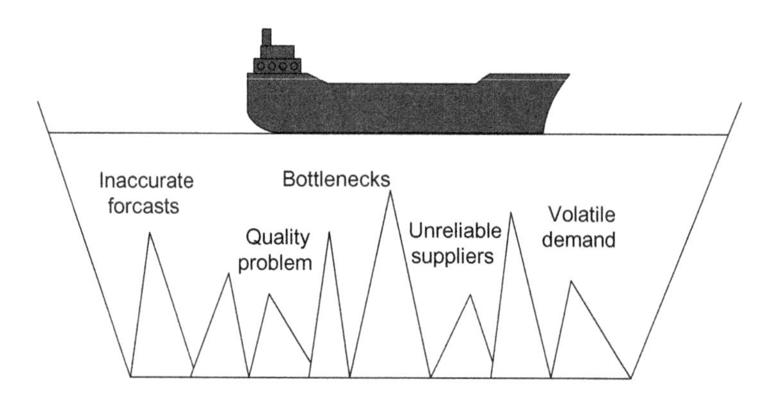

Figure 4.1 Hidden problems as a result of holding extra inventory [3].

In contrast, in the conventional *push system*, production is done according to schedules and available resources such as people, material, and machines regardless of whether or not the next step needs them at the time. In the kanban system, the aim is to meet demands right at the time they are ordered and not before. In assembly lines, stages up the chain provide needed parts. By reducing the kanban quantity (i.e., the amount demanded from each workstation) bottlenecks gradually become apparent [3]. The kanban quantity is then reduced gradually until all of the bottlenecks are revealed and removed. The aim of kanban is to reduce inventory to its minimal amount at every stage to achieve a balanced supply chain.

4.1.2 Just-in-Time Philosophy

Just-in-time (JIT) systems are the extension of kanban and link purchasing, manufacturing, and logistics [4]. The primary goals of JIT are inventory reduction, product quality, and customer-service improvement and production-efficiency maximization. In JIT systems, the main focus is on achieving continuous improvement of the process and the quality of the product and service. This is achieved by reducing inefficient and unproductive time in the production process.

JIT changes many of the principles in a firm. For example, instead of concentrating on cost, quality is considered more. 'Many suppliers' thinking is replaced by 'few suppliers with long-term open relationship' thinking. In fact, in JIT thinking, higher-quality customer service is more important than cost, which is the main issue in conventional systems.

In JIT systems, material control is based on the view that a process should operate only on a demand signal from the customer, the pull system noted earlier.

Implementing JIT has several benefits that primarily fall into four general areas [5]: inventory turnover improvement, customer-service improvement, warehouse-space reduction, and response-time improvement.

Lean Manufacturing

As a term, *lean manufacturing* was first used by Womack et al. in their book *The Machine That Changed the World* in 1990. The book is about the techniques and concepts developed by Taiichi Ohno at Toyota. The goal of lean manufacturing is reducing "waste" with the goal of total elimination. According to Russell and Taylor [6], waste is "anything other than the minimum amount of equipment, materials, parts, space, and time that are essential to add value to the product."

Lean methods target eight types of waste [7]:

1. *Defects*: money and time wasted for finding and fixing mistakes and defects
2. *Over-production*: making products faster, sooner, and more than needed
3. *Waiting*: time lost because of people, material, or machines waiting
4. *not using the talent of our people*: not using experiences and skills of those who know the processes very well
5. *Transportation*: movement of people, materials, products, and information
6. *Inventory*: raw materials, works in process (WIP), and finished goods more than the one piece required for production
7. *Motion*: Any people and machines movements that add no value to the product or service.
8. *Over-processing*: Tightening tolerances or using better materials than what are necessary.

These wastes can be seen in all logistics activities such as distribution and warehousing. In other words, waste is anything that adds no value to a product or service. Sometimes, more than 90% of a firm's overall activities are non-value added [8].

In lean manufacturing, paradigms change from conventional "batch and queue" to "one-piece flow" [9]. In fact, lean manufacturing combines best features of both mass production and craft production, which means it intends to reduce cost and improve quality while increasing diversity of production [1,10]. Lean manufacturing leads to improved product quality and production levels; reduced cycle time, WIP, inventories, and tool investment; improved on-time delivery and net income; better space, machine, and labor utilization; decreased costs; and quicker inventory investment [10].

Lean Manufacturing Tools

Various methods and tools are developed for lean implementation. We briefly describe the following five core lean methods: the kaizen rapid-improvement process, 5S, total productive maintenance (TPM), cellular manufacturing, and Six Sigma.

Kaizen means continuous improvement in Japanese. The kaizen rapid-improvement process is the foundation of lean manufacturing. It holds that by applying small but incremental changes routinely over a long period of time, a firm can realize considerable improvements. Kaizen involves workers from all levels of an organization in addressing a specific process and identifying waste in this process. After finding possible wastes, the team tries to find solutions to eliminate them and then quickly apply chosen solutions, often within 3 days. After implementing improvements, periodic events ensure that this improvement is sustained over time.

Standing for *sort*, *set in order*, *shine*, *standardize*, and *sustain*, 5S tries to reduce waste and optimize productivity by maintaining an orderly workplace [11]. *Sort* means removing every nonessential item from the workplace. *Set in order* implies that everything should have a specific place and should always be in the place unless it is being used. *Shine* means cleaning the plant and equipments. *Standardize* is the process of making people accustomed to the first three S's, and *sustain* means keeping 5S operating over time. The 5S philosophy encourages workers to maintain their workplace in good condition and ultimately leads reduced waste, downtime, and in-process inventory. The 5S implementation can also significantly reduce the space required for operations [12].

TPM reaches effective equipment operation by involving all workers in all departments. The most important concept of TPM is *autonomous maintenance*, which trains workers to be in charge of and take care of their own equipment and machines. TPM tries to eliminate breakdowns, the time spent on equipment setup or adjustment, and lost time in equipment stoppages and to minimize defects, reworks, and yield losses [10].

Cellular manufacturing is the actual practice of the pull system. The ideal cell is basically a pull system in which one piece is pulled by each machine as it needs the piece for manufacturing [11]. All of the machines needed for a process are gathered as a group into one cell. Using cellular manufacturing offers different advantages such as reduced WIP between machinery, low lead time to customers, reduced waste, and more flexibility. In a cell, when a defect occurs, only one product is defective, and it can be immediately caught. As soon as a defective part is seen, the operator starts repairing it, which leads to reduced scrap. In addition, using cells can shorten lead times. For example, if a customer's order is less than the usual company batch, the order can be delivered the moment it has been completed. In conventional manufacturing, however, the customer must wait until the company batch is completed before the order is shipped.

Developed by Motorola in the 1990s, *Six Sigma* uses statistical quality-control techniques and data-analysis methods. Six Sigma uses a set of methods that analyze processes systematically and reduce their variations, ultimately leading to continuous improvement. At the Six Sigma quality level, there will be about 3.4 defects per million, which signifies high quality and low variability of process [13].

4.1.3 Lean Principles

Five principles are basic pillars in the lean philosophy [14,15]:

1. Identifying customer value
2. Managing the value stream
3. Developing a flow production
4. Using pull techniques
5. Striving to perfection

Identifying Customer Value

Value is an important and meaningful term in the lean context, meaning something that is worth paying for in a customer's point of view [14]. Therefore, the first step in specifying this value is demonstrating a product's capabilities and its offered price.

Managing the Value Stream

Once value is identified, all required steps that create this value must be specified. Wherever possible, steps that do not add value must be eliminated.

Making Value-Adding Steps Flow

Making steps flow means specifying steps so that there is no waiting time, downtime, or other general waste within or between the steps.

Using Pull Techniques

Fulfilling customer needs means supplying a product or service only when the customer wants it. The following types of waste are eliminated by using pull techniques: Designs that are out of date before the product is completed, finished goods, inventories, and leftovers that no one wants.

Striving to Perfection

By repeatedly implementing these four steps, perfect value is ultimately created and there is no waste.

Using these principles is key in making an activity lean. These principles are also called *lean thinking*. With this thinking, any tiny change may lead to waste. A good example is changing the placement of a waste bin in a plant. After awhile workers get used to the placement of the bin and blindly throw their wastes in it without searching for it. When the bin's place is changed, however, workers have to find the bin first, which is time consuming, even if it is just a few seconds. In lean thinking, these seconds are waste. The ultimate goal is to eliminate them.

4.1.4 Lean Warehousing: Cross Docking

Warehousing has four major tasks: receiving, storage, order picking, and distribution. Among these four tasks, storage and order picking are the most costly. Storage requires inventory holding (one of the eight kinds of wastes), and order picking needs a lot of labor work hours (which is mentioned as *motion* in different kinds of wastes). To eliminate these wastes and produce a lean warehouse, the *cross-docking* concept was developed. A cross dock is just like a warehouse in which only receiving and delivering freight is being done.

Shipments will be transferred directly from incoming trucks to outgoing ones without any long-term storage. These loads are delivered to receiving doors, sorted, consolidated with other products for each destination, and loaded onto outgoing trucks at shipping doors. The whole process is done in less than 24 hours in a typical cross dock.

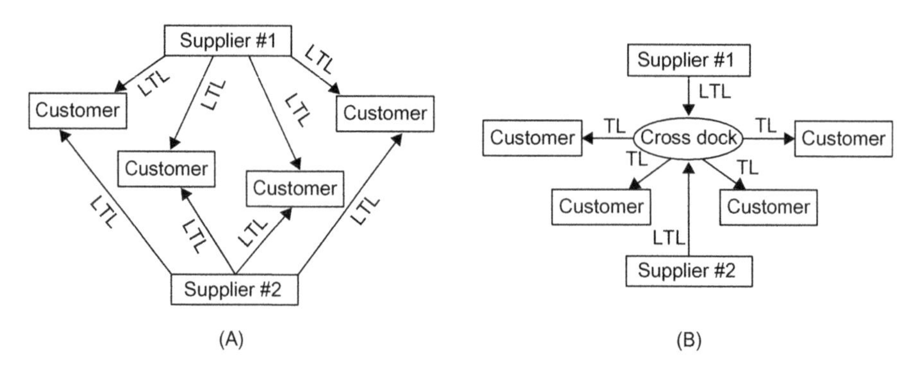

Figure 4.2 Direct shipment versus cross docking: (A) direct shipment and (B) cross dock as a consolidation center.

Companies such as American Freightways, Yellow Transport, and Old Dominion Freight Lines operate hundreds of cross docks in North America. Some retailers such as WalMart and Home Depot also operate cross docks.

Cross docks also function as a place to consolidate loads and thus reduce transportation costs. As illustrated in Figure 4.2, suppose two suppliers serve four customers. Direct-shipment suppliers may undertake extra costs as a result of a less than truckload (LTL) for each customer. Using a cross dock as a consolidation center reduces the number of LTLs and sends more truckloads (TLs) to retailers, substantially reducing transportation costs.

Types of Cross Docking

Napolitano [16] classifies the different types of cross docks as follows.

- A *manufacturing cross dock* is used to receive and consolidate supplies.
- A *distributor cross dock* consolidates products from different suppliers and delivers them to customers.
- A *transportation cross dock* consolidating LTLs from different suppliers to reach economic shipments.
- A *retail cross dock* receives, sorts, and sends products to different retail stores.

Cross docks can also be divided based on assigned or unassigned information: *predistribution* and *postdistribution*[17]. In a predistribution cross dock, the destinations are already determined and their orders are prepared by vendors for direct shipment. In this kind of cross dock, shipments that are already price tagged or labeled are transferred directly into outgoing trucks. In postdistribution operations, the cross dock must assign freights to each destination to make them ready for shipping by price tagging, labeling, and so on, which means higher labor costs and more floor space for the distributor.

Cross docks can also be classified based on the method of freight staging. According to Yang et al. [18], there are *single-stage*, *two-stage*, and *free-stage*

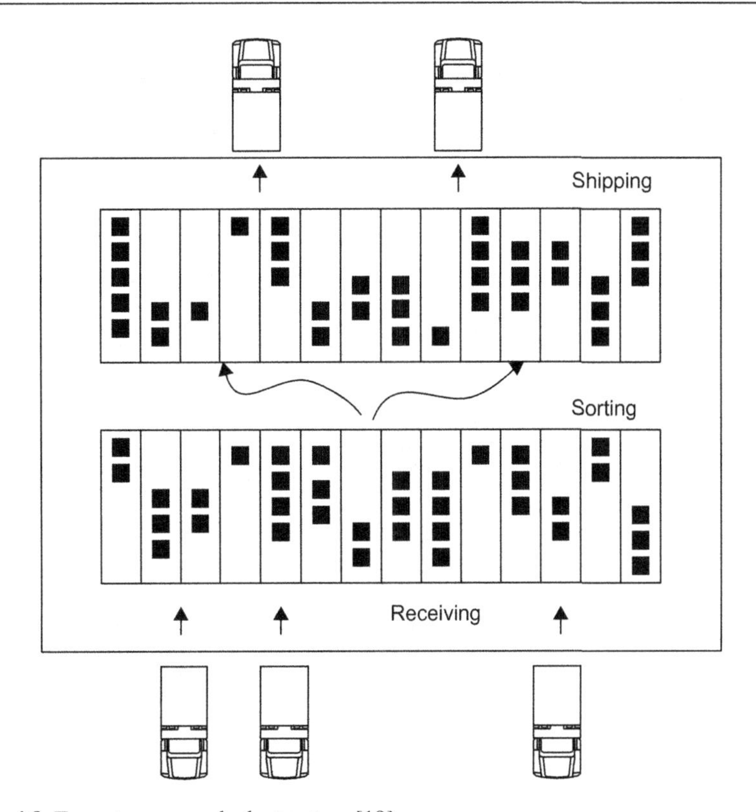

Figure 4.3 Two-stage cross-dock structure [19].

methods. In a single-stage method, one staging lane is devoted to each receiving or shipping door, and shipments are placed into these staging lanes. In a two-stage method, the shipments are unloaded in receiving doors and put directly in a receiving door's stage lines. In the center of the cross dock, shipments are resorted and repacked into the staging lines of the shipping doors. In a free-staging cross dock, there are no lanes or queues in which freights are placed and pulled to shipping doors. Instead, near the receiving or shipping doors, freights are resorted and repacked. Figure 4.3 depicts a two-stage cross dock [19].

Product Selection

Generally, products that can easily be handled with low variance and high volume are the most suitable for cross docking [17].

For supply and demand to be matched, demand for shipments must be certain or at least have low variance. If demand has high variance, then cross docking is not suitable.

As cross docking needs frequent shipments that force expenses to system, demand for shipments must be high enough to justify extra expenses.

Cross-Dock Design

There are two important parameters in cross-dock design: size and shape. The first decision is the number of doors. Generally, doors are devoted to one of the two following types of trailers [20]:

- Incoming (receiving), from which freight must be unloaded
- Outgoing (shipping), in which freight must be loaded

Typically, once the number of destinations of a cross dock is known, the number of outgoing doors needed is easily determined. The number of doors required for each destination depends on its freight flow. Destinations with a higher flow may need more than one door. The number of shipping doors thus equals the number of destinations the cross dock serves multiplied by their needed doors.

To determine the number of receiving doors needed, more issues must be addressed. In some retail cross docks, extra operations such as packaging, pricing, or labeling need the same number of doors in each side of the cross dock, with receiving doors devoted to one side and shipping doors to the other. For distributors' cross docks, which generally do no value-added processing and are just a place to consolidate freights, the number of receiving doors can be estimated by Little's law: unloading mean time multiplied by trailer throughput [21]. Unloading is relatively easier than loading. "A good rule of thumb is that it takes twice as much work to load a trailer as it does to unload it" [20]. To achieve a smooth flow without bottlenecks, there could be either twice as many outgoing doors as incoming doors or more assigned workers to load each trailer.

The next important factor is cross-dock shape. According to Bartholdi and Gue [21], most cross docks are I shaped like a long rectangle, but there are also cross docks in other shapes such as an L (as in Yellow Transportation, Chicago Ridge, IL), a U (as in Consolidated Freightways, Portland, OR), a T (American Freightways, Atlanta, GA), an H (Central Freight, Dallas, TX), and an E. Bartholdi and Gue also showed that design has an important impact on cross-dock costs. I shapes were best for cross docks with fewer than 150 doors, T shapes for cross docks with 150–250 doors, and H shapes for more than 250 doors.

The shape of a cross dock also depends on many issues such as the shape of the land on which the cross dock will be placed, the patterns of freight flows, and the material-handling systems within the facility [20].

In addition to the shape of cross dock, the location of cross dock and how it is related to its connections have a considerable effect on its success.

4.2 Agile Logistics

According to Christopher and Towill, "Agility is a business-wide capability that embraces organizational structures, information systems, logistics processes and, in particular, mindsets" [22].

The European Agile Forum (2000) defined agility as follows: "Agility is the ability of an enterprise to change and reconfigure the internal and external parts of the enterprise—strategies, organization, technologies, people, partners, suppliers, distributors, and even customers in response to change unpredictable events and uncertainty in the business environment."

Flexibility is the pivot characteristic of an agile organization [23]. In fact, agile philosophy is the extension of a flexible manufacturing system (FMS) in broader business frameworks. FMSs are systems in which scheduling, routing, controlling, and so on are mostly done by computers in order to achieve high levels of flexibility in response to market changes.

4.2.1 Agile versus Lean

Agility is different from leanness, and these two concepts should not be confused. Lean means containing little or even no fat, but agile means nimble. As mentioned before, lean focuses on eliminating waste. In fact, "it deals with doing more with less" [22], whereas agile is the ability to adapt to market changes and to keep track.

According to Naylor et al. [24], agility means "using market knowledge and a virtual corporation to exploit profitable opportunities in a volatile marketplace." They defined leanness as "developing a value stream to eliminate all kinds of wastes, including time, and to enable a level schedule."

Mason-Jones et al. [25] differentiated leanness and agility by means of two important concepts: *market qualifier* and *market winner*. To enter a competitive market, it is important and also necessary to understand what baselines—market qualifiers—this market has. The criterion for being a winner in this market is called the *market winner*. As you can see in Table 4.1, lean philosophy is most promising when the market winner is cost, but agile is appropriate when service level is the market winner.

Lean and agile have many different attributes. Agile is stronger when a market is volatile, dealing with a high variety of products with short life cycles. To apply lean, a market should be predictable and have a smaller variety of products with long life cycles. Table 4.2 compares agile and lean with their distinguishing attributes [25].

Table 4.1 Market Qualifier and Market Winner for Agile and Lean Supply [25]

	Market Qualifiers	Market Winners
Lean Supply	Quality Lead time Service level	Cost
Agile Supply	Quality Cost Lead time	Service level

Table.4.2 Comparison of Lean and Agile by Their Distinguishing Attributes [25]

Distinguishing Attributes	Lean Supply	Agile Supply
Typical products	Commodities	Fashion goods
Marketplace demand	Predictable	Volatile
Product variety	Low	High
Product life cycle	Long	Short
Customer drivers	Cost	Availability
Profit margin	Low	High
Dominant costs	Physical costs	Marketability costs
Stock-out penalties	Long-term contractual	Immediate and volatile
Purchasing policy	Buy goods	Assign capacity
Information enrichment	Highly desirable	Obligatory
Forecasting mechanism	Algorithmic	Consultative

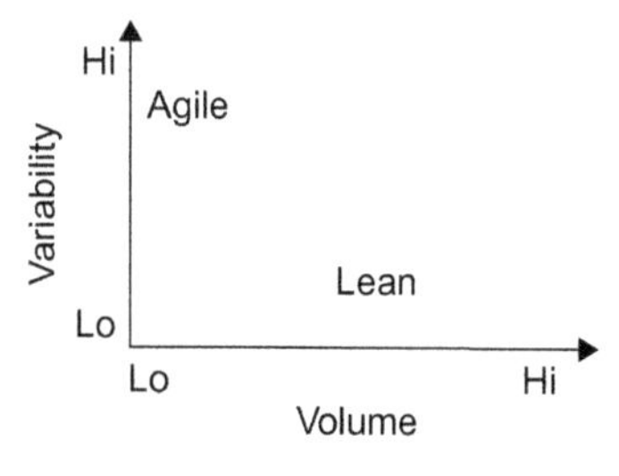

Figure 4.4 Comparison of lean and agile based on variety and volume [23].

In another comparison, Christopher [23] compares lean and agile from the points of view of volume and variety (Figure 4.4). According to him, agility works well in less-predictable environments with unstable demand and high variety. Needless to say, it is vital to be fast (or agile) in such environment in response to demand. An example of this environment is the fashion industry. The volume of fashion goods is low, but the variability is high. Lean is best in predictable environments with high volume and low diversity. Routine goods such as groceries and household stuff are good examples of merchandise when the lean philosophy is preferred.

4.2.2 Quick Response

McMichael et al. [26] define quick response (QR) as "a customer-driven business strategy of cooperative planning by supply-chain partners to ensure that the right goods are in the right place at the right time, using IT and flexible manufacturing to eliminate inefficiencies from the entire supply chain."

As a strategy in the retail sector, QR uses a number of tactics to improve inventory management and efficiency as well as speed inventory flow [4]. The fundamental idea of QR is that to exploit the advantages of markets that are time based, it is necessary to develop systems that are responsive and fast [3]. Reducing inventory levels and lead times and increasing the accuracy of forecasting are the main purposes of QR [27]. This strategy is suitable for highly engineered products and can be used by firms that produce large amounts of variable products [28].

For Christopher, "The logic behind QR is that demand must be captured in as close to real time as possible and as close to the final consumer as possible" [3], which is the most reliable information for the next logistics responses and decisions. Whole decisions are made directly based on this information.

One of the most important things that made QR possible is the development of information technology. In fact, QR keeps information instead of inventory. QR uses information technologies such as electronic data interchange (EDI), bar coding, electronic point of sale (EPOS), and laser scanners to quickly track customer sales. This information will be really helpful for manufacturers who are in charge of production scheduling and delivery. If there is a good response time by using a cross dock instead of a warehouse, then implementing QR can lead to inventory reduction.

Implementation of QR can benefit both suppliers and retailers. Giunipero et al. [29] cite cost reduction, reduced stock inventory, stock-turnover increases, and customer-service improvement as retailers benefits. Predictable production cycles, frequent orders, reduced costs, and closer relationships with retailers are supplier benefits. QR advantages are most significant in environments where a high level of service is targeted. Applying QR has a high fixed cost (because of the required information technologies), the increment of cost is low when service level improves. Figure 4.5 depicts this concept [3].

As shown, in lower levels of service, cost of implementing QR is high as a result of high-level information technology. However, by improving service level, the increment of QR cost is lower than keeping inventory. When the strategy of a market is to keep extra inventory in order to have a better service level, the amount of stored inventory must be higher, so its cost increases exponentially. After a certain level of service, QR cost is much lower than keeping inventory. It shows that, as a strategy of agile philosophy, QR concentrates on service level more than any other parameter. Whenever service level is an important key factor in a market, QR is an appropriate strategy.

Fashion and apparel industries have widely used QR. This strategy is appropriate in these fields because in these industries high service levels are demanded and inventory holding is not a suitable solution because of storage costs. Consider a hypothetical chain of fashion stores. Each of several thousand stores in the chain tracks consumer preferences daily using its point-of-sale data, which indicate

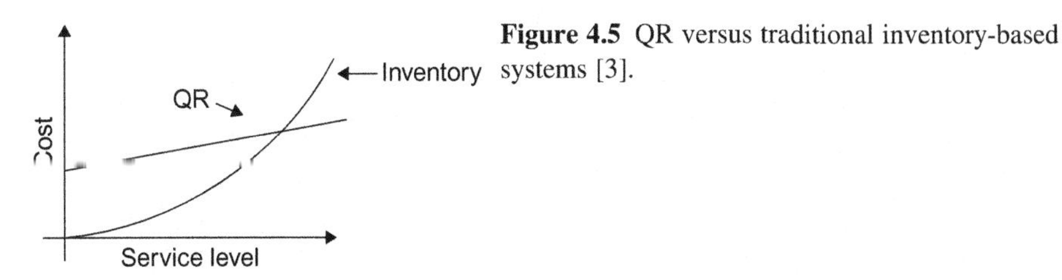

Figure 4.5 QR versus traditional inventory-based systems [3].

orders. Based on the information gathered, orders will be sent by satellite links to suppliers around the world. These orders will be consolidated in one place, and ultimately goods are flown back to the chain's distribution center. The consolidation center may even be on another continent, with shipments done by planes or ships. At the distribution center, the goods are price marked and restored for immediate delivery to the retail stores. All of this process will take place in 4–6 weeks, a time period that cannot be achieved in conventional systems.

Efficient Consumer Response

Development of the QR strategy in the fashion industry encouraged the development of the efficient consumer response (ECR) strategy, which has the same concepts as QR in the grocery industry [27]. The Europe Executive Board expresses the ECR as "working together to fulfill consumer wishes better, faster, and at less cost." ECR emphasizes cooperation between distributors and manufacturers, and its goal is to create a customer-driven system with high levels of consumer satisfaction and low levels of costs [30].

The main purpose of ECR is to provide efficient replenishment that is achieved by reducing stock holdings, making stock ranges more practical by indicating the dimensions of goods, improving space allocation, and introducing new products effectively [31].

According to Casper (1994), ECR includes the following strategies:

- EDI usage between suppliers and manufacturers, manufacturers and distributors, and distributors and customers
- Use of more accurate bar coding system and better exploitation of point-of-sale data
- Mutual close relationship between manufacturers, distributors, suppliers, and customers
- Continuous inventory replenishment
- Improved product management and development

ECR concepts and its implementation benefits have been known since the early 1990s [31,32], but it is not as widespread as expected. Although many firms in the grocery sector are using ECR-related concepts, most of them are not applying the total ECR concept, either not considering certain elements or only partly implementing them [33].

Barriers of implementing ECR are mostly organizational [30]. Haben [34] argues that these organizational obstacles could be both cultural and functional. He believes that traditional top-down organizations in which every function is operated separately, and measurement systems which assess efficiency of parts individually without attention to whole system, are the major impediments of ECR implementation and building trust between different parts.

4.2.3 Vendor-Managed Inventory

Vendor-managed inventory (VMI), also called co-managed inventory (CMI), is an agreement in which monitoring, planning, and managing inventory is done by the supplier in exchange with real-time information. In fact in VMI, retailer provides vendor with real time point of sale data and instead vendor takes the responsibility of monitoring, holding and managing inventory for retailer.

VMI was first implemented in the 1980s by WalMart and Procter & Gamble [35]; after that, many other companies from different industries used it. VMI's most important features are short replenishment lead times and frequent and punctual deliveries that optimize production and transport scheduling [36].

In traditional systems, customers place orders on their suppliers. Although this seems logical, significant inadequacies magnified the need for efficient systems.

Conventional systems are mostly based on forecasting because suppliers have no advance warning of orders; as a result, a supplier must carry unnecessary safety stocks. However, the supplier often encounters unforeseen orders, which leads to frequent changes of their production and distribution schedules [3]. Thus, customer's real-time information substitutes for orders; instead, the supplier takes the responsibility for monitoring and managing the customer's inventory.

VMI system has benefits for both supplier and customer. Benefits for customers are higher product availability and service level and diminished stock-out risk, while inventory levels and monitoring and managing costs are reduced significantly [35,37].

As suppliers have access to demand and inventory data, planning and scheduling of production, distribution, and replenishment can be done better [38], and ultimately the potential for stock outs is significantly reduced. VMI can also lead to the appropriate use of production capacity [35] and a reduction in the bullwhip effect [39,40].

References

[1] J. Womack, D.T. Jones, D. Roos, The Machine That Changed the World, Macmillan, New York, 1990.

[2] A.S. Sohal, Developing a lean production organization: an Australian case study, Int. J. Oper. Prod. Man. 16 (1996) 91–102.

[3] M. Christopher, Logistics and Supply Chain Management: Strategies for Reducing Cost and Implementing Service, second ed., Financial Times Prentice Hall, Harlow, 1998.

[4] D.M. Lambert, L.M. Ellram, J.R. Stock, Fundamentals of Logistics Management, McGraw-Hill, Boston, MA, 1998.

[5] R.C. Lieb, R.A. Miller, Why U.S. companies are embracing JIT, Traffic Manage. (1990, November) 30–39.

[6] R.S. Russell, B.W. Taylor, Operations Management, second ed., Prentice-Hall, Upper Saddle River, NJ, 1999.

[7] B.K. Brashaw, J. McCony, Lean manufacturing concepts applied to logging businesses, in Logger Conference, Minnesota Logger Education Program, 2007

[8] S. Caulkin, Waste not, want not, The Observer September 8(Sunday) (2002).

[9] Ross & Associates Environmental Consulting. Lean manufacturing and the environment. Available online at: http://www.epa.gov/lean/leanreport.pdf, 2009.

[10] S.J. Pavnaskar, J.K. Gershenson, A.B. Jambekar, Classification scheme for lean manufacturing tools, Int. J. Prod. Res. 41 (2003) 3075–3090.
[11] B. Carreira, Lean Manufacturing That Works: Powerful Tools for Dramatically Reducing Waste and Maximizing Profits, AMACOM, USA, 2004.
[12] US Environmental Protection Agency Official Website. Lean report. Available online at: http://www.epa.gov/lean/thinking/tpm.htm, 2009.
[13] M.L. George, D. Rowlands, M. Price, J. Maxey, The Lean Six Sigma Pocket Tool Book: A Quick Reference Guide to 100 Tools for Improving Quality and Speed, McGraw-Hill, New York, 2005.
[14] J. Womack, D. Jones, Lean Thinking: Banish Waste and Create Wealth in Your Corporation, Simon & Schuster, UK, 1996.
[15] B. Kollberg, J.J. Dahlgaard, P. Brehmer, Measuring lean initiatives in health care services: issues and findings, Int. J. Prod. Perform. Manag. 56 (2007) 7–24.
[16] M. Napolitano, Making the move to Cross Docking, Warehousing Education and Research Council, Illinois, US, 2000.
[17] K.R. Gue, Cross Docking: Just-In-Time for Distribution, Teaching Notes-Naval Postgraduate School, Monterey, CA, 2001.
[18] K.K. Yang, J. Balakrishnan, C.H. Cheng, An analysis of factors affecting cross docking operations, J. Bus. Logist. Available online at: http://findarticles.com/p/articles/mi_qa3705/is_201001/ai_n53508826/, 2010.
[19] J.J. Bartholdi, K.R. Gue, K. Kang, Staging freight in a cross dock, in: Proc. International Conference on Industrial Engineering and Production Management, May, Quebec City, Canada, 2001.
[20] J.J. Bartholdi, S.T. Hackman, Warehouse and Distribution Science, Available online at: http://www2.isye.gatech.edu/~jjb/wh/book/xdock/xdoc.html, 2006.
[21] J.J. Bartholdi, K.R. Gue, The best shape for a cross dock, Transport. Sci. 38 (2004) 235–244.
[22] M. Christopher, D.R. Towill, Supply chain migration from lean and functional to agile and customized, Supply Chain Manag. 5 (2000) 206–213.
[23] M. Christopher, The agile supply chain competing in volatile markets, Ind. Market. Manag. 29 (2000) 37–44.
[24] J.B. Naylor, M.M. Naim, D. Berry, Leagility: interfacing the lean and agile manufacturing paradigm in the total supply chain, Int. J. Prod. Econ. 62 (1997) 107–118.
[25] R. Mason-Jones, J.B. Naylor, D.R. Towill, Engineering the leagile supply chain, Int. J. Agile Manag. Syst. 2 (2000) 54–61.
[26] H. McMichael, D. Mackay, G. Altman, Quick response in the Australian TCF industry: a case study of supplier response, Int. J. Phys. Dist. Log. Manag. 30 (2000) 611–626.
[27] G. Birtwistle, C.M. Moore, S.S. Fiorito, Apparel quick response systems: the manufacturer perspective, Int. J. Log. Res. Appl. 9 (2006) 157–168.
[28] R. Suri, Quick response manufacturing: a competitive strategy for the 21st century, in: POLCA Implementation Workshop, 2002.
[29] L.C. Giunipero, S.S. Fiorito, D.H. Pearcy, D. Dandeo, The impact of vendor incentives on quick response, Int. Rev. Retail Distrib. Consum. Res. 11 (2001) 359–376.
[30] Kurt Salmon Associates, Inc., Efficient Consumer Response: Enhancing Consumer Value in the Grocery Industry, Food Marketing Institute, Washington, DC, 1993.
[31] J. Fernie, Quick response: an international perspective, Int. J. Retail Distrib. Manag. 24 (1994) 1–12.

[32] Kurt Salmon Associates, Efficient Consumer Response: Enhancing Consumer Value in the Grocery Industry, The research department of Food Marketing Institute, Washington, 1993.

[33] J. Aastrup, H. Koztab, D.B. Grant, C. Teller, M.A Bjerre, Model for structuring efficient consumer response measures, Int. J. Retail. Distrib. Manag. 36 (2008) 590–606.

[34] T.J. Haben, Food industry innovation: efficient consumer response, Agribusiness 14 (1998) 235–245.

[35] M.A. Waller, M.E. Johnson, T. Davis, Vendor-managed inventory in the retail supply chain, J. Bus. Logist. 20(1) (1999) 183–203.

[36] A. De Toni, E. Zamolo, From a traditional replenishment system to vendor-managed inventory: a case study from the household electrical appliances sector, Int. J. Prod. Econ. 96 (2004) 63–79.

[37] D.D. Achabal, S.H. McIntyre, S.A. Smith, K. Kalyanam, A decision support system for vendor managed inventory, J. Retail. 76 (2000) 430–454.

[38] S. Cetinkaya, C.Y. Lee, Stock replenishment and shipment for vendor managed inventory systems, Manag. Sci. 46 (2000) 217–232.

[39] H. Lee, V. Padmanabhan, S. Whang, The bullwhip effect in supply chains, Sloan Manag. Rev. 38 (1997) 93–102.

[40] S.M. Disney, D.R. Towill, Vendor-managed inventory (VMI) and bullwhip reduction in a two level supply chain, Int. J. Oper. Prod. Man. 23 (2003) 625–651.

5 Logistics Parties

Seyed-Alireza Seyed-Alagheband

Department of Industrial Engineering, Amirkabir University of Technology, Tehran, Iran

Logistics outsourcing has attracted the attention of lots of industrialists in recent years. As a result, having long-term relationships with logistics parties seems to find its undeniable place in today's growing extent of outsourcing affairs. Third-party logistics (3PL), in particular, has received substantial attention from logistics experts, leading to a great deal of research in this area. Furthermore, improved versions of logistics parties, especially fourth parties, are growing with high speed. Because of its importance, this chapter is dedicated to the introduction and general implications of logistics parties.

Logistics is basically the concept of how to deal with the movement and storage of materials or products that results in the highest consumer satisfaction [1]. The modern form of logistics concept dates back to the second half of the twentieth century. During the past several years, this field has obtained greater importance and has been theoretically and practically extended.

The logistics evolution requires that decision makers have a comprehensive and updated vision on the concept. The decision environment has become extensively complex with factors such as new management strategies and business models, global markets and sourcing, information technology (IT), new trends of customer satisfaction, and new transport-service options.

In most developed economies, the costs of logistics management are steadily growing, indicating an increasing proportion of the gross national product. Logistics costs have become an important part of the value of products, and logistics management is regarded as an important role in the international competitive market [2].

Logistics outsourcing is one of the issues that a firm has to consider about the efficiency and benefit outsourcing brings to the company. The decision to outsource logistics activities brings about the use of other companies to handle logistics affairs such as transportation and warehousing.

Logistics outsourcing is not a new trend. In the 1950s and 1960s, transportation and warehousing were commonly outsourced. This outsourcing was a pure commodity purchase, and logistics as an activity was rarely a part of a company's business strategy. By the 1970s, as companies began to emphasize cost reduction and improved productivity, they started to look for multicompetency providers for outsourcing. The long-term relationship became more common, and service providers

Logistics Operations and Management. DOI: 10.1016/B978-0-12-385202-1.00005-0

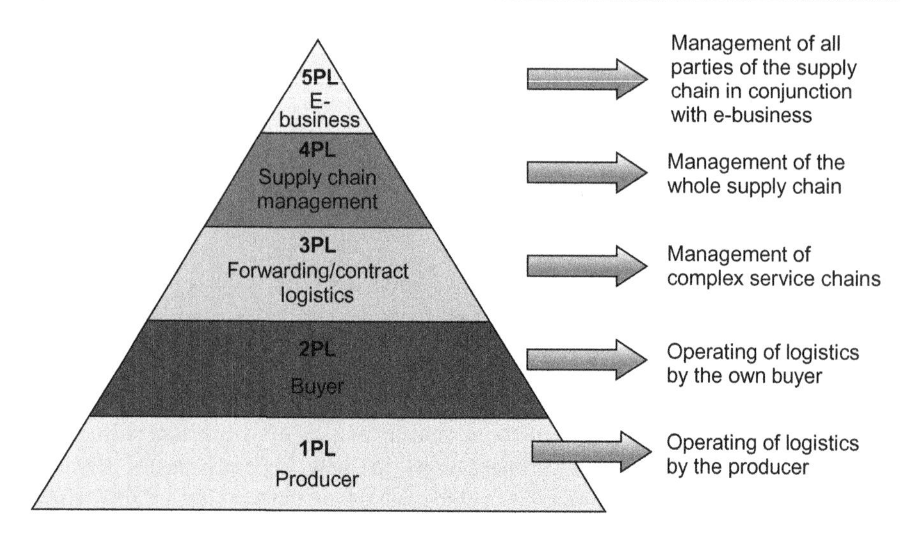

Figure 5.1 Types of logistics parties.

began to set up dedicated facilities for some of their clients. Those service providers are called *3PLs*. In the early 1980s, companies began to emphasize supply-chain optimization, but it was mostly restricted to isolated operations within their own organizations. Businesses focused on coordinating the movement of products within their facilities, integrating their financial system, ordering systems, and developing in-house inventory management. As a result, the range of services offered by logistics service providers (LSPs) also increased [3].

In the 1990s, by the advent of Internet and the emergence of global sourcing, logistics industry introduced a new generation of LSPs called *fourth-* and *fifth-party logistics* (4PLs and 5PLs). Today, virtual logistics departments, called *zero-party logistics*, are able to handle most or all of the logistics activities using an integrated information chain between buyers and carriers. Figure 5.1 demonstrates types of logistics parties.

Looking at the growth trend of logistics parties demonstrates the increasing attention and investments of firms on logistics outsourcing. Therefore, this chapter supports the decision to choose logistics activities to outsource, select suitable parties for partnership, and cooperate with logistics parties.

This chapter is organized as follows. Section 5.1 provides an overview and applications on the concept of 3PLs. In Section 5.2, new generations of logistics parties, including 4PL and 5PL, are investigated. Section 5.3 provides an in-depth study on the concept of 3PLs. Section 5.4 offers concluding remarks.

5.1 Third-Party Logistics: An Overview

5.1.1 Why 3PLs?

3PLs emerged in the early 1990s when LSPs started offering consolidated services and an increasing number of customers entered into longer business contracts with the LSPs. The total logistics market in 2003 in the United States was $910 billion, and the 3PL market was around $65 billion. The 3PL market has been increasing for the last 10 years at a rate of more than 20%. A survey of 221 US companies reported that 78% of them are using 3PLs for logistics services and spending 49% of their logistics expenditure on outsourcing [3].

The 3PL business is developing as a result of the increasing demand of advanced logistics services, including globalization, lead-time reductions, customer orientation, and outsourcing. Therefore, the role of logistics providers is changing both in content and in complexity and brings about the development of new logistics providers who offer various services to their customers [4].

The growth rate of 3PL industry and the increasing interest in outsourcing logistics activities illustrates the growing importance of the 3PL role in the industry, thus making it an interesting research field for further investigation and development.

5.1.2 Definition

Terms such as *logistics outsourcing*, *logistics alliances*, *third-party logistics*, *contract logistics*, and *contract distribution* have been used to describe the organizational practice of contracting out part of or all logistics activities that were previously performed in-house [5]. Unfortunately, no single consistent definition for the 3PL concept can be found in the 3PL literature. In some cases, 3PL is used as a label for traditional outsourcing of transportation or warehousing; in other cases, the term is used to account for the outsourcing of a wider logistics process.

In 1992, Lieb defined 3PL in the following sentences: "Third-party logistics are external companies which perform logistics functions that have traditionally been performed within an organization. The functions performed by the third party can encompass the entire logistics process or selected activities within that process" [6]. This definition suggests that all logistics activities that were previously performed in a firm can be performed in an external organization. In such broad definitions, however, some terms and conditions such as the type of logistics activities are not mentioned.

Narrower definitions express functional or interorganizational features of the logistics outsourcing. For example, in 1999, Berglund et al. [7] suggested the following definition:

> Third-party logistics are activities carried out by a logistics service provider performing at least management and execution of transportation and warehousing. In addition, other activities like inventory management, information related activities, value added activities, or supply chain management can be managed by the logistics service provider. The contract is also required to include management, analytical or design details, and the length of the cooperation should be at least one year.

Such narrower definitions appear to draw a line between 3PL and the traditional outsourcing of logistics functions, providing more distinctive features for 3PLs including the provision of a broader range of services, a long-term duration, the customization of the logistics solution, and a fair sharing of benefits and risks [8].

5.1.3 Emergence of 3PLs

According to a survey conducted by Berglund et al. in 1999 [7], there are three waves of entrants into the 3PL market. The first wave dates back to the 1980s or even earlier with the emergence of what are called *traditional LSPs*. Companies such as ASG in Sweden and Frans Maas in the Netherlands are examples of traditional 3PLs. The 3PL activities of these companies have emerged from a traditionally strong position in either transportation or warehousing.

The second wave dates from the early 1990s when a number of network players, mainly parcel and express companies such as DHL, TNT, and UPS, started their 3PL activities. Usually, the 3PL activities of these companies are based on their worldwide air-express networks and their experience with expedited freight.

The third wave dates from the late 1990s. Currently, a number of players entering the 3PL market can be seen from areas such as IT, management consultancy, and financial services. These players are working together with players from the first and second waves. In most cases, several shippers and providers are involved.

Whereas the first and second wave entrants base their strength on traditional logistics activities such as transportation, warehousing, or running a scheduled network, these new players build on very different skills such as IT, consultancy, or financial expertise. In other words, there is a gradual shift from asset-based players to skill- or systems-based players [7].

5.1.4 Activities of 3PLs

3PLs services can be relatively limited or comprise a fully integrated set of logistics activities. A 3PL can perform the following activities: transportation, warehousing, freight consolidation and distribution, product marking, labeling and packaging, inventory management, traffic management and fleet operations, freight payments and auditing, cross docking, product returns, order management, packaging, reverse logistics, carrier selection, rate negotiation, and logistics information systems [9].

Detailed 3PL activities, including global functions, include the following [10]:

- Planning functions
  - Location selection
  - Supplier selection
  - Supplier contracting
  - Scheduling
- Equipment functions
  - Selection
  - Allocation
  - Sequencing
  - Positioning
  - Inventory control
  - Ordering
  - Repair

- Terminal functions
  - Gate checks
  - Location control
- Handling functions
  - Pickup
  - Consolidation
  - Distribution
  - Expediting
  - Diversion
  - Transloading
- Administrative functions
  - Order management
  - Document preparation
  - Customs clearance
  - Invoicing
  - Inventory management
  - Performance evaluation
  - Information services
  - Communications
- Warehousing functions
  - Receiving
  - Inventory control
  - Reshipment
- Pre/Post production
  - Sequencing
  - Assorting
  - Packaging
  - Postponement
  - Marking
- Transportation functions
  - Modal coordination
  - Line-haul services
  - Tracking and tracing

According to a survey conducted by Aghazadeh in 2003, users in 2000 relied most heavily on third parties for warehousing management (56%), transportation services (49%), and shipment consolidation (43%). The use of traditional logistics services offered by 3PLs has remained relatively stable in recent years; however, there is an emerging interest from manufacturers for nontraditional applications of 3PLs [1].

5.1.5 Advantages and Disadvantages of 3PL

Firms interested in outsourcing their logistics activities should be familiar with the pros and cons of establishing relationships with 3PLs.

The 3PL can effectuate a great degree of efficiency by exploiting economies of scale among others. Capacity can be better utilized because the peaks and drops in transport quantities offered by different clients can be counterbalanced and because backhauls are often available. A 3PL can invest in specific know-how because logistics management is its core activity. The 3PL can improve the quality and flexibility of service and thus improve customer service [2].

By outsourcing logistics activities, firms can reduce the amount of capital investments. Some firms spend much on physical distribution centers or information networks, which also involves financial risks. 3PL providers can spread such risks by outsourcing to subcontractors.

Other advantages of partnership with 3PLs include expanded ranges of service, bargaining power, faster learning, networking with other providers, knowledge of various kinds, faster implementation of new systems, improved customer satisfaction, restructuring of supply chains, reduced investment base, and smoother production [4].

Despite the numerous advantages of using 3PL, there are some disadvantages as well. For instance, it is not easy to establish a reliable and cost-effective partnership between the shipper and the 3PL provider. To establish reliable partnership, efforts should be made in two directions: 3PL provider selection and contract signing [11].

If a manufacturer contracts out logistics activities, it runs the risk of losing expertise in logistics. In addition, manufacturers contracting out their logistics activities are often worried about the protection of company information because they have to share confidential data [2].

In a comprehensive manner, using 3PL services will lead an organization to the following:

- Save time
- Share responsibility
- Reengineer distribution networks
- Focus on core competencies
- Exploit external logistical expertise
- Reduction in inventory levels, order cycle times, and lead times
- Economies of scale and scope
- Improved efficiency, service, and flexibility

The following list, however, shows some possible disadvantages of cooperation with 3PLs:

- Poor searching efforts
- Poor coordination efforts
- Poor information sharing
- Loss of control
- Poor service performance
- Inadequate provider expertise
- Inadequate employee quality
- Loss of customer feedback

5.1.6 Types of 3PLs

Firms have to investigate and choose the compatible and appropriate types of 3PLs before they start a long-term relationship with them. Here a classification for 3PLs is provided.

In terms of customer adaptation, 3PLs are classified as follows [4]:

- *Standard 3PL providers* are the most basic form of 3PL provider. They perform the most basic functions of logistics such as picking and packing, warehousing, and distribution.
- *Service developers* offer advanced value-added services to their customers such as tracking and tracing, cross docking, specific packaging, and providing a unique security system.
- *Customer adapters* provide services at the request of the customer and take thorough control of the company's logistics activities. The 3PL providers improve logistics services, but do not develop a new service. The customer base for this type of 3PL provider is typically quite small.
- *Customer developers* are the highest level of 3PL provider, integrating themselves with customers and taking over entire logistics function. These providers will have few customers, but they will perform extensive and detailed tasks for them.

Third-party logistics providers can also be classified to asset-based and non-asset–based 3PLs. Asset-based 3PLs own some assets, especially transport-related assets such as trucks and warehouses; however, non-asset–based 3PLs do not own assets, and they usually work with subcontractors. This type of 3PL may possess only desks, computers, and freight industry expertise [11].

5.1.7 2009 3PLs: Results and Findings of the Fourteenth Annual Study

About the Study

The 3PL study has annually documented the growth and evolution of the 3PL industry. The 2009 3PL study includes three directions of research: a web-based survey, workshops with shippers leveraging survey content and the Capgemini Accelerated Solutions Environment (CASE), and interviews with industry executives. Respondents represent a broad range of industries and are predominantly from North America, Europe, Pacific Asia, and Latin America, in addition to other locations throughout the world such as South Africa and the Middle East.

Key Findings

According to the responses from 3PLs and shippers, key factors responsible for the success of relationships include openness, transparency, good communication, the ability to create personal relationships on an operational level, the flexibility of 3PLs to accommodate customers' needs, and the ability to achieve cost and service objectives.

The major findings of 2009's 3PL study are presented in five tracks: state of the 3PL market, economic volatility, IT capability gap, supply-chain orchestration, and strategic assessment.

Current state of the 3PL market: Relatively small percentages of shipper respondents (30%) and 3PL respondents (25%) think of the willingness of 3PLs and shippers to share risk as a success factor. However, several of the leaders interviewed and industry executives participating in the workshops and ASE sessions view the willingness to share risk as an important attribute of a successful relationship. In addition, shippers predict that the percentage of logistics budgets they devote to outsourcing will increase in the future.

Economic volatility: Unpredictable demand is the most difficult challenge to managing and operating a supply chain in an economic downturn, according to 71% of the survey respondents. For example, many consumers turn to private label goods as their confidence declines, but that trend typically reverses itself if analysts report good numbers.

IT capability gap: 3PL respondents report they are increasingly offering their IT platforms on a subscription basis as a part of their service contracts. IT services most likely to be sold this way are transportation management, warehouse and distribution center management, and visibility and customer-order management. An average of 10% of the 3PLs responding indicate that they offer all of these services today on a subscription basis.

Supply-chain orchestration: Nearly 60% of shipper respondents feel this is the time to reevaluate their relationships with their 3PLs and possibly drive these relationships deeper. A significant 19% are unsure, perhaps indicating they are somewhat confused by what the current environment means for their businesses and 3PL relationships.

Strategic assessment: Newer concepts and technologies are emerging to help both 3PLs and shippers cope with this new slower-growth world. One of them is to create horizontal, cross-company supply chains refereed by neutral third parties. This innovation is based on the concept that by clustering specific logistics activities and consolidating supply chains, significant economies of scale can be achieved in terms of efficiency (logistics cost), effectiveness (customer service), and environmental sustainability [12].

5.2 New Generations of Logistics Parties

Through the usage of 3PL providers, a firm can outsource some of its so-called traditional logistics activities to other organizations. But to ensure that the entire logistics activities are being handled in a suitable manner, a more coherent system has to be implemented. To reach this goal, it is necessary to use new versions of logistics parties.

This section provides information on new generations of logistics parties and their differences with 3PLs. In Section 5.2.1, 4PLs is explained in details. Newer versions of logistics parties and their features, including 5PLs, are presented in Section 5.2.2. Successful examples of businesses working as logistics parties are also introduced in Section 5.2.3.

5.2.1 Fourth-Party Logistics

Definition

The terms *fourth-party logistics* (4PLs) and *lead logistics provider* (LLP) were introduced in 1996 by Bob Evans of Arthur Anderson (now Accenture) and are defined as follows: "A 4PL is an integrator that assembles the resources, capabilities, and

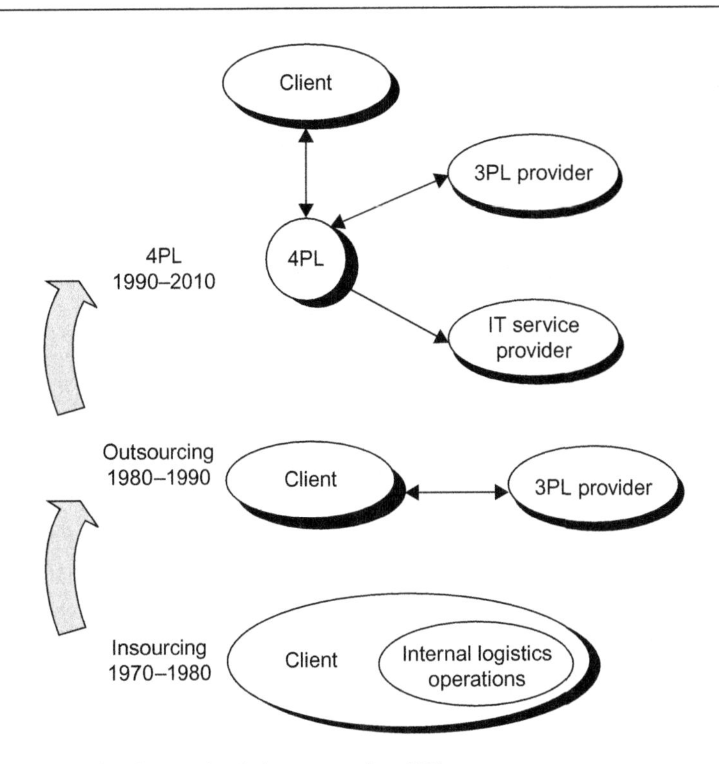

Figure 5.2 Evolution in supply-chain outsourcing [13].

technology of its own organization and other organizations to design, build, and run comprehensive supply chain solutions" [3].

Fourth-party logistics is one of the significant evolutions in supply-chain management. The convergence of technology and the rapid acceleration of e-capabilities have heightened the need for an overarching integrator for supply-chain spanning activities. 4PL is the shared sourcing of supply-chain spanning activity with a client and selected teaming partner under the direction of a 4PL integrator [13].

4PL is advertised as a refinement on the concept of 3PL, a firm that provides outsourced or 3PL services to companies for part or sometimes all of their supply-chain management functions. A 4PL uses a 3PL to supply services to customers, owning only computer systems and intellectual capital. The 3PL and 4PL relation can best be described by Figure 5.2.

What Is the Difference?

It has been argued that a 4PL is nothing but a non-asset–based 3PL. In fact, there is confusion in theory and practice about the use of the term 4PL [3]. It has been argued that any LSP that offers multiple services can be categorized under the term *3PL* and a new term is not needed for this type of LSPs.

To draw a line between 3PL and 4PL, the differences have to be clarified. According to the Council of Logistics Management, a 4PL differs from a 3PL in the following ways [3]:

1. A 4PL organization is often a separate entity established as a joint venture or long-term contract between a primary client and multiple LSPs.
2. A 4PL organization acts as a single interface between the client and multiple LSPs.
3. All aspects (ideally) of the client's supply chain are managed by the 4PL organization.
4. It is possible for a major 3PL provider to form a 4PL organization within its existing structure.

Two Key Distinctions of 4PL

Two key distinctions make the concept of 4PL apart from other 3PL outsourcing options:

1. A 4PL presents a comprehensive supply-chain solution.
2. A 4PL delivers value through the ability to have an impact on the entire supply chain.

In the former case, a 4PL has to deliver comprehensive supply-chain solutions focusing on all elements of supply-chain management in order to respond effectively to the numerous and complex needs of today's organizations. In the later case, a 4PL can impact and integrate the entire supply chain through the key drivers of shareholder value: increased revenue, operating-cost reduction, working-capital reduction, and fixed-capital reduction [13].

Operating Models of 4PL

Although 4PL solutions are usually customized for the needs of a particular client, the following models show how a 4PL generally works [13].

- *4PL–3PL partnership*: In this model, a 4PL and 3PL work together to market supply-chain solutions that fulfill the needs of both organizations.
- *Solution integrator*: In this model, the 4PL operates and manages a comprehensive supply-chain solution for a single client. This type of 4PL provider uses its resources, capabilities, and technology and also the services of other LSPs to provide a comprehensive supply-chain solution for a single organization.
- *Industry innovator*: In this model, a 4PL organization develops and runs a supply-chain solution for multiple industry players with special focus on synchronization and collaboration.

5.2.2 Fifth-Party Logistics

5PL is a new concept in outsourcing. It is the management of all parties of the supply chain in conjunction with e-business. 5PLs uses an e-logistics network focusing on global operations [14].

5PL is the discipline that bridges the gaps left between 3PL providers and 4PL providers. Where 4PLs attempt to provide supply-chain solutions with their own optimization software and the capabilities of 3PL resources, 5PLs use the buyer's (first party's) existing technology and infrastructure to optimize the supply chain by transforming into a virtual organization [15].

5.2.3 Future Trends

Since the enhancement of the new generations of logistics parties, organizations are looking forward to shape their logistics departments into a virtual format. Therefore, in a more developed way, they would use the disciplines of zero-party logistics.

Zero-party logistics is the elimination of 3PL and 4PL organizations using 5PL disciplines to transform a company into a virtual organization. In other words, the traditional logistics departments become nothing more than an integrated information chain between buyers and carriers; all planning is performed using either a company's own resources or 5PL services [15].

More recently, the concept of seventh-party logistics (7PL) has emerged, resulting from hybridizing the 3PL domain with the concept of 4PL. In fact, it is the effective fusion of physical and process expertise of 3PLs with the enhanced knowledge-based macrostrategic consulting and IT capabilities of 4PLs [16].

Although newer versions of logistics parties are not yet deeply investigated and applied, they are expected to grow and be familiarized in the near future. Of course, it requires that organizations, which are interested in outsourcing their logistics activities, get more familiar with the concept of logistics parties, especially third and fourth parties, so that they establish a well-organized framework for application of more recent and complex types of LSPs.

5.2.4 Logistics Vendors

In this section, some of the successful LSPs are introduced. These vendors are shown to have considerable global market share and high annual revenue. The selected vendors for this detailed study are United Parcel Service, Penske Logistics, and Deloitte. The selection was based on the extent of services provided by each vendor.

United Parcel Service

United Parcel Service (UPS) was found in 1907 in Seattle, WA, USA. It is now one of the world's largest package-delivery companies, delivering more than 15 million packages a day to 6.1 million customers in more than 200 countries and territories around the world. Growing into a $49.7-billion corporation, UPS employs approximately 425,300 staff, with 358,400 in the United States and 67,300 in other parts of the world.

UPS's network is based on a hub-and-spoke model. UPS uses centers that are the point of entry for parcels and send the parcels to one or more to hubs where parcels are sorted and forwarded to their destinations.

Major domestic competitors include the US Postal Service (USPS) and Federal Express (FedEx). In addition, UPS competes with a variety of international operators, including DHL, TNT NV, Royal Mail, Japan Post, and many other regional carriers, national postal services, and air-cargo handlers.

UPS has expanded its operations to include logistics and other transportation-related activities. UPS's key supply-chain solution includes logistics and distribution, transportation and freight, freight forwarding, consulting services, and industry solutions [17].

Penske Logistics

Penske Logistics is a wholly owned subsidiary of Penske Truck Leasing, which was founded in 1969. In 1988, General Electric became a limited partner in Penske Truck Leasing with Penske Corporation. Penske Logistics became a division of Penske Truck Leasing in 1995 with the acquisition of Lease Way Logistics.

As a 3PL, Penske Logistics provides the following solutions:

- Supply-chain management
- Transportation
- Distribution center management
- International transportation management

Case Study: Ford Motor Company

Ford Motor Company has worked with Penske on several Six Sigma initiatives. Penske's team of associates worked closely with Ford to organize operations and create a more centralized logistics network. In addition, Penske implemented accountability procedures and advanced logistics management technologies to gain more visibility of the overall supply network [18].

Deloitte

Deloitte Touche Tohmatsu (also branded as Deloitte) is one of the largest professional service organizations in the world. According to the organization's 2008 information, Deloitte has approximately 165,000 staff at work in 140 countries, delivering several services through its member firms.

Deloitte member firms offer services in the following functions:

- Audit and enterprise risk services
- Consulting in the areas of enterprise applications, technology integration, strategy and operations, human capital, and short-term outsourcing
- Financial advisory in the areas of dispute, personal and commercial bankruptcy, forensics, and valuation
- Tax and other services

As a 4PL organization, Deloitte serves various clients in financial services, consumer and industrial products, energy and resources, health care and life sciences, public sector, technology, telecommunications, and other industries [19].

5.3 3PLs: Theories and Conceptualizations

This section provides an in-depth investigation on the main issues of 3PLs. Some of the issues are common for other parties as well. Most of the research was descriptive in nature, simply describing trends in the industry.

5.3.1 Outsourcing Decision

The decision to outsource (or not) logistics activities depends on a multitude of variables [5]. When making an outsourcing decision, the following four categories of considerations can be distinguished [2].

1. *Economic considerations.* When a company keeps its logistics activities in-house, investments have to be made. If outsourcing logistics activities yield a higher return on investment than the investment mentioned earlier and if finance means are scarce, it can be advisable to contract out logistics management activities.
2. *Market issues*
   - *Demand fluctuations.* Most of the time, decreasing demand for one product cannot be compensated by rising demand for another product; the result is instabilities in logistics activities (i.e., capacity utilization of transport). If the peaks cannot be dealt with deploying flexible manpower, then it may be advisable to contract out logistics activities. A service provider usually serves several clients, enabling the counterbalance of a drop in one client's business with a peak in another's, particularly if the clients come from different trades.
   - *Commerce and flexibility.* In many cases, companies keep their logistics management in-house in order to maintain direct customer contact and to be able to respond to changing customer desires in a flexible manner, whereas a service provider, for reasons of efficiency, wishes to minimize deviations from schedule. Therefore, before a do-or-buy decision, the company has to select a priority between flexibility and efficiency.
3. *Availability of personnel and equipment.* A company carrying out logistics activities in-house bears the responsibility for personnel matters such as recruitment, selection, and training. Furthermore, sufficient equipment must be available to make any necessary repairs. Outsourcing logistics management can be quite a relief for a company and allows one to cut overhead, however, at the cost of loss of control on personnel and equipment.
4. *Supplier dependence.* If a company carries out logistics activities in-house, it can take action rapidly in cases of wrong deliveries and damage. If logistics has been contracted out, rapid reactions could be obstructed by the necessary consultations with the service provider and by any agreements made.

Rao and Young (1994) provided five key factors as interacting drivers in the decision of shippers to either utilize logistics parties or retain logistics activities in-house [10]:

1. Centrality of the logistics functions to core competency
2. Risk liability and control
3. Operating cost/service trade-offs
4. Information and communications systems
5. Market relationships

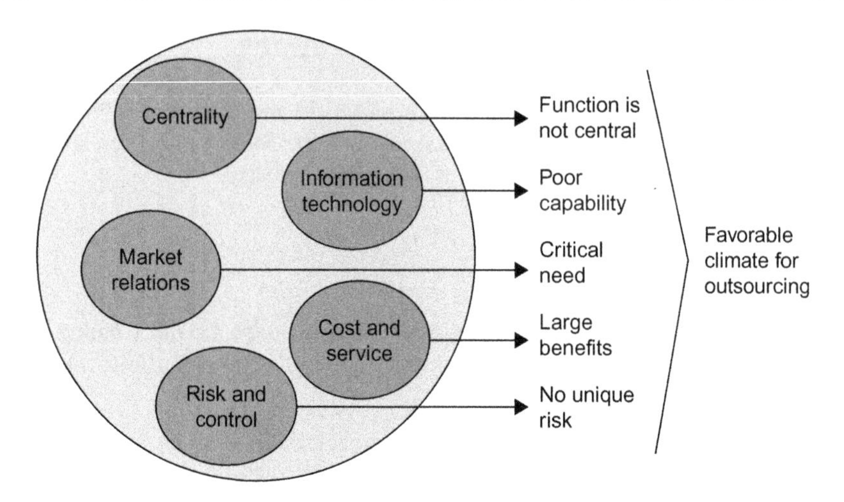

Figure 5.3 Key factors influencing outsourcing of logistics functions [10].

Figure 5.3 is one example of an appropriate climate for outsourcing.

While considering outsourcing logistics activities, companies have to proceed systematically to make a straightforward decision. By describing and prioritizing activities, products, markets, and conditions, it can be decided which activities can be or must be outsourced for which product and market combinations and under which conditions [2].

5.3.2 Selecting the Right 3PL

Before purchasing 3PL services, the client has to select from existing 3PL providers. As a result, it should first assign a selection criterion in order to choose the appropriate 3PL. Some of the criteria can be expressed as follows [2]:

- A service provider's quality could be judged based on experiences using it as a 3PL. In most cases, the service quality is even more important for the shipper than the service cost.
- The throughput rate and delivery reliability of the goods can be decisive. The 3PL must have a high degree of flexibility in place, time, volume, quantity, and product.
- The 3PL's willingness and skillfulness at having discussions with the shipper on the regular basis is of great importance. These discussions include agreements about liability, supplementary to standard transport liability for errors, negligence, and carelessness.
- A 3PL must also have a cost-control system with a clearly and logically composed tariff structure. It must be prepared to clearly state which performances and actions are covered in the tariff. The financial strength of the 3PL is an important factor for selecting the appropriate 3PL.

For the initial screening of candidate service providers, qualitative factors such as supplier reputation, references from clients, and response to information requests can be useful [20].

Table 5.1 Factors Considered in 3PL Provider Selection [21]

Factors	Consideration (%)
Price of 3PL services	87
Quality of tactical, operational services	87
Expected capability to improve service levels	67
Range of available, value-added services	62
Capable ITs	61
Expected ease of doing business	57
Availability of strategic logistics services	54
Global capabilities	51
Knowledge and advice on supply-chain innovations and improvements	44
Cultural and strategic fit with 3PL provider	42
Ability to deliver end-to-end solutions across supply-chain processes and regions	40
Coverage and experience in emerging markets	35
3PL vision and investment strategy	33

IT, services, performance metrics, and intangibles are other factors that can be used to evaluate a 3PL provider. The intangible factors include questions on the business growth of the prospective 3PL to make sure it will be conducting business for some time, including financial stability, strong profitability, experience with similar companies, and global scope [9]. Table 5.1 provides the factors to consider when selecting a 3PL and their relative considerations from the customers of 3PL services.

There are also models for 3PL selection that are further introduced in Section 5.3.5.

5.3.3 Purchasing 3PL Services

The previous section identified criteria for selecting a suitable 3PL. This complementary section provides a framework for the purchasing process of 3PL services.

In general, a buying process contains the following steps [20]:

1. Identification of the need to outsource logistics
2. Development of feasible alternatives
3. Evaluation of candidates and selection of the supplier
4. Implementation of services
5. Ongoing service evaluation

Purchasing Framework

Andersson and Norrman (2002) provided an extensive framework for the purchasing process. The proposed framework is composed of the following phases [22]:

- Define or specify the service: Identify what to define, who to define, and the nature of the factor.
- Understand the volume of services bought.

- Simplify and standardize: It is important for purchasing strategies such as reducing supplier base or buying a more standardized service.
- Market survey: It is normal when developing a bigger supplier base, especially if the strategy is to find the best price.
- Request for information (RFI).
- Request for proposal (RFP): This is sent to providers qualified from the screening process. An RFP for 3PLs, in addition to price issues and various performance factors, may include other provider characteristics such as cultural compatibility, financial issues, flexibility in meeting new requirements, and information-system capabilities.
- Negotiations.
- Contracting.

Contracting

Contracting is one of the challenging issues in the purchasing process. Contract could be used as a safeguard to minimize risks of building the cooperation; it is also regarded as an encyclopedia to describe how a logistics system is developed—i.e., include definitions of processes, activities, roles and responsibilities, incentives, and penalties [22]. Typical 3PL contract includes typical contract terms (e.g., cooperation length), costs per activity, service description, bonus payment for the best performance, penalty clauses for service failures, risks and insurance costs and contract-termination clauses [5].

There are two fundamentally opposing views on the existence of contracts: (1) Signing formal contracts is essential for defining and managing 3PL roles and relations; (2) detailed contracts can be regarded as an indication of lack of trust [5].

According to a survey carried out by Van Laarhoven et al. in 2000, written contracts, formalizing the partnership between shipper and provider, are found in almost 75% of partnerships. In addition, in just more than half the cases, the activities that are part of the partnership are specified in details. This percentage is down from 63%, meaning that providers have more flexibility in shaping the logistics activities. The inclusion of penalty clauses for providers can be seen in 40% of the contracts, up from 27% [23].

There are also a few models for the calculation and analysis of main contract features. For example, Chen et al. (2001) provided a framework for analyzing three forms of third-party warehousing contracts with space commitments and adjustment options [24].

5.3.4 Strategic Behavior of 3PLs

A survey conducted by Berglund et al. in 1999 indicates that a reasonably clear differentiation of strategies in the TPL industry exists. This strategic segmentation is found by segmenting the providers along the following two dimensions [7].

1. Providers that offer a specific service—e.g., distribution of spare parts (service providers)—versus providers that cover a complete range of services and offer their customers logistics solutions (solution providers).

2. Providers that carry out traditional transportation and warehousing activities only versus providers that offer additional activities such as value-added services.

From the survey results, it can be implied that a provider should be aware of two major issues in terms of strategic positioning. The first strategic choice that has to be made is the step from the forwarder, provider of transportation services, and so on to the provider of logistics services or solutions. The second choice is what type of logistics provider one aspires to be.

The choice of the strategy to be either service or solution has important consequences for TPL providers. For example, service providers should focus their IT efforts on developing a high-quality focused system that supports their services, whereas solution providers should have more versatile systems that can be adapted flexibly to meet the requirements of their customers' information systems. Also, when approaching their customers, providers should take notice of the fact that shippers will choose service providers when they consider logistics to be a core activity and solution providers when they do not. Finally, there are differences in skills sets required; for example, solution providers will need to develop subcontracting skills, as well as analytical and logistics design skills, to a much larger extent than service providers [7].

In 2005, Carbone and Stone investigated the strategic behavior of 20 leading European 3PLs between 1998 and 2004. The survey results show that the changing business environment caused 3PLs to alter their strategic behavior by offering different portfolios of services. The strategic behavior reveals substantial convergence, with the major 3PLs moving toward horizontal integration and business diversification, mostly by mergers and acquisitions (M&A). In addition, vertical strategic alliances between customer and logistics provider, as well as horizontal strategic alliances between logistics providers, have also been adopted [25].

5.3.5 Theoretical Models

Most of the 3PL research falls into the empirical category, suggesting an empirical and more practitioner-oriented focus on 3PL. In the 3PL literature, a few theoretical and quantitative models focus on the operational optimization and evaluation of 3PL performance. Here some of these models are presented.

There are several models for selection of the right 3PL. Jharkharia and Shankar (2007) provided an analytic network process (ANP) decision model for the selection of LSPs [26]. A similar vein for the selection of third-party reverse logistics providers is presented in the work by Meade and Sarkis (2006) [27]. Bottani and Rizzi (2006) used a multi-attribute approach based on the technique for order preference by similarity to ideal solution (TOPSIS) technique and the fuzzy set theory for the selection and ranking of the most suitable 3PL service provider [28].

The analytic hierarchy process (AHP) is also a popular technique for selecting appropriate 3PLs. For instance, Gol and Catay (2007) provided an empirical instance using AHP approach for the selection of a 3PL provider to restructure an automotive company's supply chain for export parts. In this case, several criteria were considered with respect to the general company considerations, quality, capabilities, client relationships, and labor relations, and then simple pairwise comparison judgments were used to develop overall priorities for ranking the 3PL providers. The performances of 3PL providers were evaluated using utility curves, ratings, and step functions with respect to each criterion [29].

3PLs companies that fail to sustain their economies of scale and the subsequent price leverage may not survive in today's competitive 3PL market. Therefore, the operational efficiency of 3PL providers dictates their competitiveness or survival. One way of improving the operational efficiency of 3PLs in terms of financial performance is to emulate best-practice firms that can be identified by setting a reliable financial-performance standard. Benchmarking seems to be the most effective way of setting a reliable financial standard and then measuring the operational efficiency of the 3PL because 3PL needs to measure its financial performance relative to its competitors. Min and Joo (2006) proposed a data envelopment analysis (DEA) to measure the operational efficiency of 3PLs relative to prior periods and their competitors and also help 3PLs identify potential sources of inefficiency and provide useful hindsight for the continuous improvement of operational efficiency [30].

To enhance the efficiency of warehousing operations in a 3PL, Hamdan and Rogers (2007) developed a new warehouse efficiency model using DEA to evaluate a set of homogeneous warehouses operated by a 3PL company [31]. Other factors improve 3PL operational performance. For example, Brah and Lim (2006) revealed that total quality management (TQM) technology plays important and complementing roles in improving the performance of logistics companies. Their analysis showed that both high-technology logistics firms and high-technology TQM ones perform significantly better than their low-technology peers [32].

3PL providers should have a striking distribution system to efficiently deal with various clients' service requirements. The system includes filling their customers' orders, keeping the products deliveries up to speed and reducing inventory in a dynamic and uncertain business environment. Ko et al. (2006) proposed a hybrid optimization and simulation approach to design a distribution network for 3PLs, considering the performance of the warehouses. They developed a genetic algorithm for the optimization model to determine dynamic distribution network structures, and then, they applied a simulation model to be able to consider the uncertainty in clients' demands and order-picking time [33]. Ko and Evans (2007) also presented a mathematical model for the design of a dynamic integrated distribution network to consider the integrated aspect of optimizing the forward and reverse network simultaneously [34].

5.3.6 A Framework for the Development of an Effective 3PL

To be considered as a 3PL, a logistics company should have the underlying infrastructure of a 3PL provider. In 2003, Gunasekaran and Ngai provided a framework for developing an efficient 3PL system. This framework is composed of five major dimensions [14]:

1. Strategic planning
2. Inventory management

3. Transportation
4. Capacity planning
5. Information technology

This framework is founded for the transformation of small logistics companies to 3PLs and provides managers with a straightforward approach to improve the quality of their logistics activities to the 3PL level. Table 5.2 lists some of the major underlying activities in each dimension and their corresponding strategies or techniques and technologies.

Table 5.2 A Framework for Transforming a Small Logistics Company into a Comprehensive 3PL Company [14]

Function	Activities	Strategies/Techniques	Technologies
Strategic planning	Corporate and business strategy development, resource management, budgeting, product and service selection, market-segment analysis	Forming strategic alliances, outsourcing, forecasting demand, aggregate planning, selecting criteria for partnership, gaining the support of top management, improving continuously, getting government support	Groupware, shared information systems, Internet and EDI, ERP
Inventory management	Forecasting, location analysis, network consulting, slotting and layout design, order management	Demand pull system, just-in-time, Kanban, material requirements planning, supply-chain management	MRPII, EDI, ERP, WWW, online purchasing
Transportation planning	Shipping, forwarding, De(consolidation), contract delivery, freight bill payment, load tendering and brokering	Outsourcing, forming strategic alliances, optimizing routing and scheduling, capacity management, total productive maintenance	Groupware, Internet, E-mail, intranet, extranet, linear programming
Capacity planning	Capacity of transportation vehicles, warehouse capacity, human resources, material-handling equipment	Make or buy decisions, planning aggregate capacity, minimizing costs, maximizing capacity	Linear programming, waiting-line models, scheduling optimization, MRPII, CRP, ERP
Information management	Performance measures and metrics, data collection, data processing, data reporting	Groupware, IT/IS, shareware, data mining, data warehousing, Intranet, extranet	EDI, e-commerce, Internet, AI and expert systems, ERP

5.4 Concluding Remarks

The increasing demand by a broad range of firms to outsource logistics functions has led to an increasingly developing market for logistics. This demand has resulted in considerable attention and more investigations in the important concept of logistics parties.

The 3PL industry is still growing. Among the priorities currently facing 3PL providers are regional expansion, broadening services to meet the needs of current and future customers, integrating information technologies, and developing relationships with customers and other business firms. Although most of the research on logistics parties is dedicated to 3PLs, the research on 3PLs is mostly descriptive. Therefore, it is suitable to work on the theoretical aspects of 3PLs more extensively. Organizational and technological changes, linked with globalization and information technology developments, shows that 3PL is a sector undergoing constant changes, making it necessary to study this interesting sector more theoretically [5].

However, there are also arguments that the 3PL industry has reached the mature stage of its life cycle after two decades of evolution [30]. Now it is time for other parties to grow in both theoretical and practical directions. Although rapidly developing, it is surprising to see that little research has been done on other types of logistics parties. All the 3PL issues, including selection, behavioral aspects, and operational issues, may be considered as research topics for newer generations of logistics parties. Finally, it is necessary to investigate these topics more deeply in order to appropriately utilize the talents and capabilities of parties and streamline a more integrated, cost-effective supply chain.

References

[1] S.M. Aghazadeh, How to choose an effective third party logistics provider, Manag. Res. News 26(7) (2003) 50–58.

[2] D.A. Van Damme, M.J. Van Amstel, Outsourcing logistics management activities, Int. J. Log. Manag. 7(2) (1996) 85–95.

[3] M. Ranjan, R. Tonui, Third party logistics: an analysis of the feasibility and contexts of strategic relationships, M.S. Thesis, Massachusetts Institute of Technology, 2004.

[4] S. Hertz, M. Alfredsson, Strategic development of third party logistics providers, Ind. Market. Manag. 32(2) (2003) 139–149.

[5] K. Selviaridis, M. Spring, Third party logistics: a literature review and research agenda, Int. J. Log. Manag. 18(1) (2007) 125–150.

[6] R.C. Lieb, The use of third-party logistics services by large American manufacturers, J. Bus. Logist. 13(2) (1992) 29–42.

[7] M. Berglund, P. Van Laarhoven, G. Sharman, S. Wandel, Third-party logistics: Is there a future? Int. J. Log. Manag. 10(1) (1999) 59–70.

[8] A. Marasco, Third-party logistics: a literature review, Int. J. Prod. Econ. 113(1) (2008) 127–147.

[9] G. Vaidyanathan, A framework for evaluating third-party logistics, Commun. ACM 48 (1) (2005) 89–94.

[10] K. Rao, R.R. Young, Global supply chains: factors influencing outsourcing of logistics functions, Int. J. Phys. Distrib. Logist. Manag. 24(6) (1994) 11–19.

[11] T. Nemoto, K. Tezuka, Advantage of third party logistics in supply chain management. Available online at: http://hermes-ir.lib.hit-u.ac.jp/rs/bitstream/10086/16053/1/070cm WP_72.pdf, 2004.

[12] T. Nemoto, K. Tezuka, Third-party logistics: results and findings of the 14th annual study. Available online at: http://www.3plstudy.com, 2009.

[13] D.N. Bauknight, J.R. Miller, Fourth party logistics: the evolution of supply chain outsourcing. Available online at: www.npeters.com/studium/4pl_present.pdf, 1999.

[14] A. Gunasekaran, Ngai EWT. The successful management of a small logistics company, Int. J. Oper. Prod. Manag. 33(9) (2003) 825–842.

[15] Available from: http://www.theabrahamgroup.com

[16] YCH Group, 7PL™: The definitive supply chain revolution. White Paper Version 1.2, 2002.

[17] Available from: http://www.ups.com

[18] Available from: http://www.penskelogistics.com

[19] Available from: http://en.wikipedia.org/wiki/Deloitte

[20] H.L. Sink, C.J. Langley, A managerial framework for the acquisition of third-party logistics services, J. Bus. Logist. 18(2) (1997) 163–189.

[21] H.L. Sink, C.J. Langley, Third-party logistics: results and findings of the 12th annual study. Available online at: http://www.3plstudy.com, 2007.

[22] D. Andersson, A. Norrman, Procurement of logistics services—minutes work or a multi-year project? Eur. J. Purch. Suppl. Manag. 8(1) (2002) 3–14.

[23] P. Van Laarhoven, M. Berglund, M. Peters, Third-party logistics in Europe—five years later, Int. J. Phys. Distrib. Logist. Manag. 30(5) (2000) 425–442.

[24] F.Y. Chen, S.H. Hum, J. Sun, Analysis of third-party warehousing contracts with commitments, Eur. J. Oper. Res. 131(3) (2001) 603–610.

[25] V. Carbone, M.A. Stone, Growth and relational strategies used by the European logistics service providers: rationale and outcomes, Transport. Res. E. 41(6) (2005) 495–510.

[26] S. Jharkharia, R. Shankar, Selection of logistics service provider: an analytic network process (ANP) approach, Int. J. Manag. Sci. 35(3) (2007) 274–289.

[27] L. Meade, J. Sarkis, A conceptual model for selecting and evaluating third-party reverse logistics providers, Supply Chain Manag. 7(5) (2002) 283–295.

[28] E. Bottani, A. Rizzi, A fuzzy TOPSIS methodology to support outsourcing of logistics services, Supply Chain Manag. 11(4) (2006) 294–308.

[29] H. Gol, B. Catay, Third-party logistics provider selection: insights from a Turkish automotive company, Supply Chain Manag. 12(6) (2007) 379–384.

[30] H. Min, S.J. Joo, Benchmarking the operational efficiency of third party logistics providers using data envelopment analysis, Supply Chain Manag. 11(3) (2006) 259–265.

[31] A. Hamdan, K.J. (Jamie) Rogers, Evaluating the efficiency of 3PL logistics operations, Int. J. Prod. Econ. 113(1) (2008) 234–244.

[32] S.A. Brah, H.Y. Lim, The effects of technology and TQM on the performance of logistics companies, Int. J. Phys. Distrib. Logist. Manag. 36(3) (2006) 192–209.

[33] H.J. Ko, C.S. Ko, T. Kim, A hybrid optimization/simulation approach for a distribution network design of 3PLS, Comput. Ind. Eng. 50(4) (2006) 440–449.

[34] H.J. Ko, G.W. Evans, A genetic algorithm-based heuristic for the dynamic integrated forward/reverse logistics network for 3PLs, Comput. Oper. Res. 34(2) (2007) 346–366.

6 Logistics Future Trends

Amir Zakery

Department of Industrial Engineering, Amirkabir University of Technology, Tehran, Iran

6.1 Main Influencing Issues

6.1.1 Globalization

"*Globalization* represents the cross-national functional integration and coordination of spatially dispersed economic activities" [1]. An increasing range of economic sectors are turning to globalization as a norm. So it results to complex harmonization process in coordinating of production and supply worldwide. This complexity is a result of differences in cultures, economic systems, government regulations, and also so many international rules and agreements.

The environment of business decisions is becoming fundamentally altered by globalization, which amplifies the competition between companies, so corporate decisions have to be increasingly made in this new context. Firms have to respond strategically in order to stay competitive in this situation [1].

Chatterjee and Tsai [1] have identified two primary *drivers of globalization* in recent times: (1) lowering of trade barriers including tariff and nontariff barriers and (2) considerable reduction in transportation and communication costs. These two factors have accelerated the pervasion of globalization.

The first driver explains lowered trade barriers. Trade barriers were eliminated with the free-trade regime executed by the United States, and related regulations were represented by the General Agreement on Tariffs and Trade (GATT) and the World Trade Organization (WTO). As a result, the flow of goods and products across different countries accelerated rapidly in most parts of the world.

According to Chatterjee and Tsai [1], technological change in the transportation and IT sectors make up the second driver. They emphasize that the combined effects of changes in transportation and the complementary information technologies (ITs) are visible in a different range of facilities, including traditional transportation services with more speed and reliability and lower costs, as well as the introduction of new classes of transportation services.

Chatterjee and Tsai mention that land and water transportation costs have decreased steadily in the years since World War II. But to explain the role of technology in declining costs, Chatterjee and Tsai [1] believe that the IT effect on

Logistics Operations and Management. DOI: 10.1016/B978-0-12-385202-1.00006-2

logistics' managerial and coordination functions is more influential than IT in increasing the optimal use of transportation facilities. Obviously, a lack of coordination between effective factors decreases overall efficiency and imposes different costs on the system. In supply-chain partnerships, information sharing between supply-chain members is frequently suggested as a remedy to improve collaboration. These technological improvements permit the coordination of globally dispersed sets of economic activities.

According to Kleindorfer and Visvikis [2], globalization was the major factor leading to the growth of logistics in the past decade. Smith [3] referred to trade as a fundamental factor driving economic growth. Trade means that the most cost-effective sources for product design and manufacture can be linked to end markets. It seems that in today's business world this logic has turned into the dominant theme.

In the new framework of trade policy, business has the best opportunity to grow, and logistics has become the primary "glue" for integrating the networks of intermediate and final production and service providers associated with globalization [2].

Although global trading is increasing, the nature of firms is also changing. They are shifting from being national firms to becoming international and global corporations. Becoming transnational means corporations face a more complex business environment, although they are encouraged strongly to form collaborative agreements with potential partners: suppliers, costumers, competitors, and allies. Such agreements create a network of complex business relationships. These organizational relationships have become a necessary component of globalization [4].

Different kinds of collaborations have been studied in the literature. Lemoine and Dagnæs [4], for example, studied the dynamics of a successful network of European logistics providers. Their study illustrates some major points: the complex and powerful links created between members, how they conduct their business and organize spatially to benefit from becoming global, and how firms control the transportation market and the required infrastructure. The case provides an explanation of how globalization can be attained using different organizational models and resource combinations.

In recent years, companies have tended to extend their domestic business logistics to global logistics (GLs) because of their international markets and customers. GLs cannot succeed without suitable strategies. GL strategies have more complexity and so are more difficult to develop compared with domestic logistics strategies, according to Sheu [5]. The author presents several reasons for the complexity of global strategies. Coordination of information and money flow is one important source of complexity in an international scenario. Different transfer prices, exchange rates, trade barriers, and labor costs are the other factors discussed [5].

On the other side, Sheu emphasizes that the globalization of logistical activities results in more complex business operations because international companies face more sources of uncertainty relative to domestic logistics. Greater shipment distances and longer lead times are examples of growing risks of operating globally [5]. These factors make current mathematical models that are used extensively for the design of domestic supply chains inappropriate, necessitating remodeling for newer arrangements.

As globalization matures, the number of companies operating in the global marketplace increases. Companies should take a broader perspective when operating in international scale. Before this, although some companies had a presence across a wide geographic area, their main operation was based on local or regional sourcing, manufacturing, inventorying, and distribution. But now international companies are truly global, with an organizational structure and strategy that represents a global business [6].

Typical attributes of global marketplace according to Rushton et al. include [6] inventory centralization, information centralization, and global branding, sourcing, and production.

However, global companies are capable of providing for local requirements and regional markets—e.g., electronic standards for electrical goods. It means that the transnational scope of global corporations does not mean neglecting of local customized products and services.

It seems that service companies face more demanding situations. Rushton et al. emphasize that logistics networks and operations, in order to service global markets, become far more expansive with extra complexity.

Globalization causes different changes in the logistics industry. Rushton et al. find the following implications for logistics globalization [6]:

- Supply lead times are extended.
- Transition times are extended and subject to uncertainty.
- There are multiple break-bulk and consolidation alternatives.
- There are multiple freight mode and consequently different cost alternatives.
- Production postponement is deployed.

It is obvious from above that there is a direct conflict between globalization and the move to the just-in-time (JIT) operations as a new strategy in some companies. In global companies, there is a tendency toward increasing order lead times and inventory levels because of the distances involved and the complexity of logistics. In companies moving to the JIT philosophy, the situation is reversed and there is a desire to reduce lead times and inventory as much as possible [6]. The solution is sought for trade-offs between order and inventory costs versus costs imposed by uncertainty in products' deliveries.

6.1.2 Information Technology

Many researchers have discussed IT as a means creating logistics competitiveness (e.g., [7,8]). IT in recent years has influenced the atmosphere of many economic activities, and its power is increasing because of both increasing capability and simultaneous decreasing cost. Reviewing recent publications and the overall situation of logistics reveals that IT and its derivatives in commerce, such as e-commerce, are the greatest influencing factors in today's logistics.

Closs et al. [7] identified two main streams in the literature for IT role in logistics competence. They suggest that information is a valuable logistics resource and also a means of achieving competitive advantage.

Table 6.1 Traditional and E-Logistics Characteristics [9]

	Traditional Logistics	E-Logistics
Types of load	High volumes	Parcels
Customer	Known	Unknown
Average order value ($)	>1000	<100
Destinations	Concentrated	Highly scattered
Demand trend	Regular	Lumpy

IT has encouraged several facilities in logistics business. For example, e-commerce development is a major factor in the growth of the distributive industry. However, Ghiani et al. [9] believe the rate of expansion and the extent of development of e-commerce remain uncertain. E-commerce causes a more complex organization of the whole logistics system that is called *e-logistics*. The new logistics organization should be capable of managing small and medium-sized shipments to a large number of customers in different areas (Table 6.1). As we see, IT has changed the way corporations run their business, and so changing traditional logistics system into an electronic-logistics system must be considered by logistics service providers [9].

A major role of IT through all businesses is supporting flexibility, which is necessary across the value chain in order to meet customer needs efficiently. Production phases, including product development, manufacturing, logistics, and distribution activities, should support flexibility.

Closs and Swink [8] suggest that flexible logistics programs are strongly interrelated with all performance dimensions. They focus on the "information connectivity" concept and introduce its facilitator role. Information connectivity mediates between flexible logistics and two main competencies of a firm, including asset productivity and delivery competence. They explain that information connectivity fully mediates the relationship between flexible logistics programs and asset productivity, and it partially mediates the relationship between flexible logistics programs and delivery competence.

IT is the initiator and also background for many information- and knowledge-sharing systems for supply-chain collaborations. Knowledge-sharing systems are deployed within and between organizations extensively today. These systems provide the structural (organizational), cultural, and electronic infrastructure requirements of sharing the right knowledge between the right people at the right time who need it. Electronic infrastructure mainly refers to IT facilities and capabilities, though IT provides an important and mostly unique infrastructure for knowledge-sharing systems. Other new technologies relevant to logistics management such as radio frequency identification (RFID) and global positioning system (GPS) are also dependent on IT. New technologies are discussed in the next section.

6.1.3 New Technologies

Emerging technologies are another influential trend in today's logistics. Today's forward-looking enterprises are dynamic and have collaborations with business stakeholders such as suppliers, customers, and some competitors. They use modern technologies and share information and knowledge in an effort to create a collaborative supply chain. Such a dynamic and collaborative supply chain is capable of competing if not leading a special industry. Today, managing supply-chain activities is supported by different kinds of information flows within and between parties. Examples such as material requirements planning (MRP), manufacturing resource planning (MRPII), and enterprise resource planning (ERP) are new information system developments in IT, improving logistics and supply-chain management [10].

In recent years, advanced technologies such as RFID, wireless and mobile technology, and GPS have been applied extensively in logistics systems especially in retail sectors. At the same time, they have also resulted in mixed performances in a supply chain because of large data input, analysis, and reporting in short intervals. Ketikidis et al. [10] point to the following advantages: tracking product logistics is much easier, information processing is done more efficiently, security is improved, and counterfeit reports are reduced. Fast-track ordering, improved customer relationships, and better control of supplies are other rewards. However, Ketikidis et al. emphasize that advantages have been reported more often in more-developed countries because the required infrastructure is provided there.

Electronic data interchange (EDI) has been used widely to transfer information between suppliers and customers in a supply chain. Bar coding is still extensively utilized to track products. Such long-established technologies as EDI and bar coding, although they provide lower capabilities, are not as expensive as RFID when we consider how quickly they can be implemented at any level in a supply chain. The reader standard and compatibility with suppliers' systems seems to be a constraint for RFID-integrated application in a supply chain, although RFID setup and utilization costs are decreasing [10].

Smith argued that RFID should be viewed not only as a technological innovation but also as a transformational event. It is supposed that the use of RFID-based technology by some leading and progressive enterprise (e.g., WalMart) should revolutionize and popularize inventory-tracking methods with other enterprises [11].

Like some other emergent technologies, RFID technology raises some worries from a security perspective. Kelly and Erickson [12] examined this problem and concluded that RFID technology provides enormous economic benefits for both business and consumers, while potentially being one of the most invasive surveillance technologies to threaten consumer privacy. However, most believe RFID's smart technology advantages in counteracting theft are larger than any possible threats to consumer privacy. Nonetheless, methods to resolve the security anxieties of end users should be followed to minimize the disadvantages of monitoring technologies such as RFID.

6.2 Future Trends in Some Logistics Sectors

Studying recently emerging issues in different sectors of the logistics industry is another method for studying trends in future logistics. Impacts of emerging concepts, methods, and technologies have brought changes in different sectors of the logistics industry. Major changes that are expected to become more influential over time form trends. For this purpose, three important sectors in logistics—including inventory management, transportation management, and warehousing—have been studied from the viewpoint of recent changes, today's global issues, and new strategies. This section mainly refers to Ailawadi and Santish's book *Logistics Management* [13].

6.2.1 Future Trends for Inventory Management

The global marketplace, higher product variety, shorter product life cycles, poor forecast accuracy, demand for more customized yet financially manageable products, and premium customer service are increasing logistics and supply-chain complexity. Along with these challenges, supply-chain inventory strategies are evolving to find new methods to meet customers' needs in new situations and also to benefit from new opportunities arising in the changing environment. For example, operating with minimal inventories while still meeting customer expectations is one key dimension of new inventory strategies [13], which are referred to as JIT strategies.

Postponement

Ailawadi and Santish [13] define postponement strategies as those that "combat increasing fulfillment costs associated with both geographically dispersed markets and the expansion of product variety." Postponement is conceptually based on delaying production and delivery costs until cost fulfillment is necessary. Geographic and product form postponements are increasingly applied in global supply chains.

Geographic postponement proposes to hold inventory centrally and to delay its commitment to target market areas as long as possible, often until customer orders are received. A simple example is accumulating spare parts inventories from regional warehouses into a central distribution center. However, there is a trade-off between holding inventory centrally and increasing delivery lead times. Central inventorying reduces inventory costs but increases the probability of damaging the level of customer service. Today, leading third-party logistics providers such as FedEx and UPS, with the aim of electronic order processing, provide distribution services that offer the advantages of geographic postponement and a premium delivery service [13].

The other form of postponement is product form postponement. It proposes to delay production of final goods with various appearances and functions as late as possible—until a customer's needs are known. Stocking products in their more generic form allow them to be inventoried in their least expensive and most flexible forms. The final manufacturing or assembly steps will be performed as soon as a customer's exact orders are known [13].

There are good examples for product form postponement such as stocking only white dishwashers including color panels instead of producing different colored dishwashers. Customers will select and insert their favorite color panels, so there is no need to stock different colored machines.

Quick Response

Quick response (QR) depends entirely on shortening manufacturing and distribution lead times of specific products. It acquires initial real data about market needs first and then begins the production phase. Supply chains thus should plan to enter a seasonal sales period with a small initial inventory distributed to retailers, monitor early sales patterns, generate a renewed demand distribution, and then choose the replenishment method. In this way, the manufacturer or supplier receives early sales data and utilizes them to update production schedules to better match output with demand [13].

Vendor-Managed Inventory

When logistics is studied between two or more companies, the concept of coordination becomes very crucial. Lack of coordination between partners occurs when they have incomplete or incorrect information such as what happens in the bullwhip effect. Lack of coordination between effective factors decreases overall efficiency and imposes different costs on the system.

One method gaining popularity for coordinating inventory decisions is vendor-managed inventory (VMI). This collaborative initiative authorizes suppliers to manage the buyer's inventory of stock-keeping units. It operationally integrates suppliers and buyers by using ITs. Buyers can share real-time information of sales and inventory with suppliers, and suppliers can then use this information to plan production and delivery decisions and set their inventory levels at the buyer's warehouses [14].

6.2.2 Global Transportation Issues

Today, transportation is an important part of the logistics system, and efficient supply chains rely on fast, responsive, and dependable transportation. For example, consider the following points expected in future US freight transportation [15]:

- The demand for freight transportation will nearly double by 2035. It will press the capacity of the nation's transportation system.
- There will be new pressures on freight carriers to deliver goods reliably and cost-effectively because business will move to create on-demand supply chains and replenish what customers consume as soon as it is sold.
- Businesses trade relationships with suppliers and customers will be more global, so transportation will become more necessary.
- Business will become more dependent on carefully timed and reliable freight transportation because of more intensive connectivity between members of the supply chain. When the freight transportation system fails to satisfy commitments, hundreds of members (including shippers, carriers, and worldwide markets) are affected.

Ailawadi and Santish identify two key global issues that may lengthen transit times, create more frequent and unpredictable delays, and consequently raise transportation costs: capacity constraints and increased transportation security [13].

Transportation Capacity Issues

Capacity constraints arise when demand for transportation exceeds a transport system's ability to meet demand efficiently, which results in a higher cost or a lower service level [13]. Transportation capacity constraints, particularly at domestic points of entry, now afflict several areas in the world as the practice of global sourcing increases.

Transportation capacity issues are complex because of the multiple dimensions and the integrated nature of transportation systems. In addition, the public sector has a very large and multifaceted impact on transportation capacity. The following are among the ways to address capacity problems, according to Ailawadi and Santish [13]:

- Existing transportation capacity can be used more efficiently through improved government pricing of transportation infrastructure (e.g., congestion tolls or peak and off-peak pricing differentials).
- Social regulations such as those addressing safety, labor, environment, and energy should be optimized (e.g., balanced social and economic objectives) in order to better use the existing transportation capacity.
- Technology plays a major role in alleviating capacity constraints. Two important examples are equipment and shipment tracking and the use of intelligent transportation systems (ITSs) for road traffic management.

In addition, the US Chamber of Commerce [15] emphasizes the integration of various modes into cohesive and efficient national and global networks to improve the efficacy of the transportation system, although it needs planning and funding for intermodal transportation and forms a broader geographic perspective.

Transportation Security Issues

The heightened security efforts and military actions undertaken after the terrorist attacks on the United States in 2001 have posed a potential disruption to the smooth flow of freight and have added to the congestion of transportation networks at critical points [13]. In addition, intensified security requirements in special situations have imposed huge costs on transportation.

6.2.3 Future Trends for Warehousing

Many researchers used to believe we should anticipate the demise of the warehouse because it was considered to be only a simple repository for goods. It was supposed that all companies would plan to eliminate inventory, so warehouses could not possibly play much of a role in future global supply-chain network. But as Rankin et al. [16] point out, the warehouse is still alive, playing key roles in traditional supply-chain networks as well as assuming a pivotal place in most e-commerce operations.

Rankin et al. believe we will witness a wider variety of warehouse types in the future, including (1) long and narrow warehouses with multiple dock doors to support cross-docking operations and (2) high and properly designed warehouses ready to accommodate automated storage and automated retrieval systems. Other segments of inventory management will engage in satisfying small orders of individual items required in the e-business environment. This large number of small-sized demands will present a challenging task. Warehouses will have a strategic and important role in controlling total supply-chain cost and in meeting service requirements in a dynamic environment [16].

As stated before, postponement strategies are being used more because of the emphasis put on responding to individual consumer needs in today's business atmosphere. Consequently, postponement results in significant savings in inventory, transportation, and reduced inventories of obsolete products. As a result, borders between warehousing, assembly, and retail operations are disappearing. Thus, the warehouse will be the place where final assembly, blending, labeling, and packaging, in addition to traditional stocking, will be done. Advantages for allocating this role to warehouses are closeness to markets, low labor cost, and effective systems for managing the processes of assembly, labeling, and packaging [16].

Finally, Ailawadi and Santish [13] predict the warehouse will play an important role in bringing manufacturers together to collaborate in creating consolidated shipments to major markets. Also, new technologies such as RFID will facilitate the movement and location of goods in the warehouse as soon as they become commercially viable.

6.3 Future Trends in Technical Reports

Technical reports and surveys are good resources for investigating recent changes and possible future trends in different parts of business and industry. As a living and growing subject between practitioners and industrial communities, logistics has been the focus of many technical reports. In this section, we present seven reports about recent and future trends in the logistics industry that were found through Internet searches (mostly from the Proquest database at http://www.proquest.com). Some focus on a specific region or country or on special issues of logistics, and others are more general in nature.

6.3.1 Future Trends of Logistics in the United Kingdom

In its logistics information sector, which contains an overview of UK logistics in 2005, the National Guidance Research Forum identifies the main drivers of change and accordingly lists future trends in logistics, including the following [17]:

- Globalization is affecting not only production, procurement and distribution, e-logistics and e-transport but also companies' outsourced activities.

- New technologies, including product monitoring and IT-based information systems, mostly favor large companies because of their expense. Monitoring the flow of stock is facilitated, so handling and processing of goods will speed up as a result of improvements in technology. E-commerce and direct sales require QR systems that are capable of delivering goods within very short timescales, and better IT systems will bring this capability.
- The development of e-commerce, including the choice of home shopping using the Internet, is a major driver of the distribution industry.
- National and European concerns over environmental issues will continue to intensify, and required laws and regulations will be passed. We will probably see an increase in demand for rail transportation.
- An urgent need for qualified managers has been observed in both rail and waterborne freight. Demands for expansion, privatization, and deregulation, together with increasing customer expectations, environmental standards, and safety levels, will all affect the transportation industry.

6.3.2 Thinner Margins in the Industry: A Chance to Improve for Shippers

The report *State of the Industry—Logistics* by Datamonitor predicts that not all of the companies active in the logistics sector will survive in the medium term and as the market consolidates, although the top companies have experienced revenue increases in recent years [18]. Companies can only survive if they have a good understanding of their own capabilities and can plan to meet market situations. Therefore, they should be proactive rather than reactive in identifying and exploiting market opportunities.

According to Datamonitor's report, 3PLs can exploit several trends that seem to have significant impacts on the logistics industry. Technology will play an increasingly important role. Online tracking with RFID eventually will be offered as a standard transportation service for all products. The GLs landscape is subject to change. China will still have the opportunity to be the main manufacturing region in the world, and there will be a shift in geographic focus (i.e., market share) from other areas toward China. Environmental concerns have rapidly risen up, requiring 3PLs to move toward green supply-chain options [18]. It is worth emphasizing the importance of green logistics here.

Green Logistics

Because of recent challenges and obstacles in environmental issues, companies' responsibilities about the environmental impacts of their own activities have grown, and increasing attention is being given to developing environmental-management strategies and plans for logistics in the supply chain [19]. The legal concept that "polluters should pay for pollution" is accepted in many parts of the world, and related procedures and regulations are under development and implementation. Like other sectors, the logistics industry has seen an increase in environmental concerns, and green logistics is a proper answer to this concern. Green logistics requirements may increase costs in some parts because some natural resources that had been utilized freely and without limits—for instance, pollution absorption—should be paid for in these new arrangements.

6.3.3 Third-Party Logistics Maturing Quickly

Richard Armstrong's sixth annual "Trends in 3PL/Customer Relationships" survey found that in 2006, for the first time, more than 50% of the lowest quintile of Fortune 500 companies used 3PLs to organize logistics [20]. This figure was 25% in 2001. Passing this milestone shows that the adoption of logistics outsourcing has now gone beyond leading companies and has been undertaken by a large proportion of companies. This confirms that the important psychological and performance barriers of logistics outsourcing are disappearing.

6.3.4 Strategic Shift Toward Redesigning Logistics Networks

According to the results of the survey in the *Fifteenth Annual Masters of Logistics Survey*, faced with rising transportation costs and continued service problems, shippers have begun thinking about redesigning distribution networks (33% of respondents) and forming multicompany collaborations to share transportation capacities (20%). The results also indicate that the optimization of networks is addressed on a broad scale rather than controlling costs for just one mode [21].

6.3.5 Need for Broader Range of Logistics Services

The model for providing logistics services will have to undergo major changes, according to a 2006 report from IBM Business Consulting Services in order to keep up with customers' evolving needs. This is natural: As logistics customers outsource more, they will look for a broader range of services and demand greater reliability and lower total cost.

The report predicts that a limited number of logistics providers—about 5%—will reinvent themselves in the face of these changes. They will offer end-to-end supply-chain integration and business process capabilities from the supplier management side up to the customer services side. IBM predicts that the future logistics provider will be more global, concentrated, segmented around customer type, and universally better at execution [22].

6.3.6 Five Influencing Factors in the Future of European Logistics

In another report in 2006, Datamonitor discussed the main shaping factors of future European logistics. It predicted that outsourced logistics' share of the European market in coming years will increase because of beneficial trade conditions and cost-reduction pressures. However, it is emphasized that the favorable environment for logistics outsourcing will not necessarily mean that 3PLs will automatically benefit. So the report introduces five key driving factors in the logistics industry: globalization, legislation, technology, consolidation, and alliances. These factors will help determine the success of providers and clients in the future [23].

6.3.7 Five Trends Supporting Logistics Success in China

By entering into the WTO, China agreed to open its markets and services to global competition. One of these is logistics. The road to high-performance logistics in China is not so smooth, and there are obstacles that should be considered. A managing partner of the Accenture Supply Chain Management practice in a 2006 report, counts five trends supporting logistics success in China useful for companies aiming to enter China's logistics market, as below [24].

1. *Rapid growth*: Although today's logistics indicators are not satisfactory compared to developed countries, China's logistics and distribution sector is growing rapidly. Wang et al. [25] mention that, the average annual growth rate of the China's logistics industry has been 22.2% between 1992 and 2004.
2. *Consolidation of a rather fragmented market*: Among the country's 18,000 logistics services companies, none of them offers nationwide distribution services currently (at 2006). Another guiding indicator is the largest market share possessed by a single company. This indicator is not more than 2% of the logistics market at 2006 in china. However, consolidation is expected to be happening soon, because of competitive pressures, increased service levels, and the growth of transportation to outlying destinations.
3. *Expanding markets for 3PLs*: Although the concept of outsourcing logistics functions dates back to recent years for Chinese companies, many multinational companies are accelerating outsourcing of logistics services to third parties. A report by a market analyst in 2003 predicts an annual growing rate of 25% for logistics outsourcing in China through the next decade.
4. *Greater control of downstream distribution*: More companies are imitating successful multinational companies. They are supported by strong and modern distribution networks. Guangdong Honda Automobile Co. Ltd. is a good example.
5. *Alliances result in competitive advantages*: The top companies in China plan to build competitive national distribution networks that target specific markets. Obviously, this trend is a driver force for establishing new alliances and joint ventures (JVs).

References

[1] L. Chatterjee, C. Tsai, Transportation Logistics in Global Value and Supply Chains, Center for Transportation Studies, Boston University, Boston, MA, pp. 1–17.

[2] P.R. Kleindorfer, I. Visvikis, Integration of Financial and Physical Networks in Global Logistics, Risk Management and Decision Processes Center, University of Pennsylvania, Pennsylvania, PA, 2007, pp. 1–25.

[3] A. Smith, Wealth of Nations, Prometheus Books, New York, 1776, republished in 1991.

[4] W. Lemoine, L. Dagnæs, Globalization and networking organization of European freight forwarding and logistics providers, 2nd International Conference on Co-operation & Competition, Approaches to the Organization of the Future, Vaxjo University, Sweden, 2002.

[5] J-B. Sheu, A hybrid fuzzy-based approach for identifying global logistics strategies, Transport. Res. E 40 (2004) 39–61.

[6] A. Rushton, P. Croucher, P. Baker, The Handbook of Logistics and Distribution Management, third ed., KOGAN PAGE, London, UK, 2006.
[7] D.J. Closs, T.J. Goldsby, S.R. Clinton, Information technology influences on world class logistics capability, Int. J. Phys. Distrib. Logist. Manag. 27(1) (1997) 4–17.
[8] D.J. Closs, M. Swink, The role of information connectivity in making flexible logistics programs successful, Int. J. Phys. Distrib. Logist. Manag. 35(4) (2005) 258–277.
[9] G. Ghiani, G. Laporte, R. Musmanno, Introduction to Logistics Planning and Control, John Wiley, London, UK, 2004, pp. 180–195.
[10] P.H. Ketikidis, S.C.L. Koh, N. Dimitriadisa, A. Gunasekarand, M. Kehajovae, The use of information systems for logistics and supply chain management in South East Europe: current status and future direction, Omega 36 (2008) 592–599.
[11] A.D. Smith, Exploring radio frequency identification technology and its impact on business systems, Inform. Manag. Comput. Secur. 13 (2005) 16–28.
[12] E.P. Kelly, G.S. Erickson, RFID tags: commercial applications v. privacy rights, Ind. Manag. Data Syst. 105 (2005) 703–713.
[13] S.C. Ailawadi, Logistics Management, Prentice Hall, India, 2006.
[14] Y.T. Yao, P.E. Evers, M. Dresner, Supply chain integration in vendor-managed inventory, Decis. Support Syst. 43 (2007) 663–674.
[15] Cambridge Systematics, Inc., Boston Logistics Group, Inc., Alan E. Pisarski, The Transportation Challenge: Moving the US Economy, National Chamber Foundation, Available online at: http://www.uschamber.com/reports/transportation-challenge, 2008.
[16] S.R. Rankin, T. Rogers, D. Tompkins, J. Lancioni, R. Delp, Paul, Tomorrow is here, warehousing management. Available online at: http://www.allbusiness.com/technology/software-services-applications-internet-social/7802056-1.html, 2000.
[17] S.R. Rankin, T. Rogers, D. Tompkins, J. Lancioni, R. Delp, Paul, Future trends for logistics. Available online at: http://www.guidance-research.org/future-trends/logistics/info, 2005.
[18] S.R. Rankin, T. Rogers, D. Tompkins, J. Lancioni, R. Delp, Paul. Datamonitor: 2008 set to be a challenging year for the logistics industry. Available online at: http://proquest.umi.com/pqdweb?id=1431584171&sid=1&Fmt=3&clientId=46428&RQT=309&VName=PQD, 2008.
[19] B.M. Beamon, Designing the green supply chain, Logist. Inform. Manag. 12(4) (1999) 332–342.
[20] W. Hoffman, 3PLs maturing fast. Available online at: http://proquest.umi.com/pqdweb?did=1168054231&sid=2&Fmt=3clientId=46428&RQT=309&VName=PQD, 2006.
[21] J.A. Cooke, 15th Annual Masters of Logistics Survey: Strategy Shift. Available online at: http://proquest.umi.com/pqdweb?did=1131491221&sid=2&Fmt=3&clientId=46428&RQT=309&VName=PQD, 2006.
[22] J.A. Cooke, Outsourced logistics trends creates buyer/seller needs gap. Available online at: http://findarticles.com/p/articles/mi_hb4372/is200605/ai_n18926532, 2006.
[23] J.A. Cooke, A Datamonitor report: five factors that will shape the future of European logistics, product code: BFCO0008. Available online at: www.datamonitor.com/Products/Free/Brief/BFCO0008/010BFCO0008.pdf, 2006.
[24] P.M. Byrne, Five trends support logistics success in China. Available online at: http://www.logisticsmgmt.com/article/CA6343713.html, 2006.
[25] Q. Wang, K. Zantow, F. Lai, X. Wang, Strategic postures of third-party logistics providers, in Mainland China, Int. J. Phys. Distrib. Logist. Manag. 36 (2006) 793.

Part III

Tactical and Operational Issues

7 Transportation

*Zohreh Khooban**

Department of Industrial Engineering, Amirkabir University of Technology, Tehran, Iran

7.1 Basic Aspects in Transportation Systems

Transportation systems move goods between origins and destinations using vehicles and equipment such as trucks, tractors, trailers, crews, pallets, containers, cars, and trains. Transportation represents the major role and most important element in logistics because of its considerable cost [1].

A transportation system is an organization that designs, arranges, sets up, and schedules freight-transportation orders during a given and limited time period with technical restrictions at the lowest possible cost [2].

7.1.1 The Role of Transportation in Logistics

Transportation often accounts for between one-third and two-thirds of total logistics costs—i.e., between 9% and 10% of the gross national product for the Europe economy and also between 10% and 20% of a product's price, so transportation's importance and key role is undeniable. Transportation is essential for moving any shipment in a logistic system such as raw materials from sources to manufacturer, semifinished products between plants, and final goods to retailers and customers.

These days, with the growth of science and technology, increasing consumption and global commerce highlight the role of transportation in all processes. There is a high level of competition between manufacturers and also transportation holders in the quality of their customer services. Other critical competitive factors are reducing lead times, delays, and whole transportation costs, as well as increasing efficiency, reliability, safety, and reactivity in their service systems [3].

Distribution Channels

A few manufacturers sell their products directly to end users. For most of them, bringing products to end users may be a complex process that needs sales agents or brokers who get goods from producers and distribute them to retailers. At the end

* E-mail: Faranakkhooban@aut.ac.ir

Logistics Operations and Management. DOI: 10.1016/B978-0-12-385202-1.00007-4

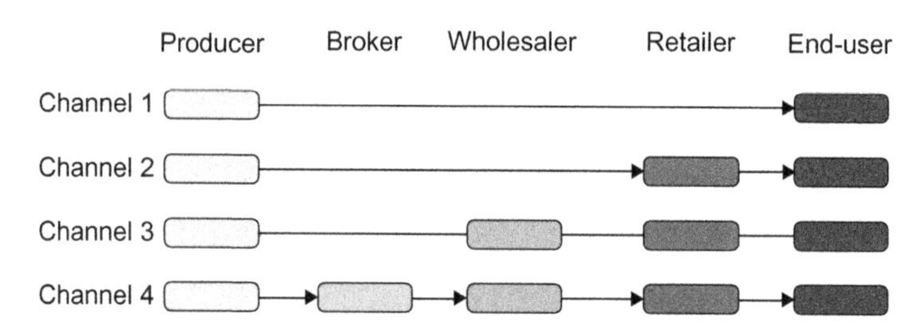

Figure 7.1 Channels of distribution [1].

of process, end users get their needs met by retailers. In this way, the cost of products may increase because of the existence of intermediaries in the product distribution process, but from a general point of view, comparing to manufacturers, intermediaries profit users by decreasing the transportation unit cost.

The path followed by a shipment from producer to customer is called the *distribution channel*. Distribution channels can be classified into four groups. In the first group, the channel has no intermediaries, so manufacturers send their products to end users directly. Some kinds of products such as cosmetics and encyclopedias sold door to door and handicrafts sold at local market are brought to the end user in this way through distribution channel 1. In the second distribution channel, retailers play intermediary roles (e.g., retailers in the tire industry buy from manufacturers and then resell products to their customers).

When manufacturers sell their products only in large quantities and retailers are not able to purchase these large quantities, wholesalers play the role of intermediary between manufacturers and retailers. Channel 3 is typical in the food industry. Channel 4 is as the same as channel 3, except that a producer contracts with a broker or sales agent who sells products to wholesalers (e.g., in the clothing industry) [1]. Figure 7.1 illustrates these four distribution channels.

7.1.2 Transportation Participants

The three basic participants in a transportation system are the shipper, the carrier, and government.

Shipper

One of the best ways to transport freight is to use shipper services. Shippers can move freights from origin to destination at the lowest cost and during a specified time period. The shipper ensures many transportation services such as particular pickup and delivery times, accurate and timely exchange of invoicing and information, zero loss and damage, and specified transit times. Some Shippers are producers of goods, this is while some others are just intermediary firms (brokers) which attribute demand to supply [4].

Carrier

Carriers tender transportation services. Railways, shipping lines, trucking companies, intermodal container services, and postal services are different kinds of carriers. Generally, carriers are classified into three main classes: common carriers, private carriers, and contract carriers. Depending on its market, any manufacturer or distributor can choose from these three choices to transport goods and products.

The best-known common carriers are public airlines, motor carriers, cruise ships, bus lines, railroads, and other freight companies. Common carriers' routes are defined and published in advance, and rate tables and time schedules for transporting people and goods require the approval of regulators (the government in most countries).

A common carrier is a business that offers its transportation services to the general public under license or the authority provided by a regulator. A common carrier holds itself out to provide transportation services to the general public without discrimination for "public convenience and necessity."

The significant problem in using common carriers is that the numbers of customers cannot be predicted in advance all the time. To get around this problem, firms may have special, private carriers to deliver their own products to end users, making shipments more predictable. For example, the Wingman's grocery store chain owns and operates its own private fleet of vehicles to deliver produce and goods to company stores; private carriers have an advantage over other carriers because of their flexibility and economy.

A contract carrier, on the other hand, is a kind of for-hire carrier agent that serves a limited number of shippers under specific contractual arrangements. According to contract, they provide a specified transportation service at specified cost; contract carriers are the same as private carriers except they do not hold serve the general public and in most instances have contract rates that are lower than those of common carriers.

Government

In most countries, public transportation systems and facilities such as rail facilities, roads, and ports are planned, constructed, and operated by governments. Governments also control the shipment of certain items (e.g., hazardous and poisonous products) and tax the transportation industry.

Governments have traditionally been more involved in the practices of carriers than in most other commercial enterprises; their regulations include restricting carriers to certain markets and regulating prices they can charge.

7.1.3 Delivery Frequency System

In an economic view, carriers must plan the frequency of service between any two points. Finding the best delivery frequency can decrease investment in equipment and facilities. Delivery frequency systems can be chosen between three approaches; they are customized transportation, consolidation transportation, and frequent operation [5].

Customized Transportation

In the customized transportation approach, truckload (TL) vehicles with a driver or driving team are dedicated to a specific customer. This transportation team starts its delivery trip when a customer asks for service. The truck is sent to the customer's origin site to begin loading. Then it moves to the customer's specified destination to unload. When driving team finish their tour, call the carrier's dispatcher to ask for next assignment if there is one. Otherwise the team should wait for next location. The customized delivery system creates a dynamic environment for TL carriers because most of transportation specifications related to customers (such as demand frequently, travel times, waiting delays at customer sites, and waiting delays of TL team until future assignment) are uncertain.

In these conditions, carriers should attempt to use their on-hand resources such as crews, fleets, vehicles, and trailers in the best possible way. To achieve this aim, developing the well-organized resource management and allocation plans should be the core of the carriers' management procedure in responding to the maximum demands of transportation [4].

Consolidation of Transportation

In the freight-transportation and logistics environment, there are many different ways to save transportation costs. One way is to consolidate transportation. In this way, it is possible to take advantage of economies of scale in transportation by substituting large shipments for small ones. Zhou et al. have delineated three general policies for transporting goods by vehicles. (i) the quantity policy, according to which the maximum capacity of a vehicle should be used by carrying the maximum number of freight quantity. (ii) the time policy, according to which the time of delivery is the most important factor and shouldn't exceed a preplaned time limits. (iii) the quantity and time policy, according to which both capacity and time are critical factors, so a vehicle is sent either when the delivery time limit arrives or when the freight quantity reaches to its maximum bound [6].

Consolidating freight is a way to cover these policies. It means consolidating demands from several points until a transporting vehicle is full. This on-demand approach has many benefits for carriers because the investment in vehicle capacity is much lower than the customized approach; as a result, lower unit transportation costs and high-capacity use are achieved. This approach may, however, be quite undesirable for customers with time-sensitive delivery requirements or who have high-value goods with high associated inventory, security, and holding costs [4].

Carriers must have accurate scheduling in order to plan their services and satisfy the expectations of the largest possible number of customers in the fewest number or series of routes. They must be able to group several services in a schedule and indicate departure and arrival times for stops along the route. To achieve this, carriers should adjust service-related characteristics such as routes, the capacities and types of vehicles and convoys, and intermediary stops and locations of different customers' origins and destinations. Carriers who use this system in the best possible way are ensured that their transport services are performed in a rational, efficient, and profitable way. In this system, carriers plan rules and policies that affect the whole system [4].

Consolidation of small shipments can occur in three ways. First, small shipments that must be transported over long distances or even short ones can be combined, just as when large shipments are transported over long distances (facility consolidation). Second, several small shipments can be replaced by a single large shipment by using an adjusted forward or backward shipment schedule (temporal consolidation). Third, when there are many pickup and delivery points, using a vehicle on a multistop route can serve less than truckload (LTL) pickup and deliveries associated with different locations (multistop consolidation) [1].

Frequent Operation

Another alternative for delivery services is frequent operations in which carriers provide fixed schedules that match their customers' shipping requirements. In this fixed schedule, delivery services are organized in advance—e.g., once a day or twice a week. In this approach, unpredictable numbers of customers in each service period cause uncertainty in shipping requirements. To cover the most possible demands, carriers need a higher-capacity investment (as compared to consolidating transport). However, predictability of operation schedules and the accuracy of anticipated shipping arrival dates are among the advantages of frequent service [5].

7.1.4 Long-Haul Consolidated Freight Transportation [4]

Freight-transportation operations over long distances and between terminals and cities may be performed by rail, truck, ship, and so on or any combination of these modes. The structure of long-haul consolidation transportation system consists of a whole network with terminals and related links. Consolidated carriers perform transportation services by using many kinds of trucks: railcars, trailers, containers, and ships, among others. Terminals with different designs and sizes play very critical roles in freight-transportation networks. They may be concentrated in specific operations, given service for only a particular kind of shipment, or presented an entire set of transportation services for any kind of shipment.

The following is a brief look at rail-transportation networks and LTL networks. Differences between these two types of networks are described in the following sections.

Railway Transportation Network

A railway network is made of single or double track lines that connect several different train yards together. When a customer calls for a service, an appropriate number of railcars at the nearest main yard are chosen, inspected, and transported to the freight pickup point. Loaded cars return to the origin yard to be ordered, grouped, consolidated, and assembled into *blocks*. A block contains a group of railcars that are considered a single unit with the same origin but perhaps different final destinations. Using blocks in train transportation systems has many economic advantages such as full train loads and the management of longer car strings in yards.

During the long-haul transportation trip, the train may travel on single-track lines, so it is common to meet trains traveling in the opposite direction. In this situation, the train with the higher priority passes first. The train may be stopped in middle train yards where cars and engines are regularly inspected and blocks are separated from one train and put on another. At the final yard, the first blocks are detached from the train and disassembled. Then cars are inspected, put in order, and moved to their final destination to be unloaded. Once a car finishes its delivery trip, it may move to a new pickup point and then be assembled in a new block or it may wait empty for a future assignment.

Managing the main yard's operations is the most complex activity in a long-haul railway transportation system.

LTL Transportation Network

In an LTL network, small vehicles pick up local traffic at origin points and deliver it to end-of-line terminals. Then local traffic from different parts of the network are grouped and consolidated into larger batches before they begin their long-haul journey. Breakbulks are terminals where arrivals from several origin points are gathered, unloaded, ordered, and consolidated for the rest of the long-haul transport [4]. Breakbulks in LTL networks are the same as main yards in rail-transportation systems. LTL carriers usually have their own terminals, but they use public transportation networks.

Railway and LTL System Differences

The structure of LTL transportation network basically is the same as a rail-transportation network but in simpler scale and with more flexibility in choosing ways to move materials to their destinations. Whereas rail-transportation links are limited, trucks may use any available link of the road and highway network while obeying weight regulations [4]. In LTL systems, terminal operations are generally simple. But in railway systems, more complicated consolidation operations are managed through grouping and consolidation of railcars into blocks and then into trains [7].

7.2 Classification of Transportation Problems

A lot of decision making problems in a logistic system are directly or indirectly related to transportation activities. In addition to standard transportation problems, handling a transportation system creates and develops many other decision making problems [1].

Freight transportation plans as decision making problems involve many different variables and constraints. Some of them can be applied for all transportation systems, whereas others are only relevant to specific modes or particular ways of system operation. For example, the vehicle and drivers scheduling problem is a common decision problem through which there is a least-cost allocation of vehicles and drivers over time. In this way, some constraints such as rules and guidelines on vehicle maintenance and crew rests may be satisfied.

7.2.1 Planning Levels [4]

Transportation systems are among the most complex organizations and involve many components such as human and material resources, complex connections, and balances between decision variables and management policies that directly or indirectly affect different components of the system. To decrease this complexity, researchers have provided a general classification for transportation problems with three planning levels: strategic (long term), tactical (medium term), and operational (short term).

Strategic (Long Term) Planning

Strategic planning involves decisions at the highest level of management and requires long-term investment. Strategic decisions develop general policies and extensively structure the functional strategies of the system. Any physical changes or development in whole network such as locating main facilities (e.g., hubs and terminals) are examples of strategic decision planning. Strategic planning takes place in international, national, and regional transportation systems.

Tactical (Medium Term) Planning

Tactical planning needs medium term investment and is not as critical as strategic planning. This class contains a well-organized allocation and operation of resources to improve system performance. Examples of this category are decision making in the design of service networks, service schedules, repositioning fleets, and traffic routing. Most carriers' decision making is at this level.

Operational (Short Term) Planning

Operational planning is short term and urgent decision making performed by local management, yard masters, and dispatchers. Decisions at this level do not need large investments. The completion and adjustment of schedules for services, crews, maintenance activities, and routing and dispatching of vehicles and crews are examples of this level.

In the next part, we introduce variants of standard transportation problems and then several important transportation problems, some of which are shipper decision problems and some of which are carrier decision problems.

7.2.2 Variants of the Standard of TPs

The time-minimization transportation problem (TMTP) minimizes the time to transport goods from m origins to n destination under some constraint of available sources and requested destinations. Such problems especially arise when perishable goods are transported or when it is required to transport essential items such as food and ammunition in the shortest possible time in a war scenario.

The fundamental difference between the cost-minimization transportation problem (CMTP) and TMTP is that the cost of transportation depends on the quantity of commodity being transported but the time involved is independent of this factor.

Many different objective function may be considered for a transportation problem such as minimization of transportation costs, minimization of labor turnover, minimization of risk to a firm or the environment, and minimization of deterioration of perishable goods [8].

Time Minimizing Solid Transportation Problem

The cost minimizing solid transportation problem (CMSTP) is [8]:

Minimize

$$z = \sum_{i=1}^{n}\sum_{j=1}^{m}\sum_{k=1}^{p} c_{ijk}x_{ijk} \tag{7.1}$$

subject to

$$\sum_{j=1}^{n} x_{ijk} = a_{jk}, \quad \sum_{j=1}^{m} x_{ijk} = b_{ki} \sum_{k=1}^{p} x_{ijk} = e_{ij} \tag{7.2}$$

$$\sum_{j=1}^{m} a_{jk} = \sum_{i=1}^{p} b_{ki}, \quad \sum_{k=1}^{p} b_{ki}, \quad = \sum_{j=1}^{m} e_{ij}, \quad \sum_{i=1}^{n} e_{ij} = \sum_{k=1}^{p} a_{jk} \tag{7.3}$$

$$\sum_{j=1}^{m}\sum_{k=1}^{p} a_{jk} = \sum_{k=1}^{p}\sum_{i=1}^{n} b_{ki} = \sum_{i=1}^{n}\sum_{j=1}^{m} e_{ij}, \quad x_{ij} \geq 0 \tag{7.4}$$

where

i is the number of origin points providing type k of goods,
j is the number of destinations,
x_{ijk} is the number of type k sent from the ith origin to the jth destination,
c_{ijk} is the cost of transporting the unit item of the kth from the ith supply point to the jth destination,
a_{jk} is the requirement at the jth destination of type k of goods,
b_{ki} is the availability of type k of goods at the ith supply point,
e_{ij} is the total quantity of goods to be sent from the ith supply point to the jth destination.

The TMTP form of this problem is as given below.
Minimize

$$[\mathrm{Max}t_{ijk} : x_{ijk} > 0] \tag{7.5}$$

subject to Equations (7.2–7.4), where t_{ijk} is the time of transportation type k of goods from the ith source to the jth destination.

It is assumed that all carriers start simultaneously. The convexity of objective function has been demonstrated in CMSTPs but not in TMTPs.

Pricing of Bottlenecks at Optimal Time in a Transportation Problem

The conventional transportation problem deals with minimizing the cost of transporting a homogeneous product from various supply points to a number of destinations without caring for the time of transportation. By increasing the bottleneck at time T, we can have a less-cost transportation schedule. However, a commissioning of the project is influenced by the bottleneck. Thus, the larger bottleneck has to be valued by comparing its compact on the functioning of the project with the saving in transportation cost. Assuming that the impact of the bottleneck flow on the functioning of the project is known, a convergent iterative procedure was proposed by Malhotra and Puri [9] which finds all various efficient pairs at time T.

Minimize

$$\sum_{i\in I}\sum_{j\in J} c_{ij}x_{ij} \tag{7.6}$$

subject to the following constraints:

$$\left.\begin{array}{l}\sum_{j\in J} x_{ij} = a_i, a_i > 0, i\in I,\\ \sum_{i\in I} x_{ij} = b_j, b_j > 0, j\in J\\ x_{ij} \geq 0, i\in I, j\in J\end{array}\right\} \tag{7.7}$$

where

i denotes the index set of supply points,
j the index set of destination,
x_{ij} the number of the product transported from the ith supply point to the jth destination,
c_{ij} the unit cost of transportation on the (i,j)th route.

Minimize

$$\sum_{i\in I}\sum_{j\in J} c'_{ij}x_{ij} \tag{7.8}$$

subject to Equation (7.7), where c'_{ij} is different for different values of t_{ij}.

$$c'_{ij} = \begin{cases} 1 \text{ if } t_{ij} = T \\ 0 \text{ if } t_{ij} < T \\ \infty \text{ if } t_{ij} > T \end{cases} \tag{7.9}$$

Bi-Criteria Transportation Problem

A bi-criteria transportation problem is a kind of problem with two linear objectives that are the minimization of the total transportation cost and minimization of the total deterioration of goods during transportation [10]. The mathematical model of problem is formulated as below.

Minimize $z = (z_1, z_2)$

$$z_1 = \sum_{i \in I} \sum_{j \in J} c_{ij} x_{ij} \tag{7.10}$$

$$z_2 = \sum_{i \in I} \sum_{j \in J} d_{ij} x_{ij} \tag{7.11}$$

subject to

$$\left. \begin{aligned} &\sum_{j \in J} x_{ij} = a_i, \quad a_i > 0, \quad i \in I \\ &\sum_{i \in I} x_{ij} = b_j, \quad b_j > 0, \quad j \in J \\ &x_{ij} \geq 0, \quad (i,j) \in I \times J \end{aligned} \right\} \tag{7.12}$$

where

i denotes the index set of supply points,
j is the index set of destination,
x_{ij} is the number of the product transported from the ith supply point to the jth destination,
c_{ij} is the unit cost of transportation on (i,j)th route,
d_{ij} is the cost of deterioration of a unit while transporting from i to j,
a_{ij} is the availability of the product at the ith supply point,
b_j is the demand at the destination j.

Observing that the nondominated set in the decision space has the larger number of extreme points compared with the extreme points of the nondominated set in the criteria space, Aneja and Nair [10] also have developed an algorithm to determine the efficient extreme points in the criteria space. They have solved the same transportation problem again and again but with unlike objectives. In continual iterations, the objective function is the positive weighted average of the two linear objectives under consideration.

7.2.3 Carrier Decision-Making Problems

A carrier decision-making problem is a problem whose objective function is defined in the same direction of maximization of a carrier's profits. Some of the carrier decision-making problems are crew-assignment problems, vehicle allocation and scheduling problems, terminal design problems, allocation and operation problems, freight-traffic assignment problems, service network design problems, and fleet-composition problems [1].

As mentioned before, most of a carrier's decision-making problems belong to the tactical level.

Dynamic Driver Assignment Problem

The first carrier decision-making problem introduced here is the dynamic driver assignment problem (DDAP) that arises in TL trucking. Crews are assigned to vehicles in order to support the planned operations [11].

In this problem, a fully loaded vehicle is assigned to a driver in a scheduled operation. It may take several days for a vehicle to be fully loaded because the customer's demands are not known in advance and are received randomly. Here each driver is supposed to be assigned to just one demand at one time [1].

The DDAP is formulated as a minimum-cost problem during which, the cost of driver's assignments (for empty moving from waiting location to pick-up point) is minimized. It is assumed that n drivers are waiting for assignment to fully loaded vehicles. Let the set of drivers be shown by $D=\{1, \ldots, n\}$ and the set of ready fully loaded vehicles by $V=\{1, \ldots, m\}$. If the maximum number of D and V are not equal ($n \neq m$), the extra number is compensated by is compensated by a dummy value 0. Once $n<m$, a dummy driver 0 is added to D, whereas if $m<n$, a dummy load 0 is added to V. In this case ($m<n$), the DDAP is formulated as follows.

Minimize

$$\sum_{i \in D} \sum_{j \in L} c_{ij} x_{ij} \tag{7.13}$$

subject to

$$\begin{aligned} &\sum_{i \in D} x_{ij} = 1, \quad j \in L \backslash \{0\} \\ &\sum_{j \in L} x_{ij} = 1, \quad i \in D \\ &x_{ij} \in \{0,1\}, \quad i \in D, \quad j \in L \end{aligned} \tag{7.14}$$

where

c_{ij} is the predefined cost for driver i who drives vehicle j,

x_{ij} is a binary variable equal to 1 if driver i is assigned to vehicle j and 0 otherwise.

There are also many other known problems in the crew-scheduling management category, such as optimizing the scheduling of terminal employees or the quantity of the reserve crew [12].

Fleet-Composition Problem

Carriers always try to decrease the investment in their own crew. In other words, paying attention to the variety of demands over a year, they avoid having their own maximum number of vehicles needed in peak periods during a year. They usually have a base number of their own vehicles to answer their usual demands, and they hire additional required vehicles in peak periods. In this way, carriers can save money by creating a balance point at which the total cost of their own and hired vehicles are minimized. For this problem, similarity of vehicles is an assumption [1].

Let a year include n time periods; the number of own vehicles is shown by v. If $i = \{1, \ldots, n\}$ is the set of time periods, then v_i is the number of needed vehicles during the time period t. The main variable m is the number of time periods per year in which carriers need to hire vehicles ($v_i > v$). There are two types of costs for carriers' own vehicles during a time period: fixed cost (c_f) and variable cost (c_v), respectively, while c_h is the cost of a hired vehicle in the same time period. This problem is formulated in a simple minimization model without any constraint as follows.

Minimize

$$c_i(v) = c_\mathrm{f}(v) + c_\mathrm{v}\min(v_i, v) + c_\mathrm{h}(v_i - v) \tag{7.15}$$

To find the annual cost, we just need to add up the costs of different periods of a year. The formulation is changed in Equation (7.16).

Minimize

$$\sum_{i=1}^{n} c_i(v) = \sum_{i=1}^{n} c_f(v) + \sum_{i=1}^{n} c_v\min(v_i, v) + \sum_{i=v_i>v} c_h(v_i - v) \tag{7.16}$$

Let C_f be the annual fixed cost, while C_v and C_h are the annual variable costs for own and hired vehicles, respectively. Now we can formulate this problem in a simpler form in Equation (7.17).

Minimize

$$C(v) = nC_\mathrm{f}v + C_\mathrm{v}\sum_{i=1}^{n}\min(v_i, v) + C_\mathrm{h}\sum_{i:v_t>v}(v_i - v) \tag{7.17}$$

Because this transportation cost problem has a simple linear formulation without any constraint, the optimal annual cost is found by setting the derivation of $c(v)$

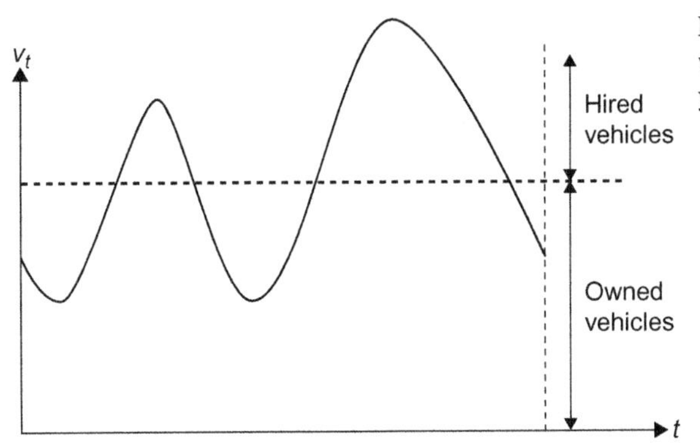

Figure 7.2 Fleet compositions when demand varies over the year [1].

(with regard to v) equal to 0 (Equation 7.18). Derivative of $c(v)$ is found in the following equation, so the best value of m is obtained in Equation (7.19).

$$nC_\mathrm{f} + C_\mathrm{v}m - C_\mathrm{h}m = 0 \tag{7.18}$$

Figure 7.2 demonstrates the variety of demands over the year. Minimum annual cost is obtained when the area below and above the line $v_t = v$ be equal. This condition is achieved if the Equation (7.18) is satisfied.

$$m = n\frac{C_\mathrm{f}}{C_\mathrm{h} - C_\mathrm{v}} \tag{7.19}$$

Vehicle Allocation and Scheduling Problem

Carriers can best use their vehicles by applying an optimum allocated schedule to respond the maximum number of demands. A vehicle-allocation problem is a kind of carrier's decision problem formulated as a minimum-cost flow problem in which carriers decide which demand should be responded to and which one should be rejected, which vehicle should be moved to a new pickup point, and which one should wait for a future assignment [1].

In this problem, it is supposed that all demands have been known in advance and all vehicles are the same in type, size, and capacity. Let $t = \{1, \ldots, T\}$ be the set of time periods assumed to divide the planning horizon and N be the set of demand pickup or delivery points. For this problem, there are several parameters: d_{ijt}, $i \in N$, $j \in N$, $t = 1, \ldots, T$, is an available demand delivered from pickup point i to destination j during time period t; τ_{ij}, $i \in N$, $j \in N$ is the travel time from point i to point j; c_{ij}, $i \in N$, $j \in N$, is the cost of moving an empty vehicle from pickup point i to point j; p_{ij}, $i \in N$, $j \in N$, is the profit obtained by delivering a shipment from origin i to destination j; and m_{it}, $i \in N$, $t = 1, \ldots, T$, is the available number of vehicles ready to be moved from point i in time period t.

There are two types of decision variables in this problem: x_{ijt}, $i \in N$, $j \in N$, $t = 1, \ldots, T$, denotes the number of vehicles starting their delivery services from origin i and ending at destination j in time period t; and y_{ijt}, $i \in N$, $j \in N$, $t = 1, \ldots, T$, indicating the number of vehicles moving empty from point i to point j at time period t. This minimum-cost flow problem can be formulated as a maximum-profit problem.

Maximize

$$\sum_{t=1}^{T} \sum_{i \in N} \sum_{j \in N, j \neq i} (p_{ij} x_{ijt} - c_{ij} y_{ijt}) \tag{7.20}$$

subject to

$$\sum_{j \in N} (x_{ijt} + y_{ijt}) - \sum_{k \in N, k \neq i: t > \tau_{ki}} \left(x_{ki(t-\tau_{ki})} + y_{ki(t-\tau_{ki})} \right) - y_{iit} - 1 = m_{it}, i \in V, t \in \{1, \ldots, T\} \tag{7.21}$$

$$x_{ijt} \leq d_{ijt}, \quad i \in N, \quad j \in N, \quad t \in \{1, \ldots, T\} \tag{7.22}$$

$$x_{ijt} \geq 0, \quad i \in N, \quad j \in N, \quad t \in \{1, \ldots, T\} \tag{7.23}$$

$$y_{ijt} \geq 0, \quad i \in N, \quad j \in N, \quad t \in \{1, \ldots, T\} \tag{7.24}$$

where the objective function (Equation 7.20) is the total profit over the whole planning horizon derived by revenues minus costs.

Constraint (7.21) confirms that the number of entered vehicles at point i during time period t must equal the exact value m_{it}. Constraint (7.22) states that the number of vehicles moved from point i to point j in time period t should not be more than the number of demands between these points during this period. Regarding this, $d_{ijt} - x_{ijt}$ represent loads should be rejected.

7.2.4 Shipper Decision-Making Problems

A shipper decision-making problem is a problem with an objective function defined in the same direction as the maximization of a shipper's profits. Some of the shipper decision-making problems are transportation mode selection, a shipment consolidation and dispatching problem, a commodities load planning and packing problem, and a carrier-type decision problem [1].

Shipment Consolidation and Dispatching

Shipment consolidation and dispatching problem is a kind of shipper decision-making problem often faced by producers (if they do their delivery activities themselves) and contracted shippers. The manufacturer or shipper has to choose the best way for timely delivery of orders to customers during a time horizon divided by T periods. They should find the most suitable transportation mode for each shipment. They also should choose the best design for consolidation of shipments and estimate the start time of dispatching. In this way, any related scheduling critical factor should be considered [1].

Shipment consolidation and dispatching problem is formulated as a minimization model. Herein there are several parameters for each vehicle i: a destination $d_i \in N$, a weight $w_i \geq 0$, a ready time r_i (the period in which vehicle i is ready for delivery), and a deadline k_i (the period in which vehicle i should be reached to destination d_i).

A shipper may use LTL carriers or hire one-way truck trips. For rented trucks, consider a set of route R with the following characteristics: S_r is the set of cessations during route r; f_r is the fixed cost for using route r; q_r is the available capacity of route r. Let τ_{ir}, $i \in I$, $r \in R$, be the number of time periods taken to deliver shipment i on route r. By using LTL carriers, g_i is the cost of delivery for shipment i, and τ'_i is the number of time periods this delivery takes.

This problem has three kinds of binary variables: (1) x_{irt}, $i \in I$, $r \in R$, $t = 1, \ldots, T$, is equal to 1 if shipment i starts its delivery trip on time period t during route r and 0 otherwise; (2) y_{rt}, $r \in R$, $t = 1, \ldots, T$, is equal to 1 if route r is used in time period t and 0 otherwise; and (3) w_i is equal to 1 if shipment i is delivered by LTL carriers and 0 otherwise (this variable is definable only when $r_i + \tau'_i < k_i$).

Minimize

$$\sum_{r \in R} \sum_{t=1}^{T} f_r y_{rt} + \sum_{i \in I} g_i w_i \tag{7.25}$$

subject to

$$\sum_{i: r_i \leq t \leq d_i - \tau_{ir}, s_i \in S_r} w_i x_{irt} \leq q_r y_{rt} r \in R, \quad t = 1, \ldots, T \tag{7.26}$$

$$\sum_{r: s_i \in S_r} \sum_{t: r_i \leq t \leq k_i - \tau_{ir}} x_{irt} + w_i = 1 \; i \in I \tag{7.27}$$

$$x_{irt} \in \{0,1\}, \quad i \in I, \quad r \in R, \quad t = 1, \ldots, T \tag{7.28}$$

$$y_{rt} \in \{0,1\}, \quad r \in R, \quad t = 1, \ldots, T \tag{7.29}$$

$$w_i \in \{0,1\}, \quad i \in I \tag{7.30}$$

In objective function (Equation 7.25), the total cost for delivering shipments is minimized. Constraint (7.26) states that for each route r and time period $t = 1, \ldots, T$, if LTL carriers are used, the total weight carried on route r during time period t should not be more than capacity q_r. Constraints (7.27) confirm that delivery of each shipment should be assigned to only one of transportation operators (LTL carriers or rented trucks).

7.3 Case Study: An Application of Cost Analyses for Different Transportation Modes in Turkey

In this study, a cost analysis is conducted and costs are compared by using data concerning Turkey for different modes of transportation to calculate the transportation cost of a unit of cargo for each mode. Therefore, current data for investment costs, operational and maintenance costs, and fuel and external costs for each transportation mode are collected for the proposed cost-analysis method. In this study, different modes of truck, rail, and ship are compared with each other [13].

Studies and data on external cost estimations of different transportation modes on country basis are not satisfactory. Therefore, taking into consideration the available data for Turkey and the results of different international analyses such as those carried out by [14–18], estimations are made for the specific external cost data for different transportation modes.

Standard vehicle types that can be suitably used in this country are selected for different transportation modes. The selected vehicles are a general cargo ship with a capacity of 3300 deadweight tonnage for the seaway, a freight train with a capacity of 700 tons for the railroad, and a truck with a capacity of 20 tons for the road.

The mathematical formulation for calculating the transportation cost of a unit of cargo derived in this research is:

$$U_{\mathrm{T}} = \frac{\sum_{t=1}^{n} C_{\mathrm{k}}(t) + C_{\mathrm{f}}(t) + C_{\mathrm{m}}(t) + C_{\mathrm{ex}}(t)}{\sum_{t=1}^{n} Y_{\mathrm{s}}(t)} \tag{7.31}$$

where

$C_{\mathrm{k}}(t)$ denotes the investment cost per unit of cargo or passenger,
$C_{\mathrm{f}}(t)$ is the fuel and lubricant costs per unit of cargo or passenger,
$C_{\mathrm{m}}(t)$ is the operational and maintenance costs per unit of cargo or passenger,
$C_{\mathrm{ex}}(t)$ is the external costs per unit of cargo or passenger,
$Y_{\mathrm{s}}(t)$ is the number of annual cargoes and passengers or the number of cars that can be carried in a ferry,
U_{T} is the total cargo or passenger cost per unit.

Results of the Study

The total cost of a unit of cargo in sea transportation consists of 26% investment cost, 32% fuel cost, 35% operational and maintenance cost, and 7% external cost. These percentages for railroad transportation are 22% for investment, 46% for fuel,

30% for operations and maintenance, and 2% for external costs. For road transportation, these values are 14%, 60%, 17%, and 9%, respectively.

An analysis of the figures for sea transportation revealed the following points: It is considered that a fullness ratio of 60% is the lower limit for sea-cargo transportation, which may change according to route lengths. Thus, the fleet size and the optimal vessel capacity can be determined by taking into consideration the annual cargo potential on a given sea transportation route.

By looking at the figures for road transportation, we can conclude that a fullness ratio of 80% was chosen as the lower limit for road-cargo transportation. Determining the fleet size for this mode of transportation based on this fullness ratio may provide an important reduction in transport costs. The proportion of road-cargo transport costs to sea-cargo transport costs is 7. This ratio changes a little with route length.

The evaluation of the figures for railroad transportation reveals that if the route length becomes greater than 350 kilometers, railroad transportation becomes more economical than road transport. Within the range of 350–1000 kilometers, the economic advantage of the railroad transportation changes between 0% and 20%. This advantage does not change with the fullness ratio.

References

[1] G. Ghiani, G. Laporte, R. Musmanno (Eds.), Introduction to Logistics Systems Planning and Control, John Wiley & Sons, New York, 2004.

[2] T. Gudehus, K. Herbert (Eds.), Comprehensive Logistics, Springer, Berlin Heidelberg, 2009.

[3] A. Hoff, H. Andersson, M. Christiansen, G. Hasle, A. Løkketangen, Industrial aspects and literature survey: fleet composition and routing, Comput. Oper. Res. 37 (2010) 2041–2061.

[4] T.G. Crainic, W.H. Randolph, Long-haul freight transportation, in: W.H. Randolph (Ed.), Hand Book of Transportation Science, second ed., Kluwer Academic publishers, New York, 2003.

[5] G.D Taylor (Ed.), Logistics Engineering Handbook, Taylor & Francis, London and New York, 2008.

[6] G. Zhou, Y.V. Hui, L. Liang, Strategic alliance in freight consolidation, Transp. Res. Part E. 47 (2011) 18–29.

[7] C. Barnhart, G. Laporte (Eds.), Handbook in Operation Research and Management Science, vol. 14, Elsevier, Amsterdam, 2007, p. 783.

[8] R. Malhotra, S.S Lalitha, P. Gupta, A. Mehra, R. Sonia, Combinatorial Optimization: Some Aspects, Narosa, India, 2007.

[9] R. Malhotra, M.C Puri, Pricing of bottlenecks at optimal time in a transportation problem, in: N.K. Jaiswal (Ed.), Scientific Management of Transport Systems, North Holland Publishing Company, Amsterdam, 1981, pp. 234–261.

[10] Y.P. Aneja, K.P.K. Nair, Bi-criteria transportation problem, Manag. Sci. 25 (1979) 73–78.

[11] C. Barnhart, E.L. Johnson, G.N. Nemhauser, P. Vance, Airline crew scheduling, in: W.H. Randolph (Ed.), Hand Book of Transportation Science, second ed., Kluwer Academic publishers, New York, 2003.
[12] Y. Nobert, J. Roy, Freight handling personnel scheduling at air cargo terminals, Transport. Sci. (1998) 32295–32301.
[13] B. Sahin, H. Yilmaza, Y. Usta, A.F. Gunerib, B. Gulsunb, An approach for analyzing transportation costs and a case study, Eur. J. Oper. Res. 193 (2007) 1–11.
[14] D.J. Forkenbrock, External costs of intercity truck freight transportation, Transp. Res Part A. 33 (1999) 505–526.
[15] D.J. Forkenbrock, Comparison of external costs of rail and truck freight transportation, Transp. Res Part A. 35 (2001) 321–337.
[16] M. Beuthe, F. Degrandsart, J.F. Geerts, B. Jourquin, External costs of the Belgian inter urban freight traffic: a network analysis of their internalization, Transp. Res Part E. 7 (2002) 285–301.
[17] E. Quinet (Ed.), Internalising the Social Costs of Transport, OECD/ECMT, Paris, 1994, Chapter 2.
[18] E. Quinet, A meta-analysis of western European external costs estimates, Transport. Transp. Res Part D. 9 (2004) 465–476.

8 The Vehicle-Routing Problem

Farzaneh Daneshzand

Department of Industrial Engineering, Amirkabir University of Technology, Tehran, Iran

In brief, the solution of vehicle-routing problem (VRP) determines a set of routes that starts and ends at its own depot, each performed by a single vehicle in a way that minimizes the global transportation cost and fulfills the demands of the customers and operational constraints (Figure 8.1) [1].

8.1 Definitions and Applications

The pioneers of the VRP were Dantzig and Ramser [1]. They proposed the first mathematical programming formulation and algorithmic approach of VRP in a real-world application in 1959. A few years later, Clark and Wright [2] improved the Dantzig–Ramser approach by proposing a heuristic. Following these two papers, many researchers studied algorithms and models for different versions of the VRP.

Researchers are interested in studying the VRP for two reasons: its practical relevance and its difficulty. Lenstra and Rinnooy Kan [3] have analyzed the complexity of the VRP and concluded that practically all of the VRP problems are nondeterministic polynomial-time hard (NP-hard).

VRP has many applications in real-world cases. Some applications are solid-waste collection, street cleaning, school bus routing, routing of salespeople and maintenance units, transportation of handicapped people, and so forth.

8.2 Basic VRP Variants

Fundamental components of the VRP are road network, customers, depots, vehicles, and drivers. To make different versions of VRP, different constraints and situations can be imposed on each component, and each of them can be supposed to achieve particular objectives [4].

Basic variants are capacitated VRP (CVRP), distance-constrained and capacitated VRP (DCVRP), VRP with time window (VRPTW), VRP with backhauls (VRPB), VRP with pickup and delivery (VRPPD), and any combination of these variants.

Logistics Operations and Management. DOI: 10.1016/B978-0-12-385202-1.00008-6

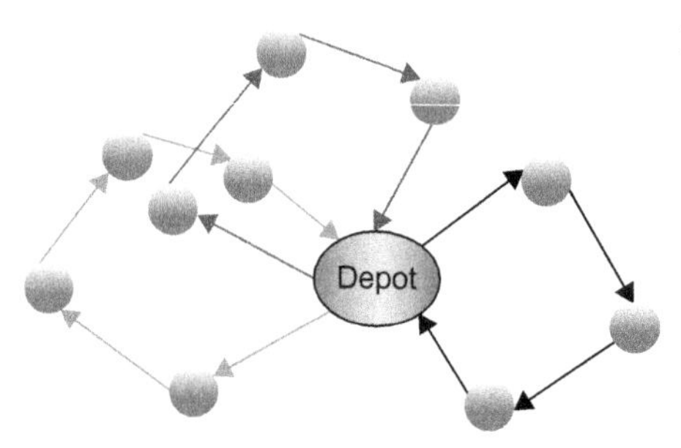

Figure 8.1 The scheme of the VRP [2].

Other VRP variants that have been studied recently in the literature are open VRP (OVRP), multidepot VRP (MDVRP), mix fleet VRP (MFVRP), split-delivery VRP (SDVRP), periodic VRP (PVRP), stochastic VRP (SVRP), and fuzzy VRP (VRPF). In the following sections, we will discuss these variants and their most important formulations.

8.2.1 The Capacitated VRP

The basic version of VRP is CVRP. In this problem, each vehicle has a capacity that is known in advance, so loading the vehicle more than its capacity is not allowed. There are two versions of CVRP: ACVRP, when the cost matrix is asymmetric, and SCVRP, when the cost matrix is symmetric.

The integer linear programming formulation of ACVRP proposed by Toth and Vigo [4] is presented as follows.

Model Assumptions

- The demands are deterministic.
- The demands may not be split.
- The vehicles are identical.
- The vehicles are based at a single central depot.
- The capacity restrictions for the vehicles are imposed.

Model Inputs

$G=(V, A)$: A complete graph
$V=\{0, \ldots, n\}$: The vertex set
A: The arc set
d_j: The demand of each customer ($d_0=0$)
c_{ij}: The nonnegative travel cost spent to go from vertex i to vertex j
$S \subseteq V$: The customer set
$d(S)=\sum d_i$: The total demand of the set
K: The number of identical vehicles
C: The capacity of each vehicle
$K_{\min}$, $r(S)$: The minimum number of vehicles needed to serve all customers

Model Output

$x_{ij} = 1$ if arc $(i, j) \in A$ belongs to the optimal solution and 0 otherwise.

Objective Function and Its Constraints

$$\min \sum_{i \in V} \sum_{j \in V} c_{ij} x_{ij} \tag{8.1}$$

$$\sum_{j \in V} x_{ij} = 1 \quad \forall i \in V/\{0\} \tag{8.2}$$

$$\sum_{j \in V} x_{ij} = 1 \quad \forall i \in V/\{0\} \tag{8.3}$$

$$\sum_{i \in V} x_{i0} = K \tag{8.4}$$

$$\sum_{j \in V} x_{0j} = K \tag{8.5}$$

$$\sum_{i \notin S} \sum_{j \in S} x_{ij} \geq r(s) \;\; \forall S \subseteq V/\{0\},\; S \neq \phi \tag{8.6}$$

$$x_{ij} = \{0, 1\} \quad \forall i, j \in V \tag{8.7}$$

Equations (8.2) and (8.3) are indegree and outdegree constraints, respectively. Constraints (8.4) and (8.5) impose the degree requirement for the depot vertex. Inequalities (Eqn (8.6)) are called capacity cut constraints (CCCs), and they impose vehicle capacity requirements while ensuring the connectivity of the solution. In fact, they stipulate that each cut $(V/S, S)$ defined by a customer set S is crossed by a number of arcs not smaller than $r(S)$.

An alternative formulation may be obtained by transferring the CCCs into subtour elimination constraints (SECs):

$$\sum_{i \in S} \sum_{j \in S} x_{ij} \leq |S| - r(S) \tag{8.8}$$

This constraint indicates that at least $r(S)$ arcs leave each customer set S. Both families of constraints (8.6) and (8.8) grow exponentially with n. It means that it is practically impossible to solve the linear programming relaxation of the problem directly (Eqns (8.1–8.7)). A possible way to solve these problems is to consider only some of these constraints and to add the remaining ones only if needed.

8.2.2 Distance-Constrained and Capacitated VRP

The DCVRP is a variant of CVRP on which both vehicle capacity and maximum distance constraints are imposed. In such problems, each tour length should not exceed the quantity known before. The symmetric DCVRP model of Laporte et al. [5] is presented as follows.

Model Assumptions (Other Than Assumptions of CVRP)

- Distance restrictions are imposed.

Model Inputs (Other Than Inputs of CVRP)

$r'(S)$: Given a subset S of customer vertices, the quantity $r'(S)$ represents the minimum number of vehicles needed to serve all the customers in S.

Model Output

$x_{ij} = 1$ if arc $(i, j) \in A$ belongs to the optimal solution and 0 otherwise.

Objective Function andits Constraints

$$\min \sum_{i \in V/\{n\}} \sum_{j>i} c_{ij} x_{ij} \tag{8.9}$$

$$\sum_{h<i} x_{hi} + \sum_{j>i} x_{ij} = 2 \quad \forall i \in V/\{0\} \tag{8.10}$$

$$\sum_{j \in V/\{0\}} x_{0j} = 2K \tag{8.11}$$

$$\sum_{i \in S} \sum_{\substack{j>i \\ j \in S}} x_{ij} \leq |S| - r'(S) \quad \forall S \subseteq V/\{0\},\ S \neq \phi \tag{8.12}$$

$$x_{ij} \in \{0, 1\} \quad \forall i, j \in V/\{0\},\ i<j \tag{8.13}$$

$$x_{0j} \in \{0, 1, 2\} \quad \forall i \in V/\{0\} \tag{8.14}$$

Constraints (8.10) and (8.11) are the degree constraints. Inequality (Eqn (8.12)) is an SEC that imposes the connectivity of solution, the vehicle capacity, and the maximum route length requirements by forcing a sufficient number of edges to leave each subset of vertices.

8.2.3 VRP with Time Windows

The VRPTW is the extension of CVRP where the service at each customer must start within a specified time window and the vehicle must remain at the customer's location during service. The model of Toth and Vigo [4] for VRPTW is presented here.

Model Assumptions (Other Than Assumptions of CVRP)

- For each customer i, the service starts within the time window, $[a_i, b_i]$, and the vehicle stops for s_i time instants.

Model Inputs (Other Than Inputs of CVRP)

E: The earliest possible departure from the depot
L: The latest possible arrival at the depot
$[a_0, b_0] = [a_{n+1}, b_{n+1}] = [E, L]$: The time window associated with node 0, $n + 1$
$\Delta^+ (i)$: The vertices that are directly reachable from i, the forward start of i
$\Delta^- (i)$: The vertices from which i is directly reachable, the backward start of i
S_i: The service time for customer i
t_{ij}: The travel time for each arc $(i, j) \in A$
w_{ik}: The start of service at node i when serviced by vehicle k.

Model Outputs

$x_{ijk} = 1$ if arc (i, j) is used by vehicle k, $(i, j) \in \mathrm{A}$, $k \in K$.
Note that the depot is presented by two nodes: 0, $n + 1$.

Objective Function and its Constraints

$$\min \sum_{k \in K} \sum_{(i,j) \in A} c_{ij} x_{ijk} \tag{8.15}$$

$$\sum_{k \in K} \sum_{j \in \nabla^+ (i)} x_{ijk} = 1 \quad \forall i \in N \tag{8.16}$$

$$\sum_{j \in \nabla^+ (0)} x_{0jk} = 1 \quad \forall k \in K \tag{8.17}$$

$$\sum_{i \in \nabla^- (j)} x_{ijk} - \sum_{i \in \nabla^+ (j)} x_{jik} = 0 \quad \forall k \in K, \; j \in N \tag{8.18}$$

$$\sum_{i \in \nabla^-(n+1)} x_{i,n+1,k} = 1 \quad \forall k \in K \tag{8.19}$$

$$x_{ijk}(w_{ik} + S_i + t_{ij} - w_{ik}) \leq 0 \quad \forall k \in K, \ (i,j) \in A \tag{8.20}$$

$$a_i \sum_{j \in \nabla^+(i)} x_{ijk} \leq w_{ik} \leq b_i \sum_{j \in \nabla^+(i)} x_{ijk} \ \forall k \in K, \ i \in N \tag{8.21}$$

$$E \leq w_{ik} \leq L \quad \forall k \in K, \ i \in \{0, n+1\} \tag{8.22}$$

$$\sum_{i \in N} d_i \sum_{j \in \nabla^+(i)} x_{ijk} \leq C \quad \forall k \in K \tag{8.23}$$

$$x_{ijk} \geq 0 \quad \forall k \in K, \ (i,j) \in A \tag{8.24}$$

$$x_{ijk} \in \{0, 1\} \quad \forall k \in K, \ (i,j) \in A \tag{8.25}$$

The objective function (Eqn (8.15)) expresses the total cost. Constraint (8.16) restricts the assignment of each customer to exactly one vehicle route. Constraints (8.17–8.19) characterize the flow on the path to be followed by vehicle k. Constraints (8.20–8.23) guarantee schedule feasibility according to time and capacity considerations, respectively, and the last constraint imposes binary conditions on flow variables.

8.2.4 VRP with Backhauls

In VRPB, customers can demand or return some commodities. In fact, it is an extension of the CVRP in which the customers are partitioned into two subsets: line-haul and back-haul customers. Each line-haul customer requires a given quantity to be delivered while a given quantity of products must be picked up from back-haul customers.

This kind of mixed distribution causes a significant saving in transportation costs, because one is able to visit back-haul customers while delivering the products to the line-haul customers. The assumption is that on each route, all deliveries are made before any pickups [6].

Here, we present the formulation of Toth and Vigo [4] for VRPB as an asymmetric problem.

Model Assumptions (Other Than Assumptions of CVRP)

- The sum of demands of the line-haul and back-haul vertices visited by a circuit does not exceed separately the vehicle capacity, C.
- In each circuit, the line-haul vertices precede the back-haul vertices, if any.

Model Inputs (Other Than Inputs of CVRP)

$L = \{1, \ldots, n\}$: Line-haul customer subset

$B = \{n+1, \ldots, n+m\}$: Back-haul customer subset

$F = L \cup B$

C_{ij}: The nonnegative cost associated with each arc $(i, j) \in A$

$L_0 = L \cup \{0\}$

$B_0 = B \cup \{0\}$

$\overline{G} = (\overline{V}, \overline{A})$: A directed graph obtained from G by defining $\overline{V} = V$

$$\overline{A} = (A_1 \cup A_2 \cup A_3)$$

$A_1 = \{(i, j) \in A: i \in L_0, j \in L\}$

$A_2 = \{(i, j) \in A: i \in B, j \in B_0\}$

$A_3 = \{(i, j) \in A: i \in L, j \in B_0\}$

Model Outputs

$x_{ij} = 1$ if and only if arc (i, j) is in the optimal solution and 0 otherwise.

Objective Function and Its Constraints

$$\min \sum_{(i,j)\in\overline{A}} c_{ij}x_{ij} \tag{8.26}$$

$$\sum_{i\in\nabla_j^-} x_{ij} = 1 \quad \forall j \in \overline{V}/\{0\} \tag{8.27}$$

$$\sum_{i\in\nabla_i^+} x_{ij} = 1 \quad \forall i \in \overline{V}/\{0\} \tag{8.28}$$

$$\sum_{i\in\nabla_0^-} x_{i0} = K \tag{8.29}$$

$$\sum_{i\in\nabla_0^+} x_{0j} = K \tag{8.30}$$

$$\sum_{j \in S} \sum_{i \in \nabla_j^- / S} x_{ij} \geq r(S) \quad \forall S \in F \tag{8.31}$$

$$\sum_{i \in S} \sum_{i \in \nabla_j^+ / S} x_{ij} \geq r(S) \quad \forall S \in F \tag{8.32}$$

$$x_{ij} \in \{0, 1\} \quad \forall (i,j) \in \overline{A} \tag{8.33}$$

The objective function minimizes the total cost. Equations (8.27–8.30) impose indegree and outdegree constraints for the customer and the depot vertices, respectively. Constraints (8.31) and (8.32) are CCCs and impose the connectivity and the capacity constraints. Because of the degree constraints, for any given S, the left-hand side of Eqns (8.31) and (8.32) are equal. Hence, if constraint (8.31) is imposed, constraint (8.32) is redundant and vice versa.

8.2.5 VRP with Pickup and Delivery

In VRPPD, the vehicles have two sets of tasks, one delivering goods to customers and the other picking goods up at customer locations.

In the VRPPD, a heterogeneous vehicle fleet must satisfy a set of transportation requests. Each request is defined by a pickup point, a corresponding delivery point, and a demand to be transported between these locations.

VRPPD can be formulated as a mixed-integer linear programming model. The integer linear model proposed by Hoff, Gribkovskaia, Laporte, and Lokketangen is presented [7].

Model Assumptions (Other Than Assumptions of CVRP)

- There are n customers; i represents two vertices i and $n+i$. It means that vertex i is used to perform a delivery, and vertex $i+n$ is used to perform a pickup.
- $p_i = d_{i+n}$ for $i = 1, 2, \ldots n$
- Visiting i, $i+n$ in succession by the same vehicle is, in fact, making a simultaneous pickup and delivery operation at customer i. Otherwise, the two operations are performed separately by the same vehicle or by two different vehicles.

Model Inputs (Other Than Inputs of CVRP)

The extended cost matrix $\overline{C} = (\overline{c}_{ij})_{(2n+1)(2n+1)}$

$$\overline{c}_{ij} = \begin{cases} c_{ij} & \text{if } i \leq n \text{ and } j \leq n \\ c_{i-n,j} & \text{if } i \geq n+1 \text{ and } j \leq n,\ j \neq i-n \\ c_{i,j-n} & \text{if } i \leq n,\ j \geq n+1,\ i \neq j-n \\ c_{i-n,j-n} & \text{if } i \geq n+1,\ j \geq n+1 \\ 0 & \text{if } j = i-n \text{ or } i = j-n \end{cases}$$

u_{ik}: An upper bound on the total pickup demand accumulated in vehicle k on leaving vertex i ($i = 0, 1, \ldots, 2n$; $k = 1, \ldots, m$)

v_{ik}: An upper bound on the total delivery demand remaining in vehicle k on leaving vertex i

q_{ij}: The distance between customer i and j

d_i: The demand of customer i

p_i: The supply of customer i

D: The maximum distance that the vehicles may cover in a tour

C: The maximum capacity of a vehicle.

Model Outputs

x_{ijk}: 1 if vehicle k travels directly from vertex i to vertex j ($i, j = 0, \ldots, 2n$, $i \neq j$, $k = 1, 2, \ldots, m$) and 0 otherwise.

Y_{ik}: 1 if vehicle k performs a delivery at vertex i ($i = 1, 2, \ldots, n$, $k = 1, 2, \ldots, m$).

Z_{ik} is 1 if vehicle k performs a pickup at vertex i ($i = 1 + n, \ldots, 2n$, $k = 1, 2, \ldots, m$).

Objective Function and its Constraints

$$\min \sum_{k=1}^{m} \sum_{i=0}^{2n} \sum_{j=0}^{2n} \bar{c}_{ij} x_{ijk} \tag{8.34}$$

$$\sum_{j=0}^{2n} x_{ojk} = 1 \quad k = 1, 2, \ldots, m \tag{8.35}$$

$$\sum_{j=0}^{2n} x_{ijk} = \sum_{j=0}^{2n} x_{ijik} \quad (i = 0, \ldots, 2n; \;\; k = 1, 2 \ldots, m) \tag{8.36}$$

$$\sum_{k=1}^{m} \sum_{j=0}^{2n} x_{ijk} = 1 \quad (i = 0, \ldots, 2n) \tag{8.37}$$

$$u_{0k} = 0 \quad (k = 1, 2, \ldots, m) \tag{8.38}$$

$$v_{0k} = \sum_{i=1}^{n} d_i y_{ik} \quad (k = 1, 2, ..., m) \tag{8.39}$$

$$0 \leq u_{ik} + v_{ik} \leq Q_k \quad (i = 0, \ldots, 2n; \;\; k = 1, 2, \ldots, m) \tag{8.40}$$

$$u_{jk} \geq u_{ik} + p_j z_{jk} - (1 - x_{ijk})Q_k \quad (i = 0, \ldots, 2n;\ j = 1, \ldots, 2n;\ k = 1, 2, \ldots, m) \tag{8.41}$$

$$v_{jk} \geq v_{ik} - d_j y_{jk} - (1 - x_{ijk})Q_k \quad (i = 0, \ldots, 2n;\ j = 1, \ldots, 2n;\ k = 1, 2, \ldots, m) \tag{8.42}$$

$$x_{ijk} \leq y_{ik} + z_{ik} \quad (i = 1, \ldots, 2n;\ j = 0, \ldots, 2n;\ k = 1, 2, \ldots, m) \tag{8.43}$$

$$x_{ijk} \in \{0, 1\} \quad (i,\ j = 0, \ldots, 2n;\ i \neq j;\ k = 1, \ldots, m) \tag{8.44}$$

$$y_{ijk} \in \{0, 1\} \quad (i = 1, \ldots, n;\ k = 1, \ldots, m) \tag{8.45}$$

$$z_{ik} \in \{0, 1\} \quad (i = n + 1, \ldots, 2n;\ k = 1, 2, \ldots, m) \tag{8.46}$$

Constraint (8.35) implies that m vehicles leave the depot. Constraint (8.36) ensures that the incoming flow at each customer vertex is equal to the outgoing flow and that the same vehicle enters and leaves the vertex. Constraint (8.37) means that each vertex is visited exactly once, and constraints (8.38) and (8.39) are used to initialize the pickup and delivery demands in the vehicles. Constraint (8.40) guarantees that the vehicle load never exceeds the vehicle capacity. Inequalities (Eqns (8.41 and 8.42)) control the upper bounds on the amounts of pickup and delivery demands in the vehicle on leaving each vertex. These constraints are, in fact, SECs. Constraint (8.43) states that if a vehicle performs no delivery and no pickup at vertex i, then it does not travel along any arc (i, j). Because constraint (8.37) forces each vertex to be visited by exactly one vehicle, there necessarily exists an index k, for which both sides of Eqn (8.43) will be equal to 1.

Constraints (8.38–8.42) ensure that the u_{ik} and v_{ik} variables are nonnegative.

8.3 Solution Techniques for Basic VRP Variants

Different solving methods were proposed for VRP variants, including exact algorithms, heuristics, and metaheuristics. When n (the number of vertices) increases, connectivity or CCC causes a dramatic increase in computation time and the exact algorithms will not be effective.

Exact approaches for the CVRP are mainly branch-and-bound and branch-and-cut algorithms and set-covering-based solution methods. Lots of heuristics and metaheuristics were proposed for solving the CVRP. Some of them are simulated annealing, deterministic annealing, Tabu search, genetic algorithms, ant algorithms, and neural networks. A good reference for studying CVRP and DCVRP is Laporte [8].

Exact algorithms used for VRPTW are mostly branch-and-bound and branch-and-cut. Also, heuristics and metaheuristics have led the way in generating near-optimal solutions and are much faster than exact algorithms. A comprehensive literature review on VRPTW can be found in Cordeau et al. [9].

Like other variants of VRP, VRPB is known to be NP-hard in the strong sense, and many heuristic algorithms were proposed for the approximate solution of the problem with symmetric or Euclidean cost matrices. Some exact algorithms are set-covering-based and branch-and-bound algorithms.

Based on classical procedures such as insertion methods, edge and vertex exchanges, and customer relocations, heuristic algorithms were developed for VRPPD. Some metaheuristic algorithms and optimization-based approaches such as Benders' decomposition, dynamic programming, polyhedral approach, and column generation were proposed. For more information about solving methods, refer to Berbeglia et al. [10] and Parragh et al. [11] parts 1 and 2 for classification of different exact, heuristic, and metaheuristic methods.

8.4 Other Variants of VRP

8.4.1 Open VRP

One of the variants of VRP is the OVRP. The important feature of the OVRP is that the vehicles are not obliged to return to the depot.

This kind of problem appears for the companies that do not own a vehicle fleet at all or a vehicle fleet that is inadequate to the demand of all customers. Therefore, the company is obliged to contract all or part of the product distribution to external couriers. The hired vehicles will be assigned to routes and do not have to return to the company's distribution center (depot).

The problem solution will provide the company with the minimum number of vehicles that must be hired in order to serve the customers and the set of routes that minimizes the traveling cost. Furthermore, in the situation in which the company has its own vehicle fleet and customer demand varies significantly over time, the solution will provide the proper combination of owned and hired vehicles [12].

A practical example of this kind of problem is the delivery and collection of mail in which, after delivery, the vehicles start collecting new mail and return it to the sorting office. Air-courier companies also have to determine such routes for fast and efficient service.

The elimination of vehicle return to the depot, which is a constraint in the VRP, does not actually lead to an easier and less complex problem in the OVRP [13], yet OVRP remains NP-hard. Because of its resolution, it is necessary to find the best Hamiltonian path for each set of customers assigned to a vehicle [14]. Therefore, any exact algorithm for solving the OVRP will certainly have the inefficiency of the exact algorithms for the VRP.

The OVRP received very little attention from the early 1980s to the late 1990s. However, since 2000, several researches have used Tabu search, deterministic annealing, and large neighborhood search to solve the OVRP with some success [12].

Table 8.1 OVRP Literature

Author(s)	Year	Type	Algorithm
Repoussis et al. [16]	2007	OVRPTW	
Letchford et al. [17]	2007	COVRP	Branch-and-cut algorithm
Aksen et al. [18]	2007	COVRPTW	Modified Clarke–Wright parallel savings algorithm, a nearest insertion algorithm, and a Tabu search heuristic
Fleszar et al. [19]	2008	OVRPTW, DCOVRP	Variable neighborhood search algorithm
Derigs and Reuter [20]	2008	OVRP	Attribute-based hill climber

COVRP, capacitated OVRP; OVRPTW, OVRP with time window; COVRPTW, capacitated OVRP with time window; DCOVRP, distance-constrained OVRP.

Li et al. [15] provided an extensive review of the literature on OVRP until 2007. Other studies in this area are reported in Table 8.1.

8.4.2 Multidepot VRP

In classical VRP, there is only one depot and all vehicles start and end their routes in that depot. In the MDVRP, more than one depot exists. In this problem, every customer is visited by a vehicle based at one of the several depots.

The MDVRP can be viewed as a clustering problem, in the sense that the output is a set of vehicle schedules clustered by depot. Therefore, the MDVRP can be solved in two stages: first, customers must be allocated (assigned) to depots; second, customers assigned to the same depot must be linked together through routes. Ideally, it is more efficient to deal with the two steps simultaneously. When faced with larger problems, however, a reasonable approach would be to divide the problem into as many subproblems as there are depots and to solve each subproblem separately [21].

Crevier et al. [22] summarized the works on VRP with multiple depots from 1969 to 2002. Other papers since that period are listed in Table 8.2.

MDVRP with interdepot routes is an extension of the MDVRP in which depots can act as intermediate replenishment facilities along the route of a vehicle. This problem is referred to as the VRP with intermediate facilities (VRP-IF). In a distribution system, these facilities are warehouses; in a collection system, these facilities represent the sites where the vehicles are unloaded.

There are two common points among these proposed methodologies. First, the MDVRP was decomposed; second, the subproblems were solved sequentially and iteratively.

8.4.3 Mix Fleet VRP

MFVRP is a different kind of VRP that differs from the classical one in that it deals with a heterogeneous fleet of vehicles having various capacities and fixed and variable costs. The routing cost is the sum of the fixed and variable cost wherein the variable cost is in proportion to the travel distance.

Table 8.2 MDVRP Literature

Author(s)	Year	Type	Algorithm
Giosa et al. [21]	2002	MDVRPTW	Designing six heuristics for assigning customers to depots and the same VRP heuristic for each depot
Angelelli and Speranza [23]	2002	VRP-IF	A Tabu search-based algorithm
Wasner and Zapfel [24]	2004	MDLRP for planning parcel service	Heuristic based on local search
Polacek et al. [25]	2004	MDVRPTW	Variable neighborhood search
Nagi and Salhi [26]	2005	VRPPD, MDVRPPD	Several heuristic methods for VRPPD can be modified to tackle MDVRPPD
Ho et al. [27]	2008	MDVRP	Hybrid genetic algorithm
Crevier et al. [22]	2007	MDVRP with interdepot routes	Heuristic combining adaptive memory principle, Tabu search for solution of subproblems, and integer programming
Chunyu and Xiaobo [28]	2009	MDVRPB	Hybrid genetic algorithm
Yu et al. [29]	2010	MDVRP	Ant colony metaheuristic
Sombuntham and Kachitvichayanukul [30]	2010	MDVRPPD in time window	Particle swarm optimization algorithm

MDVRP, multidepot VRP; MDVRPPD, MDVRP with pickup and delivery; MDVRPTW, MDVRP with time window; MDLRP, multidepot location routing problem; MDVRPB, multidepot VRP with backhauls.

There are three kinds of MFVRP in the literature. Introduced by Golden et al. in 1984 [34], the first one uses the same value for the variable costs regardless of the vehicle type and has an unlimited number of vehicles of each type. It is regarded as the vehicle fleet mix (VFM), the fleet size and mixed VRP, and the fleet size and composition VRP.

The second type, proposed by Salhi et al. [31], considers different variable costs dependent on the vehicle type and also has an unlimited number of vehicles of each type. It is the heterogeneous VRP (HVRP), VFM with variable unit running costs, and MFVRP.

Taillard [32] introduced the last type that differs from the second type in which there are restrictions on the number of available different vehicles of each type.

Because of the complexity of the MFVRP, some attempts have been made to formulate it using mixed-integer linear programming, but no exact algorithm for MFVRP has ever been developed [33].

Here, the literature of MFVRP is categorized in Table 8.3.

8.4.4 Split-Delivery VRP

SDVRP is a relaxation of the VRP in which the same customer can be served by different vehicles if it reduces overall costs. This problem was first introduced by

Table 8.3 MFVRP Literature

Author(s)	Year	Type	Algorithm
Golden et al. [34]	1984	FSMVRP	Saving heuristic based on Clarke–Wright method, two-step procedures
Gheysens et al. [35]	1984	FSMVRP	Penalty function approach
Gheysens et al. [36]	1986	FSMVRP	Two-stage method
Salhi et al. [31]	1992	VFM	Route perturbation (RPERT) procedure for different variable costs
Salhi and Rand [37]	1993	FSMVRP	Extension of RPERT procedure
Rochat and Semet [38]	1994	Heterogeneous fixed fleet	Tabu search
Osman and Salhi [39]	1996	VFM	Modified version of RPERT, called MRPERT, allowing search process to restart several times to produce several solutions
Taillard [32]	1999	VFM	Heuristic column-generation method
Liu and Shen [40]	1999	FSMVRPTW	Insertion-based savings heuristics
Renaud and Boctor [41]	2002	FSMVRP	Sweep-based algorithm
Wassan and Osman [33]	2002	FSMVRPTW	New variants of Tabu search mixed with reactive Tabu search concepts, variable neighborhoods, special data-memory structures, and hashing functions
Dullaert et al. [42]	2002	MFVRPTW	Sequential insertion heuristic for the FM
Tarantilis et al. [43]	2004	HFM	Backtracking adaptative threshold accepting
Dell'Amico et al. [44]	2006	FSMVRPTW	Constructive insertion heuristic and metaheuristics algorithm
Belfiore and Favero [45]	2007	FSMVRPTW	Scatter search approach
Braysy et al. [46]	2008	FSMVRPTW	Multirestart deterministic annealing metaheuristic
Lee et al. [47]	2008	VFM	Tabu search and set partitioning
Brandao [48]	2009	VFM	Deterministic Tabu search algorithm
Braysy et al. [49]	2009	FSMVRPTW	Three-phase metaheuristic
Liu et al. [50]	2009	FSMVRP	Genetic algorithm-based heuristic
Repoussis and Tarantilis [51]	2009	FSMVRP	Adaptive memory programming solution approach

FSMVRP, fleet size and mixed VRP; VFM, vehicle fleet mix; FSMVRPTW, FSMVRP with time window.

Dror and Trudeau [52]. They proved that split deliveries result in savings, both in the total distance traveled and the number of vehicles utilized.

An example in Gendreau et al. [53] shows that when the number of demand points goes to $+\infty$, the ratio of the optimal value of the SDVRP over that of the corresponding CVRP approaches 1/2. These savings are more significant when

Table 8.4 SDVRP Literature

Author(s)	Year	Type	Algorithm
Song et al. [56]	2002	SDVRP (newspaper logistic problem)	
Ho et al. [55]	2004	VRPSDTW	Tabu search
Archetti et al. [57]	2006	SDVRP	Tabu search
Gendreau et al. [58]	2006	SDVRTW	
Lee et al. [59]	2006	SDVRP (split pickups)	Exact algorithm based on shortest path
Campos et al. [60]	2007	SDVRP	Scatter search
Chen et al. [61]	2007	SDVRP	Heuristic combining mixed-integer program and record-to-record travel algorithm
Nakao and Nagamochi [62]	2007	SDVRP	Dynamic programming
Archetti et al. [63]	2008	SDVRP	Method based on Tabu search and using integer programming
Suthikarnnarunai [64]	2008	SDVFM	Sweep heuristic method
Derings et al. [65]	2009	SDVRP	Local search-based metaheuristic
Bolduc et al. [66]	2010	SDVRP	Tabu search
Aleman et al. [67]	2010	SDVRP	Adaptive memory algorithm
Moreno et al. [68]	2010	SDVRP	Proposing an algorithm to obtain lower bounds
Gulczynski et al. [69]	2010	SDVRP-MDA	Heuristic applying modified Clarke–Wright saving algorithm

SDVRPTW, SDVRP with time window; SDVFM, split-delivery vehicle fleet mix; SDVRP-MDA, SDVRP with minimum delivery amounts.
The split-pickup VRP is the same as SDVRP, but products have to be picked up from suppliers.

average customer demand is more than 10% of the vehicle capacity [54]. Ho and Haugland [55] gave a survey on SDVRP up to 2002. Other works since then are reported in Table 8.4.

8.4.5 Periodic VRP

In the PVRP, a set of customers has to be visited on a given time horizon one or more times. Different customers usually require different numbers of visits in that certain time horizon. PVRP with service choice (PVRP-SC) is a variant of the PVRP in which the visit frequency to nodes is a decision variable of the model. This can result in more efficient vehicle tours or greater service benefit to customers [70].

Solving the problem requires assigning a visiting schedule to each customer. For each day of the time horizon, the routes of the vehicles must be defined in such a way that all customers whose assigned schedules include that day are served.

Table 8.5 PVRP Literature

Author(s)	Year	Type	Algorithm
Delgado et al. [72]	2005	Minimizing labor requirements in a PVRP	Two-level heuristic
Francis and Smilowitz [70]	2005	PVRP-SC	Continuous approximation model
Belanger et al. [73]	2006	PVRPTW	Nonlinear integer multicommodity network flow formulation and new branch-and-bound strategies in branch and price
Alegre et al. [74]	2007	PVRP (pickups)	Scatter search
Alonso et al. [75]	2008	Site-dependent multitrip PVRP	Tabu-based algorithm
Pirkwieser and Raidl [76]	2008	PVRPTW	Variable neighborhood search
Pirkwieser and Raidl [77]	2009	PVRPTW	Column-generation method
Pirkwieser and Raidl [78]	2009	PVRPTW	Integer linear programming solver with variable neighborhood search

MDPVRP, multidepot PVRP.
In site-dependent multitrip PVRP, a vehicle can have multiple trips during the day and they are site-dependent—i.e., not every vehicle can visit every customer.

Therefore, a VRP has to be solved for each day of the planning horizon. In such a case, the choice of the visiting schedules and the definition of the routes are two interrelated problems. This feature is essential in some applications such as waste-collection problems in which each customer has to be served in a given period (e.g., twice a week).

Published papers on this subject from 1974 to 2005 are reported in Hemmelmayr et al. [71], and the summary of other works since then are reviewed briefly in Table 8.5.

8.4.6 Stochastic VRP

The deterministic CVRP has been widely studied in the literature. In the classic definition of VRP, the associated parameters such as cost, customer demands, and vehicle travel times are deterministic. SVRPs arise when some elements of the problem are random.

Common types are the following:

- In the VRP with stochastic demands (VRPSD), each customer's demand is assumed to follow a given probability distribution instead of having a single known value. The actual customer demand is known only on arrival at the customer's location.
- In the VRP with stochastic travel time, the matrix of travel time is not deterministic.
- In the VRP with stochastic customers, the set of customers is not known with certainty and each customer has a probability of being present.
- In the VRP with stochastic service time, the service time of each customer is not deterministic.

SVRPs differ from their deterministic ones in several aspects. Solution methodologies are more intricate and combine the characteristics of stochastic and integer programs. SVRPs are often computationally intractable; therefore, only relatively small instances can be solved to optimality, and good heuristics are hard to design and assess [79]. SVRPs can be cast within the framework of stochastic programming.

SVRPs are usually modeled using mixed- or pure-integer stochastic programs or as Markov decision processes. All known exact algorithms belong to the first category.

Tillman in 1969 [80] addressed the CVRPSD for the first time. He considered a multidepot variant of the CVRP with Poisson-distributed demands. The model considered a cost trade-off between exceeding the vehicle capacity and finishing the route with excess capacity [81].

The published papers covering all kinds of SVRP are categorized in Table 8.6 by specifying the parameters that have been assumed stochastic.

8.4.7 Fuzzy VRP

There is widespread evidence that the exact values of the mean demands, travel times, numbers and locations of customers, and so on that follow probability distributions are very difficult to obtain. In some new systems, it is also hard to describe the parameters of the problem as random variables because of insufficient data to analyze the distribution. Using methods from fuzzy sets theory makes it possible to successfully model problems that contain an element of uncertainty, subjectivity, ambiguity, and vagueness.

Fuzzy logic was used by Teodorovic and Pavkovic [130] in VRP when the demands were uncertain. The model was based on the heuristic sweeping algorithm, rules of fuzzy arithmetic, and fuzzy logic.

Cheng and Gen [131] introduced the concept of fuzzy die-time in the vehicle-routing and scheduling context. They represented the fuzzy time window in two types: the tolerable interval of service time and the desirable time for service. Usual approaches consider the tolerable interval of service time without minding customers' desired time. Their fuzzy approach can handle both kinds of customers' preferences simultaneously. Table 8.7 categorizes the papers on VRPF.

8.5 Case Studies

8.5.1 The Product Distribution of a Dairy and Construction Company

Distribution of materials is a challenging problem faced by major industrial and construction companies. Therefore, there are a lot of case studies that apply VRP methods to solve such real problems. One of them is the case studied by Tarantilis and Kiranoudis [140] in which they investigated two real-life distribution problems faced by a dairy and by a construction company.

Table 8.6 VRPSD Literature

Author(s)	Year	Type	Author(s)	Year	Type
Tillman [80]	1969	MDVRPSD	Golden and Stewart [82]	1978	VRPSD
Golden and Yee [83]	1979	VRPSD	Yee and Golden [84]	1980	VRPSD
Stewart and Golden [85]	1980	VRPSD	Stewart [86]	1981	VRPSD
Bodin et al. [87]	1983	VRPSD	Stewart and Golden [88]	1983	VRPSD
Jezequel [89]	1985	VRPSD, VRPSC, VRPSDC	Dror and Trudeau [90]	1986	VRPSD
Jaillet [91]	1987	VRPSD, VRPSC, VRPSDC	Bertsimas [92]	1988	VRPSD, VRPSC
Jaillet and Odoni [93]	1988	VRPSD, VRPSC, VRPSDC	Laporte et al. [94]	1989	VRPSD
Waters [95]	1989	VRPSD VRPSC	Dror et al. [96]	1989	VRPSD
Laporte and Louveaux [97]	1990	VRPSD	Bertsimas et al. [98]	1990	VRPSD
Bastian and Rinnooy Kan [99]	1992	VRPSD	Benton and Rosset [100]	1992	VRPSCD
Dror [101]	1992	VRPSD, VRPSC	Laporte et al. [102]	1992	VRPSD
Trudeau and Dror [103]	1992	VRPSCD	Bertsimas [104]	1992	VRPSD, VRPSCD
Dror et al. [105]	1993	VRPSD	Bouzdiene-Ayari et al. [106]	1993	SDVRPSD
Laporte and Louveaux [107]	1993	VRPSD	Gendreau et al. [79]	1995	VRPSD, VRPSCD
Gendreau et al. [58]	1996	VRPSD, VRPSCD	Secomandi [108]	1998	VRPSD
Hjorring and Holt [109]	1999	VRPSD	Yang et al. [110]	2000	SVRP
Secomandi [111]	2001	VRPSD	Laporte et al. [112]	2002	CVRPSD
Markovic et al. [113]	2005	VRPSD	Chepuri et al. [114]	2005	VRPSD
Dessouky et al. [115]	2006	VRPSD	Novoa et al. [116]	2006	VRPSD
Ak and Erera [117]	2007	VRPSD	Tan et al. [118]	2007	VRPSD
Haugland et al. [119]	2007	VRPSD	Sungur et al. [120]	2008	CVRPSD
Jula et al. [121]	2008	SVRPT	Shen et al. [122]	2009	VRPSDT
Christiansen and Lysgaard [123]	2009	VRPSD	Secomandi and Margot [124]	2009	VRPSD
Novoa and Storer [125]	2009	VRPSD	Li et al. [126]	2010	VRPSTST
Smith et al. [127]	2010	VRPSD	Rei et al. [128]	2010	VRPSD
Mendoza et al. [129]	2010	VRPSD			

MDVRPSD, multidepot VRPSD; VRPSC, VRP with stochastic customers; VRPSDC, VRP with stochastic demand and customer; VRPSTST, VRP stochastic travel time and service time.

Table 8.7 FVRP Literature

Author(s)	Year	Type	Algorithm
Cheng and Gen [131]	1995	VRPFTW	Genetic algorithm
Teodorovic and Pavkovic [130]	1996	VRPFD	Heuristic sweeping algorithm, rules of fuzzy arithmetic and fuzzy logic
Werners and Drawe [132]	2003	VRPF	Fuzzy modeling based on mixed-integer linear programming
Kuo et al. [133]	2004	VRPFT	Ant colony optimization
Sheng et al. [134]	2005	FVRP	Compares fuzzy measure method with other programming methods
He and Xu [135]	2005	VRPFD	Genetic algorithm
Zheng and Liu [136]	2006	VRPFT	Fuzzy simulation, genetic algorithm
Lin [137]	2008	VRPFTW	Genetic algorithm (multiobjective)
Erbao and Mingyong [138]	2009	VRPFD	Stochastic simulation
Erbao and Mingyong [139]	2010	OVRPFD	Stochastic simulation

VRPFTW, VRPF with fuzzy time window; VRPFD, VRP with fuzzy demand; VRPFT, VRP with fuzzy travel time; OVRPFD, OVRP with fuzzy demand.

The first case study considers a central warehouse of a dairy company that hosts a heterogeneous fleet of vehicles and stores perishable foods. The foods have to be delivered to a set of customers through daily deliveries. The distance between each pair of customers and also the central warehouse and each customer's location is known. After deliveries, each vehicle route ends at the central warehouse.

The second actual case study considers a distribution center of a concrete company in which a heterogeneous fleet of concrete-mixer trucks load ready-to-pour concrete for delivery to a set of construction sites. Every construction site requires a specific type of concrete-mixer truck of different capacity that can carry different blends of concrete. Concrete-mixer trucks return to the distribution center after unloading the demands.

Tarantilis and Kiranoudis [140] formulated these problems as HVRP with different fixed and variable cost for each vehicle and solved them by a heuristic based on adaptive memory. Computational outcome on the first case results in substantially improving on the current practice of the company by using 24 vehicles instead of 27 and saving at least 28.23% of the total duration of the trips. The same results are attained for the next case: the number of concrete-mixer trucks reduces from 13 to 10 and a 49.66% saving in total duration of trips.

8.5.2 The Collection of Urban Recyclable Waste

The next case is about the collection of recyclable waste in Portugal's central coastal region. There are two central depots in which the waste of 1642 distinct collection sites is unloaded by five vehicles. Three types of waste must be carried separately. Because 70% of the operational cost is dedicated to the transportation, creating the best collection routes minimizes the total distance of vehicles with the restrictions in the vehicle's capacity and route duration that must be managed in one work shift.

The problem is modeled as a PVRP and develops routes for every day of each month. This model is repeated in each month with 20 workdays and two work shifts in each day. It is noted that because the vehicles are busy about half the time with the exit and return trips to depots, a single route is created in one shift.

The problem is solved in three phases using heuristic algorithms. For each zone and for each work shift of each day, the decision variables are the type of waste, the sites, and the routes in which the waste must be collected [141].

References

[1] G.B. Dantzig, J.H. Ramser, The truck dispatching problem, Manag. Sci. 6 (1959) 80–91.

[2] G. Clark, J.V. Wright, Scheduling of vehicles from a central depot to a number of delivery points, Oper. Res. 12 (1964) 568–581.

[3] J.K. Lenstra, A.H.G. Rinnooy Kan, Complexity of vehicle and scheduling problems, Networks 11 (2) (1981) 221–227.

[4] P. Toth, D. Vigo, The vehicle routing problem, SIAM Monographs on Discrete Mathematics and Applications, Philadelphia, 2002.

[5] G. Laporte, Y. Nobert, M. Desrochers, Optimal routing under capacity and distance restrictions, Oper. Res. 33 (1985) 1050–1073.

[6] P. Toth, D. Vigo, A heuristic algorithm for the symmetric and asymmetric vehicle routing problems with backhauls, Eur. J. Oper. Res. 113 (1999) 528–543.

[7] A. Hoff, I. Gribkovskaia, G. Laporte, A. Lokketangen, Lasso solution strategies for the vehicle routing problem with pickups and deliveries, Eur. J. Oper. Res. 192 (3) (2007) 755–766.

[8] G. Laporte, The vehicle routing problem: an overview of exact and approximate algorithms, Eur. J. Oper. Res. 59 (1992) 345–358.

[9] G.F. Cordeau, G. Laporte, A. Mercier, A unified Tabu search heuristic for vehicle routing problems with time windows, J. Oper. Res. Soc. 52 (2001) 928–936.

[10] G. Berbeglia, J.F. Cordeau, I. Gribkovskaia, G. Laporte, Static pickup and delivery problems: a classification scheme and survey, J. Spanish Soc. Stat. Oper. Res. 15 (2007) 1–31.

[11] S.N. Parragh, K.F. Doerner, R.F. Hartl, A survey on pickup and delivery problems. Part I: transportation between customers and depot, J. Betriebswirtschaft 58 (1) (2008) 21–51.

[12] D. Sariklis, S. Powell, A heuristic method for the open vehicle routing problem, J. Oper. Res. Soc. 51 (5) (2000) 564–573.

[13] J. Brandao, A Tabu search algorithm for the open vehicle routing problem, Eur. J. Oper. Res. 157 (2004) 552–564.

[14] M. Syslo, N. Deo, J. Kowaklik, Discrete Optimization Algorithms with Pascal Programs, Prentice Hall, Englewood Cliffs, NJ, 1983.

[15] F. Li, B. Golden, E. Wasil, The open vehicle routing problem: algorithms, large scale test problems, and computational results, Comput. Oper. Res. 34 (10) (2007) 2918–2930.

[16] P.P. Repoussis, C.D. Tarantilis, G. Ioannou, The open vehicle routing problem with time windows, J. Oper. Res. Soc. 58 (3) (2007) 355–367.

[17] A.N. Letchford, J. Lysgaard, R.W. Eglese, A branch and cut algorithm for the capacitated open vehicle routing problem, J. Oper. Res. Soc. 58 (12) (2007) 1642–1651.

[18] D. Aksen, Z. Ozyurt, N. Aras, The open vehicle routing problem with driver nodes and time deadlines, J. Oper. Res. 132 (2007) 760–768.

[19] K. Fleszar, I.H. Osman, K.S. Hindi, A variable neighborhood search algorithm for the open vehicle routing problem, Eur. J. Oper. Res. 195 (3) (2009) 803–809.

[20] U. Derigs, K. Reuter, A simple and efficient tabu search heuristic for solving the open vehicle routing problem, J. Oper. Res. Soc. 60 (12) (2008) 1658–1669.

[21] I.D. Giosa, I.L. Tansini, I.O. Viera, New assignment algorithms for the multi-depot vehicle routing problem, J. Oper. Res. Soc. 53 (2002) 977–984.

[22] B. Crevier, J.F. Cordeau, G. Laporte, The multi-depot vehicle routing problem with inter-depot routes, Eur. J. Oper. Res. 176 (2) (2007) 756–773.

[23] E. Angelelli, M.G. Speranza, The periodic vehicle routing problem with intermediate facilities, Eur. J. Oper. Res. 137 (2002) 233–247.

[24] M. Wasner, G. Zapfel, An integrated multi-depot hub-location vehicle routing model for network planning of parcel service, Int. J. Prod. Econ. 90 (2004) 403–419.

[25] M. Polacek, R.F. Hartl, K.F. Doerner, M. Reimann, A variable neighborhood search for the multi depot vehicle routing problem with time windows, J. Heurist. 10 (2004) 613–627.

[26] G. Nagi, S. Salhi, Heuristic algorithms for the single and multiple depot vehicle routing problems with pickups and deliveries, Eur. J. Oper. Res. 162 (2005) 126–141.

[27] W. Ho, G.T.S. Ho, P. Ji, H.C.W. Lau, A hybrid genetic algorithm for the multi-depot vehicle routing problem, Eng. Appl. Artif. Intell. 21 (4) (2008) 548–557.

[28] R. Chunyu, W. Xiaobo, Study on improved hybrid genetic algorithm for multi-depot vehicle routing problem with backhauls, AICI 2 (2009) 347–350.

[29] B. Yu, Z.Z. Yang, J.X. Xie, A parallel improved ant colony optimization for multi-depot vehicle routing problem, J. Oper. Res. Soc. 62 (1) (2011) 183–188.

[30] P. Sombuntham, V. Kachitvichayanukul, A particle swarm optimization algorithm for multi-depot vehicle routing problem with pickup and delivery requests, Lecture Notes in Engineering and Computer Science 2182 (2010) 1998–2003.

[31] S. Salhi, M. Sari, D. Saidi, N. Touati, Adaptation of some vehicle fleet mix heuristics, Omega 20 (1992) 653–660.

[32] E.D. Taillard, A heuristic column generation method for the heterogeneous fleet VRP, RAIRO Oper. Res. 33 (1999) 1–14.

[33] N.A. Wassan, I.H. Osman, Tabu search variants for the mix fleet vehicle routing problem, J. Oper. Res. Soc. 53 (2002) 768–782.

[34] B. Golden, S. Assad, L. Levy, F. Gheysens, The fleet size and mix vehicle routing problem, Comput. Oper. Res. 11 (1) (1984) 49–66.

[35] F. Gheysens, B. Golden, A. Assad, A comparison of techniques for solving the fleet size and mix vehicle routing problems, OR Spectrum 6 (1984) 207–216.

[36] F. Gheysens, B. Golden, A. Assad, A new heuristic for determining fleet size and composition, Math. Program. 42 (1986) 234–242.

[37] S. Salhi, G.K. Rand, Incorporating vehicle routing into the vehicle fleet composition problem, Eur. J. Oper. Res. 66 (1993) 313–330.

[38] Y. Rochat, F. Semet, A tabu search approach for delivering pet food and flour in Switzerland, J. Oper. Res. Soc. 45 (1994) 1233–1246.
[39] I. Osman, S. Salhi, Local search strategies for the vehicle fleet mix problem, in: V.J. Rayward-Smith, I.H. Osman, C.R. Reeves, G.D. Smith (Eds.), Modern Heuristic Search Methods, Wiley, New York, 1996, pp. 131–153.
[40] F.H. Liu, S.Y. Shen, The fleet size and mix vehicle routing problem with time windows, J. Oper. Res. 50 (7) (1999) 721–732.
[41] J. Renaud, F. Boctor, A sweep-based algorithm for the fleet size and mix vehicle routing problem, Eur. J. Oper. Res. 140 (2002) 618–628.
[42] W. Dullaert, G.K. Janssens, K. Sorensen, B. Vernimmen, New heuristics for the fleet size and mix vehicle routing problem with time windows, J. Oper. Res. Soc. 53 (2002) 1232–1238.
[43] C.D. Tarantilis, D. Diakoulaki, C.T. Kiranoudis, Combination of geographical information system and efficient routing algorithms for real life distribution operations, Eur. J. Oper. Res. 152 (2004) 437–453.
[44] Monaci M. Dell'Amico, C. Pagani, D. Vigo, Heuristic approaches for the fleet size and mix vehicle routing problem with time windows, Transport. Sci. 4 (2006) 516–526.
[45] P. Belfiore, L. Favero, Scatter search for the fleet size and mix vehicle routing problem with time windows, Cent. Eur. J. Oper. Res. 15 (4) (2007) 351.
[46] O. Braysy, W. Dullaert, G. Hasle, D. Mester, M. Gendreau, An effective multi-restart deterministic annealing metaheuristic for the fleet size and mix vehicle routing problem with time windows, Transport. Sci. 42 (3) (2008) 371–386.
[47] Y.H. Lee, J.I. Kim, K.H. Kang, K.H. Kim, A heuristic for vehicle fleet mix problem using tabu search and set partitioning, J. Oper. Res. Soc. 59 (6) (2008) 833–841.
[48] J. Brandao, A deterministic tabu search algorithm for the fleet size and mix vehicle routing problem, Eur. J. Oper. Res. 195 (3) (2009) 716–728.
[49] O. Braysy, P.P. Porkka, W. Dullaert, P.P. Repoussis, C.D. Tarantilis, A well-scalable metaheuristic for the fleet size and mix vehicle routing problem with time windows, Expert Syst. Appl. 36 (4) (2009) 8460–8475.
[50] S. Liu, M. Haung, H. Ma, An effective genetic algorithm for the fleet size and mix vehicle routing problems, Transport. Res. E Logist. Transport. Rev. 45 (2009) 434–445.
[51] P.P. Repoussis, C.D. Tarantilis, Solving the fleet size and mix vehicle routing problem with time windows via adaptive memory programming, Transport. Res. C Emerg. Tech. 8 (5) (2009) 695–712.
[52] M. Dror, P. Trudeau, Savings by split delivery routing, Transport. Sci. 23 (1989) 141–145.
[53] M. Gendreau, M. Feillet, P. Dejax, C. Gueguen, Vehicle routing with time windows and split deliveries, Technical Paper 2006–851, Laboratoire d'Informatique d'Avignon, 2006.
[54] M. Jin, K. Liu, R.O. Bowden, A two-stage algorithm with valid inequalities for the split delivery vehicle routing problem, Int. J. Prod. Econ. 105 (2007) 228–242.
[55] S.C. Ho, D. Haugland, A Tabu search heuristic for the vehicle routing problem with time windows and split deliveries, Oper. Res. 31 (2004) 1947–1964.
[56] S. Song, K. Lee, G. Kim, A practical approach to solving a newspaper logistic problem using a digital map, Comput. Ind. Eng. 43 (2002) 315–330.
[57] C. Archetti, A. Hertz, M.G. Speranza, A Tabu search algorithm for the split delivery vehicle routing problem, Transport. Sci. 40 (2006) 64–73.

[58] M. Gendreau, G. Laporte, R. Seguin, Stochastic vehicle routing, Eur. J. Oper. Res. 88 (1) (1996) 3–12.
[59] C.G. Lee, M. Epelman, C.C. White, A shortest path approach to the multiple-vehicle routing with split picks-up, Transport. Res. B Meth. 40 (2006) 265–284.
[60] V. Campos, A. Corberan, E. Mota, A scatter search algorithm for the split delivery vehicle routing problem, in: C. Cotta, J van Hemert (Eds.), Evolutionary Computation in Combinatorial Optimization, Lecture Notes in Computer Science, 4446, Springer, Berlin, 2007, pp. 121–129.
[61] S. Chen, B. Golden, E. Wasil, The split delivery vehicle routing problem: applications, algorithms, test problems and computational results, Networks 49 (2007) 318–328.
[62] Y. Nakao, H. Nagamochi, A DP-based heuristic algorithm for the discrete split delivery vehicle routing problem, J. Adv. Mech. Des. Syst. Manuf. 1 (2) (2007) 217–226.
[63] C. Archetti, M.W.P. Savelsbergh, M.G. Speranza, An optimization-based heuristic for the split delivery vehicle routing problem, Transport. Sci. 42 (2008) 22–31.
[64] N. Suthikarnnarunai, A sweep algorithm for the mix fleet vehicle routing problem, Proceedings of the International Multi Conference of Engineers and Computer Scientists, Hong Kong, 2008.
[65] U. Derings, B. Li, U. Vogel, Local search-based metaheuristics for the split delivery vehicle routing problem, J. Oper. Res. Soc. 61 (9) (2009) 1356–1364.
[66] M.C. Bolduc, G. Laporte, J. Renaud, F.F. Boctor, A Tabu search heuristic for the split delivery vehicle routing problem with production and demand calendars, Eur. J. Oper. Res. 202 (2010) 122–130.
[67] R.F. Aleman, X. Zhang, R.R. Hill, An adaptive memory algorithm for the split delivery vehicle routing problem, J. Heurist. 16 (2010) 441–473.
[68] L. Moreno, M.P. Aragao, E. Uchoa, Improved lower bounds for the split delivery vehicle routing problem, Oper. Res. Lett. 38 (4) (2010) 302–306.
[69] D. Gulczynski, B. Golden, E. Wasil, The split delivery vehicle routing problem with minimum delivery amounts, Transport. Res. E Logist. Transport. Rev. 46 (2010) 612–626.
[70] P. Francis, K. Smilowitz, Modeling techniques for periodic vehicle routing problems, Transport. Res. B 40(10) (2006) 872–884.
[71] V.C. Hemmelmayr, K.F. Doerner, R.F. Hartl, A variable neighborhood search heuristic for periodic routing problems, Eur. J. Oper. Res. 195 (3) (2009) 791–802.
[72] C. Delgado, M. Laguna, J. Pacheco, Minimizing labor requirements in a periodic vehicle loading problem, Comput. Optim. Appl. 32 (3) (2005) 299–320.
[73] N. Belanger, G. Desaulniers, F. Soumis, J. Desrosiers, Periodic airline fleet assignment with time windows, spacing constraints, and time dependent revenues, Eur. J. Oper. Res. 175 (3) (2006) 1754–1766.
[74] J. Alegre, M. Laguna, J. Pacheco, Optimizing the periodic pick-up of raw materials for a manufacturer of auto parts, Eur. J. Oper. Res. 179 (2007) 736–746.
[75] F. Alonso, M.J. Alvarez, J.E. Beasley, A tabu search algorithm for the periodic vehicle routing problem with multiple vehicle trips and accessibility restrictions, J. Oper. Res. Soc. 59 (2008) 963–976.
[76] S. Pirkwieser, G.R. Raidl, A variable neighborhood search for the periodic vehicle routing problem with time windows, Proceedings of the 9th EU/Meeting on Metaheuristics for Logistics and Vehicle Routing, Troyes, France, 2008.
[77] S. Pirkwieser, G.R. Raidl, A column generation approach for the periodic vehicle routing problem with time windows, Proceedings of the International Network Optimization Conference, Pisa, Italy, 2009.

[78] S. Pirkwieser, R. Gunther, Boosting a variable neighborhood search for the periodic vehicle routing problem with time windows by ILP techniques, Proceedings of the 8th Metaheuristic International Conference (MIC 2009), Hamburg, Germany, 2009.
[79] M. Gendreau, G. Laporte, R. Segiun, An exact algorithm for the vehicle routing problem with stochastic demands and customers, Transport. Sci. 29 (1995) 143–155.
[80] F.A. Tillman, The multiple terminal delivery problem with probabilistic demands, Transport. Sci. 3 (1969) 192–204.
[81] C.H. Christiansen, J. Lysgaard, A branch-and-price algorithm for the capacitated vehicle routing problem with stochastic demands, Oper. Res. Lett. 35 (2009) 773–781.
[82] B.L. Golden, W.R. Stewart Jr., Vehicle routing with probabilistic demands, in: D. Hogben, D. Fife (Eds.), Computer Science and Statistics: Tenth Annual Symposium on the Interface, NBS Special Publication 503, National Bureau of Standards, Washington, DC, 1978.
[83] B.L. Golden, J.R. Yee, A framework for probabilistic vehicle routing, Am. Inst. Ind. Eng. Trans. 11 (1979) 109–112.
[84] J.R. Yee, B.L. Golden, A note on determining operating strategies for probabilistic vehicle routing, Nav. Res. Logist. Q. 27 (1980) 159–163.
[85] W.R. Stewart Jr., B.L. Golden, A chance-constrained approach to the stochastic vehicle routing problem, in: G. Boseman, J. Vora, Proceedings of 1980 Northeast AIDS Conference, Philadelphia, 1980, pp. 33–35.
[86] W.R. Stewart, New algorithms for deterministic and stochastic vehicle routing problems, Ph.D. Thesis, Report No. 81-009, University of Maryland at College Park, MD, 1981.
[87] L.D. Bodin, B.L. Golden, A.A. Assad, M.O. Ball, Routing and scheduling of vehicles and crews: the state of the art, Comput. Oper. Res. 10 (1983) 69–211.
[88] W.R. Stewart, B.L. Golden, Stochastic vehicle routing: a comprehensive approach, Eur. J. Oper. Res. 14 (1983) 371–385.
[89] A. Jezequel, Probabilistic vehicle routing problems, M.Sc. Dissertation, Department of Civil Engineering, Massachusetts Institute of Technology, Cambridge, MA, 1985.
[90] M. Dror, P. Trudeau, Stochastic vehicle routing with modified savings algorithm, Eur. J. Oper. Res. 23 (1986) 228–235.
[91] P. Jaillet, Stochastic routing problems, in: G. Andreatta, F. Mason, P. Serafini (Eds.), Stochastics in Combinatorial Optimization, World Scientific, New Jersey, 1987.
[92] D.J. Bertsimas, Probabilistic combinatorial optimization problems. Ph.D. Thesis, Report No. 193, Operations Research Center, Massachusetts Institute of Technology, Cambridge, MA, 1988.
[93] P. Jaillet, A.R. Odoni, The probabilistic vehicle routing problem, in: B.L. Golden, A.A. Assad (Eds.), Vehicle Routing: Methods and Studies, North-Holland, Amsterdam, 1988.
[94] G. Laporte, F.V. Louveattx, H. Mercure, Models and exact solutions for a class of stochastic location-routing problems, Eur. J. Oper. Res. 39 (1989) 71–78.
[95] C.D.J. Waters, Vehicle-scheduling problems with uncertainty and omitted customers, J. Oper. Res. Soc. 40 (1989) 1099–1108.
[96] M. Dror, G. Laporte, P. Trudeau, Vehicle routing with stochastic demands: properties and solution frameworks, Transport. Sci. 23 (1989) 166–176.
[97] G. Laporte, F.V. Louveaux, Formulations and bounds for the stochastic capacitated vehicle routing problem with uncertain supplies, Economic Decision-Making: Games, Econometrics and Optimization. Contribution in Honour of Dreze JH, in: J. Gabzewicz, J.F. Richard, L. Wolsey (Eds.), Elsevier Science, North Holland, Amsterdam, 1990, pp. 443–455.

[98] D.J. Bertsimas, P. Jaillet, A.R. Odoni, A priori optimization, Oper. Res. 38 (1990) 1019–1033.

[99] C. Bastian, A.H.G. Rinnooy Kan, The stochastic vehicle routing problem revisited, Eur. J. Oper. Res. 56 (1992) 407–412.

[100] W.C. Benton, M.D. Rossetti, The vehicle scheduling problem with intermittent customer demands, Comput. Oper. Res. 19 (1992) 521–531.

[101] M. Dror, Vehicle routing with uncertain demands: stochastic programming and its corresponding TSP solution, Eur. J. Oper. Res. 64 (1992) 432–441.

[102] G. Laporte, F.V. Louveaux, H. Mercure, The vehicle routing problem with stochastic travel times, Transport. Sci. 26 (1992) 161–170.

[103] P. Trudeau, M. Dror, Stochastic inventory routing: route design with stock outs and route failure, Transport. Sci. 26 (1992) 171–184.

[104] D.J. Bertsimas, A vehicle routing problem with stochastic demand, Oper. Res. 40 (1992) 574–585.

[105] M. Dror, G. Laporte, F.V. Louveaux, Vehicle routing with stochastic demands and restricted failures, Math. Method Oper. Res. 37 (3) (1993) 273–283.

[106] B. Bouzaiene-Ayari, M. Dror, G. Laporte, Vehicle routing with stochastic demands and split deliveries, Found. Comput. Decision Sci. 18 (1993) 63–69.

[107] G. Laporte, F.V. Louveaux, The integer L-shaped method for stochastic integer programs with complete recourse, Oper. Res. Lett. 13 (1993) 133–142.

[108] N. Secomandi, Exact and heuristic dynamic programming algorithms for the vehicle routing problem with stochastic demands. Ph.D. thesis, Department of Decision and Information Sciences, University of Houston, Houston, TX, 1998.

[109] C. Hjorring, J. Holt, New optimality cuts for a single-vehicle stochastic routing problem, Ann. Oper. Res. 86 (1999) 569–584.

[110] W. Yang, K. Mathur, R.H. Ballou, Stochastic vehicle routing problem with restocking, Transport. Sci. 34(1) (2000) 99–112.

[111] N. Secomandi, A rollout policy for the vehicle routing problem with stochastic demands, Oper. Res. 49(5) (2001) 796–802.

[112] G. Laporte, F. Louveaux, L.V. Hamme, An integer L-shaped algorithm for the capacitated vehicle routing problem with stochastic demands, Oper. Res. 50 (2002) 415–423.

[113] H. Markovic, I. Cavar, T. Caric, Using data mining to forecast uncertain demands in stochastic vehicle routing problem, 13th International Symposium on Electronics in Transport (ISEP), Slovenia, 2005, pp. 1–6.

[114] K. Chepuri, T. Homem de Mello, Solving the vehicle routing problem with stochastic demands using the cross entropy method, Ann. Oper. Res. 134 (2005) 153–181.

[115] M. Dessouky, F. Ordonez, H. Jia, Z. Shen, Rapid distribution of medical supplies, Int. Ser. Oper. Res. Manag. Sci. 91 (2006) 309–338.

[116] C. Novoa, R. Berger, J. Linderoth, R. Storer, A Set-Partitioning-Based Model for the Stochastic Vehicle Routing Problem, Technical Report 06T-008, Industrial and Systems Engineering, Lehigh University, Bethlehem, PA, 2006.

[117] A. Ak, A.L. Erera, A paired-vehicle recourse strategy for the vehicle-routing problem with stochastic demands, Transport. Sci. 41(2) (2007) 222–237.

[118] K.C. Tan, C.Y. Cheong, C.K. Goh, Solving multi objective vehicle routing problem with stochastic demand via evolutionary computation, Eur. J. Oper. Res. 177 (2007) 813–839.

[119] D. Haugland, S.C. Ho, G. Laporte, Designing delivery districts for the vehicle routing problem with stochastic demands, Eur. J. Oper. Res. 180 (2007) 997–1010.

[120] F. Sungur, F. Ordonez, M.M. Dessouky, A robust optimization approach for the capacitated vehicle routing problem with demand uncertainty, Inst. Ind. Eng. Trans. 40 (5) (2008) 509–523.
[121] H. Jula, M.M. Dessouky, P. Ioannou, Truck route planning in non-stationary stochastic networks with time-windows at customer locations, IEEE Trans. Intell. Transport Syst. 9 (2008) 97–110.
[122] Z. Shen, F. Ordonez, M.M. Dessouky, The stochastic vehicle routing problem for minimum unmet demand, Optim. Appl. 30 (4) (2009) 349–371.
[123] C.H. Christiansen, J. Lysgaard, A branch-and-price algorithm for the capacitated vehicle routing problem with stochastic demands, Oper. Res. Lett. 35 (2009) 773–781.
[124] N. Secomandi, F. Margot, Reoptimization approaches for the vehicle-routing problem with stochastic demands, Oper. Res. 57 (1) (2009) 214–230.
[125] C. Novoa, R. Storer, An approximate dynamic programming approach for the vehicle routing problem with stochastic demands, Eur. J. Oper. Res. 196 (2009) 509–515.
[126] X. Li, P. Tian, S.C.H. Leung, Vehicle routing problems with time windows and stochastic travel and service times: models and algorithm, Int. J. Prod. Econ. 125 (2010) 137–145.
[127] S.L. Smith, M. Pavone, F. Bullo, E. Frazzoli, Dynamic vehicle routing with priority classes of stochastic demands, SIAM J. Contr. Optim. 48 (5) (2010) 3224–3245.
[128] W. Rei, M. Gendreau, P. Soriano, A hybrid Monte Carlo local branching algorithm for the single vehicle routing problem with stochastic demands, Transport. Sci. 44 (1) (2010) 136–146.
[129] J.E. Mendoza, B. Castanier, C. Gueret, A.L. Medaglia, N. Velasco, A memetic algorithm for the multi-compartment vehicle routing problem with stochastic demands, Comput. Oper. Res. 37 (2010) 1886–1898.
[130] C. Teodorovic, G. Pavkovic, The fuzzy set theory approach to the vehicle routing problem when demand at nodes is uncertain, Fuzzy Sets Syst. 3 (1996) 307–317.
[131] R. Cheng, M. Gen, Vehicle routing problem with fuzzy due-time using genetic algorithms, J. Fuzzy Theory Sys. 7(5) (1995) 1050–1061.
[132] B. Werners, M. Drawe, Capacitated vehicle routing problem with fuzzy demand, Studies Fuzziness Soft Comput. 126 (2003) 317–336.
[133] R.J. Kue, C.Y. Chiu, Y.J. Lin, Integration of fuzzy theory and ant algorithm for vehicle routing problem with time window, Proceedings of the IEEE Annual Meeting of the Fuzzy Information[C] 2, 2004, pp. 925–930.
[134] H.M. Sheng, J.C. Wang, H.H. Haung, D.C. Yen, Fuzzy measure on vehicle routing problem of hospital materials, Expert Syst. Appl. 30(2) (2005) 367–377.
[135] Y. He, J. Xu, A class of random fuzzy programming model and its application to vehicle routing problem, World J. Model. Simulation 1 (2005) 3–11.
[136] Y.S. Zheng, B.D. Liu, Fuzzy vehicle routing model with credibility measure and its hybrid intelligent algorithm, Appl. Math. Comput. 176 (2006) 673–683.
[137] J.J. Lin, A GA-based multi-objective decision making for optimal vehicle transportation, J. Inf. Sci. Eng. 24 (2008) 237–260.
[138] C. Erbao, L. Mingyong, A hybrid differential evolution algorithm to vehicle routing problem with fuzzy demands, J. Comput. Appl. Math. 231 (2009) 302–310.
[139] C. Erbao, L. Mingyong, The open vehicle routing problem with fuzzy demands, Expert Syst. Appl. 37 (2010) 2405–2411.

[140] C.D. Tarantilis, C.T. Kiranoudis, A flexible adaptive memory-based algorithm for real-life transportation operations: two case studies from dairy and construction sector, Eur. J. Oper. Res. 179 (2007) 806–822.

[141] J. Teixeira, A.P. Antunes, J.P. de Sousa, Recyclable waste collection planning—a case study, Eur. J. Oper. Res. 158 (2004) 543–554.

9 Packaging and Material Handling

Mahsa Parvini

Faculty of Industrial Engineering, Amirkabir University, Tehran, Iran

---

Learning Objectives in Material Handling and Packaging

- To learn material-handling (MH) principles
- To identify MH equipment
- To know the utilization of unit loads role in MH
- To know the MH designing systems process
- To identify the functions performed by packaging
- To identify labeling importance
- To know how packaging affects logistics activities

9.1 Material Handling

9.1.1 History

MH is not a new subject. Human beings who first inhabited Earth were faced with the problem of moving things. They needed to transport both themselves and the materials they needed for their existence.

History has recorded continual progress in MH. Probably one of the greatest achievements in the ancient world was the construction of the pre-Inca temple near Cuzco, Peru. Stones weighing as much as 20 tons were quarried at the bottom of a valley and moved more than 2000 feet up to the temple site. In 1913, the Ford Motor Company instituted the first mechanized progressive-assembly line. World War II stimulated the implementation of MH mechanization. Companies that had government cost-plus contracts were encouraged to make capital expenditures for MH equipment.

Progress in current modern facilities is evident in the use of both mechanized and automated MH equipment to provide desired efficiencies [1].

9.1.2 Definition

One idea of how the concepts of material management, physical distribution management, and business logistics are related is that they overlap in MH, which can be described as the systematic physical movement of materials. The areas of

Logistics Operations and Management. DOI: 10.1016/B978-0-12-385202-1.00009-8

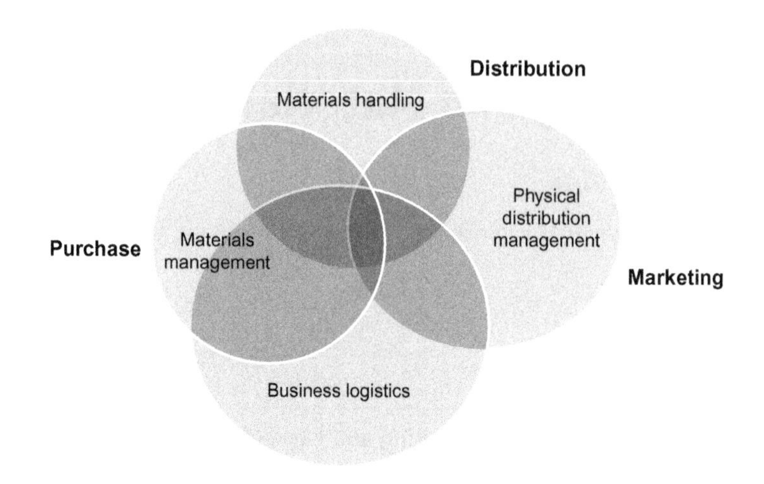

Figure 9.1 The movement of goods [2].

overlap present management with its most serious MH problems. The most important areas that are influenced by MH are shown in Figure 9.1.

The following are some of the definitions of MH.

1. In Ballou's definition, MH is physically moving objects or goods in small quantities over relatively short distances [3].
2. The way materials and products are handled physically is the subject of MH movement. To this point, the emphasis has been on the movement of products that are packaged in customer-sized boxes [4].
3. For Magad and Amos, MH is the art and science of moving, storing, protecting, and controlling materials [1].
4. MH means providing the right amount of the right material, in the right condition, at the right place, in the right position, in the right sequence, for the right cost by the right methods [5].

The first definition conveys the fact that MH is a physical movement between short distances. It is an activity that takes place in warehouses, production facilities, and retail stores and also between transportation modes, so it must be repeated many times [6]. In the second definition, the emphasis is on the concept of building blocks: MH is moving products as building blocks such as boxes, bottles, and cans [4].

The first and second definitions regard MH as a science that studies the movement of physical materials, whereas the third and fourth definitions consider MH also to be an art. The third definition conveys the fact that the MH design process is both a science and an art, and that MH function involves moving, storing, protecting, and controlling materials. It is a science-based discipline involving many areas of engineering, so engineering design methods must be applied. Thus, MH design process involves defining the problem, collecting and analyzing data, generating alternative solutions, evaluating alternatives, selecting and implementing preferred alternatives, and performing periodic reviews. It is an art because MH systems cannot be explicitly designed based solely on scientific formulas or mathematical models. As mentioned in the fourth definition, MH requires knowledge and appreciation of right and wrong, which is based on significant practical experience in the field [6].

The fourth definition exactly explains the abstract of the MH functions. The *right amount* refers to the problem of how much material is needed. The *right material* refers to the fact that an accurate identification system is needed. The *right condition* is the state in which the customer desires to receive the material. The *right sequence* of activities affects the efficiency of a manufacturing or distribution operation in MH. The *right place* addresses both transportation and storage. The *right time* means on-time delivery. The *right cost* does not mean the lowest cost. Minimizing cost is solely the wrong objective in MH system design. The more appropriate goal is to design the most efficient MH systems at the most reasonable costs [6].

9.1.3 MH Principles

No mathematical model can provide extensive solutions to overall MH problems. Applying experience is an important key in managing the MH processes. MH principles are the essence of practical experience. Condensed from decades of expert MH experience, these principles provide guidance and perspective to those who design MH systems.

These are some of the MH principles that have been developed by the College Industry Council on Material Handling Education after designing and testing MH systems through rigorous engineering analysis. Some of the principles are the results of Eastman's experiences in practice [4]. The principles are more important when laying out the intended design or when troubleshooting to discover why a system is not performing well. The principles are as follows [1].

- *Orientation principle*: Look at the entire system and study it first to learn how it operates. Identify the system components and their relationships. Also, look at relationships to other systems to find physical limitations.
- *Planning principle*: Prepare a plan to meet the basic requirements. In a reasonable form, an MH plan identifies the material (what), the moves (when and where), and the method (how and who).
- *Systems principle*: Integrate the handling, packaging, and storage activities that make up a coordinated system.
- *Unit-load principle*: Pick up products as a unit.
- *Space utilization principle*: Optimize the utilization of all space.
- *Standardization principle*: Standardize the methods and equipment employed. Reduce customization.
- *Ergonomic principle*: Adapt working conditions to workers' needs and abilities.

- *Energy principle*: Reduce energy consumption by the MH activities.
- *Ecology principle*: Minimize adverse effects on the environment when selecting MH system components.
- *Mechanization principle*: Use machines, where they can be justified, to replace human effort.
- *Flexibility principle*: Use methods and components that can work with reasonable tolerance and can perform a variety of tasks.
- *Simplification principle*: Change handling procedures by eliminating, decreasing, or combining unnecessary movements or equipment.
- *Gravity principle*: Rely on gravity to move materials easily wherever possible.
- *Safety principle*: Provide safe MH system components to handle the entire system.
- *Computerization principle*: Use computers to operate both individual pieces of equipment and massive supply chains spread across several continents.
- *Systems flow principle*: Integrate data flow with the physical material flow in handling to make a coordinated system.
- *Layout principle*: Organize an operation sequence and equipment layout for all variable system solutions.
- *Cost principle*: Recognize that all MH alternatives have associated costs and that these costs must be carefully considered as the system is devised. Investment proposals must be presented to top management for approval.
- *Maintenance principle*: Schedule a plan for maintenance on MH equipment.
- *Obsolescence principle*: Establish a long-term and economical program to replace obsolete equipment and methods, paying special consideration to after-tax life-cycle costs.
- *Automation principle*: Apply electronics and computer-based systems to operate and control the entire system activities.
- *The team-solution principle*: Collaborate with MH team members to devise the best system.
- *The just-in-time principle*: Hold products that are not moved until needed.
- *Minimum travel principle*: Systems should be set up so that loads move the shortest distances.
- *Using the right equipment*: Use equipment that is needed for MH.
- *Designing capacity for present and future*: Consider the development of MH systems in future system design.
- *Developing technological assessments*: Prepare assessments that make operations simple with using technological facilitates.
- *Using the systematical approach*: Consider the components and their relationships as an integrated system to unify them and increase efficiency.

9.1.4 MH Equipment

MH equipment and systems often represent a major capital expenditure for an organization to set up. The decisions related to the MH can affect many aspects of the organization operations.

Equipment analysis is an important part of analyzing an MH system. A reasonable solution often requires more than one individual piece of equipment. The various pieces of equipment that comprise an integrated system are needed. Materials management personnel who are involved in developing solutions must become acquainted with the diverse types of MH equipments and their applications [2].

The following are some of the criteria for selecting MH equipment.

- Cost
- Reliability and maintainability
- Service facilities
- Operating characteristics
- Safety and environmental characteristics
- Compatibility and the system's concepts

MH equipment is classified as *continuous* (e.g., conveyors), *discontinuous* (e.g., cranes and industrial trucks), or *potential movement* (e.g., unit-load equipment, pallets, and containers) [2].

In some studies such as Stock and Lambert's research on logistics management issues, automation is the classification base in the first level. In another point of view, applications are used for the classification base [6]. This list includes almost all MH equipment as follows.

1. Containers and unitizing equipment
    A. Containers
        1. Pallets
        2. Skids and skid boxes
        3. Tote pans
    B. Unitizers
        1. Stretch wrap
        2. Palletizers
2. Material-transport equipment
    A. Conveyors
        1. Chute conveyor
        2. Belt conveyor
        3. Roller conveyor
        4. Wheel conveyor
        5. Slat conveyor
        6. Chain conveyor
        7. Tow-line conveyor
        8. Trolley conveyor
        9. Power and free conveyor
        10. Cart-on-truck conveyor
        11. Storing conveyor
    B. Industrial vehicles
        1. Walking
        2. Riding
        3. Automated
            – Automated guided vehicles
            – Automated electrified monorail
            – Sorting transfer vehicles
    C. Monorails, hoists, and cranes
        1. Monorail
        2. Hoist
        3. Cranes

3. Storage and retrieval equipment
   A. Unit-load storage and retrieval
      1. Unit-load storage equipment
         - Pallet-stacking frame
         - Single deep selective rack
         - Drive-in rack
         - Mobile rack
      2. Unit-load retrieval equipment
         - Walkie stacker
         - Counterbalance lift truck
         - Narrow-aisle vehicle
         - Automated storage (AS) retrieval machines
   B. Small-load storage and retrieval
      1. Operator-to-stock storage equipment
         - Bin shelving
         - Modular storage drawers
      2. Operator-to-stock retrieval equipment
         - Picking cart
         - Order-picker truck
      3. Stock-to-operator equipment
         - Carousels
         - Vertical lift module
         - Automated dispenser
4. Automatic data-collection and communication equipment
   A. Automatic identification and recognition
      1. Bar coding
      2. Optical character recognition
      3. Magnetic strip
      4. Machine vision
   B. Automatic paperless communication
      1. Radiofrequency data terminal
      2. Voice headset
      3. Light and computer aids
      4. Smart card

9.1.5 Unit-Load Design

Baily and Framer define a *unit load* as a standardized combination of a number of items into an integrated one that can be handled as a single item. Reasonable reasons for designing unit loading include making the MH easier, reducing costs, and increasing transportation security. The elementary principle behind the unit load is making smaller units more convenient, economical, and easier to handle, transport, and store [2].

Unit load is an extension of the building-block concept to large quantities. Based on that concept, unit loading involves securing boxes to a pallet; the boxes or containers secured to a pallet are a unit load. The term *unitization* describes this kind of handling [3].

The high cost of manual labor has made the individual handling of small packages and items prohibitively expensive. If a number of items can be handled as a unit, then MH costs are reduced by moving larger loads, which eliminates unloading and reloading, cuts travel time, uses space more efficiently, reduces inventories, and facilitates shipping, transport, and receiving [6].

The size and type of unit load depend on a whole range of factors, the most important being the goods or materials to be handled; the size, weight, strength, and shape of intermediate packs; the type of storage required; the type of transport required; the type of handling equipment that may be available; and the quantities of goods and materials to be handled [2].

The unit load has several advantages. First, it adds protection to the cargo because the pallets are secured by straps, shrink-wrapping, or some other bonding device. Second, because removing a single package or its contents is difficult, pilferage is discouraged. Third, the unit load enables mechanical devices to be substituted for hand labor. Many machines have been devised that can quickly build up or tear down a pallet load of materials. Robots can be used when more sophisticated integrated movements are needed for loading or unloading. Ballou [3] presented an example of robot-assisted palletizing and depalletizing in the printing industry in which bundles of printed pages must be stacked in a specified order.

Unit-Load Criteria

Two important limits that help determine the unit-load design are those for size and weight. The unit-load size must be standardized so it can be handled easily and economically with modern equipment. Some unit loads may be bigger or smaller to meet other criteria and/or product characteristics.

The weight of the unit load must be kept within the capacities of the MH equipment and the storage facilities. Unit loads of some high-density materials such as steel, flour, and stone are smaller than optimal size in order to keep the unit-load weight within limits [3].

The unit load calls for a standard base or container. Among the possibilities are:

- Pallets
- Stillages
- Skids
- Slip sheets
- Containers
- Self-contained cartons
- Intermediate bulk containers

9.1.6 Designing MH Systems

Design is the most important step in operating an MH system that will accomplish its objectives. From a variety of possible alternatives, the designer selects one set that will result in an efficient and economic system.

The design process is not a well-defined algorithm that the designer can follow through to a successful outcome. It is both a procedure and an art based on empirical experience, engineering knowledge, and ingenuity.

The MH systems design process involves the following six-step engineering design process [5]:

1. Define the problem and identify the system scope and objectives of the MH system.
2. Identify the requirements and analyze them for moving, protecting, and controlling materials.
3. Generate alternative designs for satisfying the MH system requirements.
4. Evaluate alternative MH system designs.
5. Select the preferred design for moving, protecting, and controlling materials.
6. Implement the selected design, including selecting suppliers and equipment, training personnel, installing equipment, and periodically auditing system performance.

Problem Definition

MH problems must be first identified clearly so that a solution can be achieved. Existing operations should be reviewed, beginning with receiving activities and continuing all the way to final shipment.

The problem may arise from a critical incident, management's perception that improvement is needed, competitive pressures, or even other parts of company such as warehouses.

The objectives of MH systems are the end results that the system is expected to accomplish. As Stock and Lambert mentioned in their research, a typical objective is cost reduction. Others may be better customer service, greater space productivity, increased efficiency, or decreased accidents and damages. Those criteria which measure the extent to which the systems design is expected to meet the objectives [6].

Analyzing the Requirements

The next step is to analyze the problem and related information gathered so far based on the problem definition. The designer reviews what has been learned and what there is to work with. A major result of this analysis is limiting the number of alternatives to be investigated based on the objectives that are identified in the previous step. Careful selection of the most promising choices to investigate is the key to efficient use of engineering resources and to a successful design outcome. Another outcome is determining any additional data that must be collected and any additional changes that must be made in the problem statement, objectives, and constraints [6].

Analytical techniques can provide valuable information for the design and decision-making process in this step. Some of these techniques are as follows [1].

From–to chart: A from–to chart is a matrix used to summarize information regarding material movement between related predefined nodes. It can be used to prepare material flow patterns, compare alternative, identify bottlenecks, and determine candidates for mechanization and automation.

Flow-process chart: The flow-process chart is a step-by-step record of activities performed to accomplish a task. The flow-process chart is useful for analysis and for determining improvements, such as combining operations, eliminating unnecessary handling, simplifying a method, or changing a sequence or routing.

Flow diagram: The flow diagram is a graphical outline of the steps in a process similar to a flow-process chart. It is valuable for obtaining a macroperspective on the entire activity.

Product quantity (PQ) chart: The PQ chart is a graphical record of various products, parts, or materials produced or used for a particular time period. Quantities should be related to standard unit loads.

Simulation and waiting line analysis: These two analytical techniques help designers to simplify the problem analysis.

Developing Alternatives

To help develop alternative MH system designs, the MH system equation may be useful. The equation gives the key for identifying solutions to MH problems. It determines three important system components: *What* defines the type of materials moved, *where* and *when* identify the place and time requirements, and *how* and *who* point to the MH methods. These questions all lead us to the system.

The MH system equation is given by [7]:

$$\text{Materials} + \text{Moves} + \text{Methods} = \text{System}$$

Evaluating Alternatives

The objective of this step is to determine the value of alternative MH systems so that the designer can find the optimum solution. Economic analysis is an obligation for determining the best solution to a problem. It is also important in preparing a justification of capital expenditures for consideration and approval by upper management. Basic methods of cost comparison are discussed below.

Payback period: The most commonly used method for economic analysis—payback period—is the easiest method to compute and understand. It computes the time period required for estimated project savings to equal the investment. A serious shortcoming is the assumption that one alternative is better than another because it pays for itself more rapidly.

Return on investment (ROI): Unlike the payback period method, the ROI method takes into consideration the equipment's useful life. Normally, it relates net profit after taxes and depreciation to the total investment, thus indicating what each alternative will earn with respect to the investment.

Discounted cash flow (DCF): DCF computes the total present worth of cash flow over the project's life, using an interest rate equal to the company's minimum required rate of ROI. This method considers the present value of money after interest payments have been added to it over a period of time. Basically, it finds the interest rate that discounts future earning of the project alternative down to a present value equal to the project cost.

Some noneconomic factors can help the designer evaluate alternatives:

- Capacity
- Ability to handle the product
- Maintainability
- Reliability
- Damage and safety
- Compatibility
- Installation and lead time

Selecting the Preferred Design

In this step, all of the design alternatives that were evaluated in the previous steps are compared with each other, and the one that satisfies the objective is selected. If the MH system follows more than one objective, then the multicriteria decision methods will help the designer select the right alternative that will satisfy the majority of objectives.

Implementing the System

Implementation means to give practical effect to and ensure actual fulfillment by concrete measures. In this step, the approved MH project is implemented into a physical operating system which moves materials. The effectiveness with which implementation is carried out will determine the degree of success attained by the MH system.

This step includes the following tasks.

Organize for implementation: The quality of the design, the smoothness of installation, and the efficiency of the resulting system all depend on good organization and competent personnel.

Determine roles in implementation: The MH system designer works with many other departments and individuals in the design and installation of an MH system. Some of these have the authority to require changes in design and operation. Others may be specialists who can furnish valuable advice on some facets of the system design. The system designer must be able to work with all those involved to secure the best results.

Determine the implementation procedure: After the MH system design has been approved, the system designer has to coordinate carrying out the plan and installing the system. This requires considerable efforts, good technical abilities, and personnel skills.

Train personnel: It is not a necessary part of project implementation to train operating and maintenance personnel in new methods and equipment, but this may be a major effort as in a factory introducing new developed equipment for the first time. On the other hand, little new training may be required if the system is similar to existing ones [6].

9.1.7 MH Costs

MH represents a major portion of total costs for almost every type of business. Depending on the nature of industry and the type of facility, MH may include 10–80% of total costs. MH adds cost, but not value; hence, companies try to reduce their MH costs. Reducing their total MH costs allows a profit-oriented business to maintain a competitive edge over its competition.

To control MH costs, an analyst should review possible cost-reduction projects in order to identify general sources of potential savings. According to Magad and Amos [1], the majority of MH costs is related to space, labor, inventory, equipment, and waste. These factors and their effects on costs are explained as follows.

Space: Improvement in space unitization can reduce costs. One of the most effective options is to make use of air right or cube space. For instance, by increasing the storage height from 12 to 16 feet, one company increased its space unitization by 33%.
Labor: Automated and mechanized systems can considerably reduce labor costs. For example, the US Postal Service installed automatic mail-handling and -processing equipment to reduce labor costs and to improve flow. It reduced the error rate from 4% to 1% and increased letter sorting from 600 to 35,000 letters per hour.
Inventory: One objective of most businesses is to minimize inventories, because inventory reduction offers tremendous cost-reduction opportunities. Storing inventories increases holding cost for the company. Material-management programs that have been instituted to accomplish this objective include kanban, just-in-time, and material requirement planning (MRP). For example, the Schwitzer Turbochargers company implemented an MRP system that called for the installation of two AS and automated retrieval (AR) systems. The equipment was justified by a 30% inventory reduction.
Equipment: Developments in MH equipment can reduce costs. Numerous illustrations can be found in the transportation industry. For example, in cargo ship loading and unloading, every day a ship is in port means increased costs for the operating company, so it is critical that loading and unloading be performed expeditiously. Container ships, which are designed to transport containers that hold large amount of materials, are being used because they can be loaded and unloaded in record time.
Waste: Optimized materials management systems require good-quality materials. Concentrated efforts to eliminate the waste of damaged materials through best handling techniques and personnel-training programs will reduce costs by decreasing damage to materials.

Estimating MH Cost

Studies show that MH in a typical industrial firm accounts for 25% of all employees, 55% of all factory space, and 87% of production time. Another study shows that as much as 60% of total production cost refers to MH costs. In another point of view, 20–30% of direct labor costs and 50–70% of indirect labor costs refer to MH costs, so MH is one activity where many improvements can be achieved, resulting in significant cost saving.

Logistics-related costs are dynamic and do not readily fit with traditional accounting methods. The accounting difficulties become more pronounced when management tries to determine costs for a particular operation or a particular customer or to evaluate, outsource, or gain shared opportunities [8]. As an element of logistics, MH is no exception. Estimating the cost of MH alternatives and components is not a trivial task. At one end of the spectrum is a "rough-cut" method that uses standard data and rules of thumb. As an example of the use of rules-of-thumb data, one can get a reasonably accurate estimate of the purchase cost of specifying several walkie pallet jacks by using the unit cost of each walkie pallet jack. Use rules of thumb with caution.

At the other end of the MH cost-estimation spectrum is the use of detailed cost-estimation models based on engineering information. For instance, to estimate a MH equipment cost, a cost model must be presented based on different equipment size and alternative components.

For the design of more accurate MH systems, the interactions among the various components of the MH system design must be verified. More systematic approaches such as simulation analysis are needed to perform these verifications [5].

9.1.8 MH System Models

The importance of material-handling system selection (MHSS) and facility design in terms of production is widely recognized. Various approaches to the problem have been developed. However, the complexity of the involved issues in developing mathematical modeling-based approaches increase difficulties.

Attempts have been made to optimize the material flow system design but not the overall manufacturing operations. It is necessary to improve overall manufacturing operations and to integrate material flow design in the manufacturing system.

In recent years, there have been efforts to integrate MHSS problems. Chittratanawat and Noble [9] presented an integrated model for solving both MH and operation allocation (OA) problems. The integrated model was formulated as a nonlinear mixed-integer program. The model simultaneously determined the facility locations, pickup and dropoff points, and equipment. The approach integrated many of the significant factors in the facilities design while minimizing the overall facility design costs [10]. Then Paulo [11] and Paulo et al. [12] presented a 0−1 integer-programming formulation consisting of OA and MHSS submodels. The objective of the MHSS submodel was to maximize the compatibility between the MH equipment and the part types, as well as the ability of the MH equipment to perform the required tasks.

Lashkari [13] presented an integrated model of OA and MH selection in cellular manufacturing systems based on the work reported by Paulo and Paulo et al.

In this section, we will consider the mathematical model modified and extended by Lashkari as an example of mathematical models that were recently studied.

MHSS Mathematical Model

The 0−1 integer-programming model of MHSS is explained in the following. In this model, the decision variables were modified to include the index p.

Consider the set of n part types and the set of m machines as in the OA model. A part type i can be described in terms of five key product variables—complexity, precision, lot or batch size, diversity, and mass or linear dimension—that define the choices of manufacturing technology associated with its conception; these are labeled by indices $t = 1 \ldots 5$.

h: operation, $h = 1 \ldots H$.

$\hat{h}$: suboperation, $\hat{h} = 1 \ldots \hat{H}$.

$h\hat{h}$: operation and suboperation combination, as $e = 1 \ldots E$.

$Y_{h\hat{h}ejs}(ip)$: 0−1 decision variable, $Y_{h\hat{h}ejs}(ip) = 1$ if $(h\hat{h})$ requires MH equipment e at machine j where manufacturing operation s of (ip) is performed.

The objective is to generate the "most compatible" MH selection for a given mix of part types and MH requirements. Thus, the objective function is to maximize the following index, which represents the overall compatibility of the MH equipment and the part types:

$$\sum_{e=1}^{E} \sum_{h=1}^{H} \sum_{\hat{h}=1}^{\hat{H}} W_{h\hat{h}e} \sum_{i=1}^{n} C_{ei} \sum_{p=1}^{P(i)} \sum_{s=1}^{S(ip)} \sum_{j \in J_{ips}} A_{sj}(ip)\alpha_{h\hat{h}sjip} Y_{h\hat{h}ejs}(ip)$$

where

$$C_{ei} = 1 - \frac{\sum_{t=1}^{T=5} \left| W_{et} - \hat{W}_{it} \right|}{5T}$$

The parameter C_{ei} is a measure of the compatibility of a piece of equipment and a part type, and it is experimentally constructed to evaluate a value between 0 and 1, where 0 indicates incompatibility and 1 indicates complete compatibility.

The three rating factors (W_{hhe}, W_{et}, and $\hat{W}_{it}$) are largely subjective and represent an attempt to capture the relationships between the ability of MH equipment to perform various MH operations and the part types and their technological characteristics.

$\alpha_{h\hat{h}sjip} = 1$ if operation s of (ip), to be performed at machine j, requires $(h\hat{h})$, and 0 otherwise.

The following constraint sets are needed to ensure that the respective conditions are met.

1. Only one type of MH equipment is chosen to perform $(h\hat{h})$, associated with operation s of (ip) at machine j:

$$\sum_{e \in E_{ips}} Y_{h\hat{h}ejs}(ip) = A_{sj}(ip)\alpha_{h\hat{h}sjip} \quad \forall \ s,(ip),j,h,\hat{h}$$

2. The selection of a piece of MH equipment e may depend on the prior selection of another piece of MH equipment $\hat{e}$. This set of constraints is provided to allow precedence relationships that may exist in the assignment of MH equipment:

$$D_e \leq D_{\hat{e}} \quad \forall \ e,\hat{e}$$

where $D_e = 1$ if a piece of equipment e has been chosen and 0 otherwise, and similarly for $D_{\hat{e}}$.

3. If a piece of MH equipment is chosen, then at least one $(h\hat{h})$ must be assigned to that equipment:

$$\sum_{i=1}^{n}\sum_{h=1}^{H}\sum_{\hat{h}=1}^{\hat{H}}\sum_{p=1}^{P(i)}\sum_{s=1}^{S(ip)}\sum_{j\in J_{ips}} A_{sj}(ip)\alpha_{h\hat{h}sjip}Y_{h\hat{h}ejs}(ip) \geq D_e \quad \forall\ e$$

4. The total time for all the jobs assigned to a piece of MH equipment does not exceed the time available on it. It is given by:

$$\sum_{i=1}^{n} \mathrm{di}\sum_{h=1}^{H}\sum_{\hat{h}=1}^{\hat{H}}\sum_{p=1}^{P(i)}\sum_{s=1}^{S(ip)}\sum_{j\in J_{ips}} t_{h\hat{h}e}A_{sj}(ip)\alpha_{h\hat{h}sjip}\mathrm{Y}_{h\hat{h}ejs}(ip) \leq T_e D_e \quad \forall\ e$$

where $t_{h\hat{h}e}$ is the time required by MH equipment e to perform $(h\hat{h})$, and T_e is the time available on the MH equipment e.

Assembling the above, we get the following complete statement of our 0−1 integer-programming model of the MHSS, which is designated as P(MHSS):

$$\text{Maximize} \sum_{e=1}^{E}\sum_{h=1}^{H}\sum_{\hat{h}=1}^{\hat{H}} W_{h\hat{h}e}\sum_{i=1}^{n} C_{ei}\sum_{p=1}^{P(i)}\sum_{s=1}^{S(ip)}\sum_{j\in J_{ips}} A_{sj}(ip)\alpha_{h\hat{h}sjip}Y_{h\hat{h}ejs}(ip)$$

where

$$C_{ei} = 1 - \frac{\sum_{t=1}^{T=5}\left|W_{et} - \hat{\mathrm{W}}_{it}\right|}{5T}$$

Subject to

$$\sum_{e\in E_{ips}} \mathrm{Y}_{h\hat{h}ejs}(ip) = A_{sj}(ip)\alpha_{h\hat{h}sjip} \quad \forall\ s,(ip),j,h,\hat{h}$$

$$D_e \leq D_{\hat{e}} \quad \forall\ e,\hat{e}$$

$$\sum_{i=1}^{n}\sum_{h=1}^{H}\sum_{\hat{h}=1}^{\hat{H}}\sum_{p=1}^{P(i)}\sum_{s=1}^{S(ip)}\sum_{j\in J_{ips}} A_{sj}(ip)\alpha_{h\hat{h}sjip}Y_{h\hat{h}ejs}(ip) \geq D_e \quad \forall\ e$$

$$\sum_{i=1}^{n} di\sum_{h=1}^{H}\sum_{\hat{h}=1}^{\hat{H}}\sum_{p=1}^{P(i)}\sum_{s=1}^{S(ip)}\sum_{j\in J_{ips}} t_{h\hat{h}e}A_{sj}(ip)\alpha_{h\hat{h}sjip}Y_{h\hat{h}ejs}(ip) \leq T_e D_e \quad \forall\ e$$

$$[Y_{hh^{\wedge}ejs}(ip), D_e, D_{\hat{e}}] \in \{0,1\} \quad \forall\ i,p,s,j,h,\hat{h} \in J_{ips}$$

The assignments determined by the model (i.e., the $Y_{h\hat{h}ejs}(ip)$) are then summarized in the matrix $B_{sje}(ip)$ where an element $\{b_{sje}(ip)\}$ is equal to 1 if operation s of (ip) is performed at machine j using MH equipment e to transport the part to the next machine, and 0 otherwise. The matrix B provides the necessary information about the assignment of the MH equipment to perform various operations of (ip) so as to enable P(OA) to compute the MH costs in its objective function. At the same time, it reduces the choices of MH equipment in the set E_{ips} as iterations continue. The matrix $B_{sje}(ip)$ forms the feedback link between P(MHSS) and P(OA), thus completing the loop.

9.2 Packaging

9.2.1 *History*

Iron and tin-plated steel were used to make cans in the early nineteenth century. Paperboard cartons and fiberboard boxes were first presented in the late nineteenth century [14].

Packaging advances in the early twentieth century included Bakelite closures on bottles, transparent cellophane overwraps and panels on cartons, and increased processing efficiency, much of it helping to improve food safety. As additional materials such as aluminum and several types of plastic were developed, they were incorporated into packages to improve performance and functionality [15].

9.2.2 *Definition*

Packaging is an important warehousing and material-management concern, one that is closely tied to warehouse efficiency and effectiveness. An appropriate packaging increases service, decreases cost, and makes for better handling. Good packaging can have a positive impact on layout, design, and overall warehouse productivity [7].

Packaging has also been defined as the science, art, and technology of enclosing products for distribution, storage, sale, and use. Packaging also refers to the process of design, evaluation, and production of packages [15].

9.2.3 *Functions of Packaging*

In terms of the different requirements to which packaging are subjected, García-Arca et al. [16] associate packaging with three large functions: marketing, logistics, and environmental.

In its marketing function, the package presents customers with information about the product and promotes the product through the use of color and shape.

The central purpose of the environmental function is to optimize packaging while minimizing packaging waste wherever appropriate and to reuse or recycle.

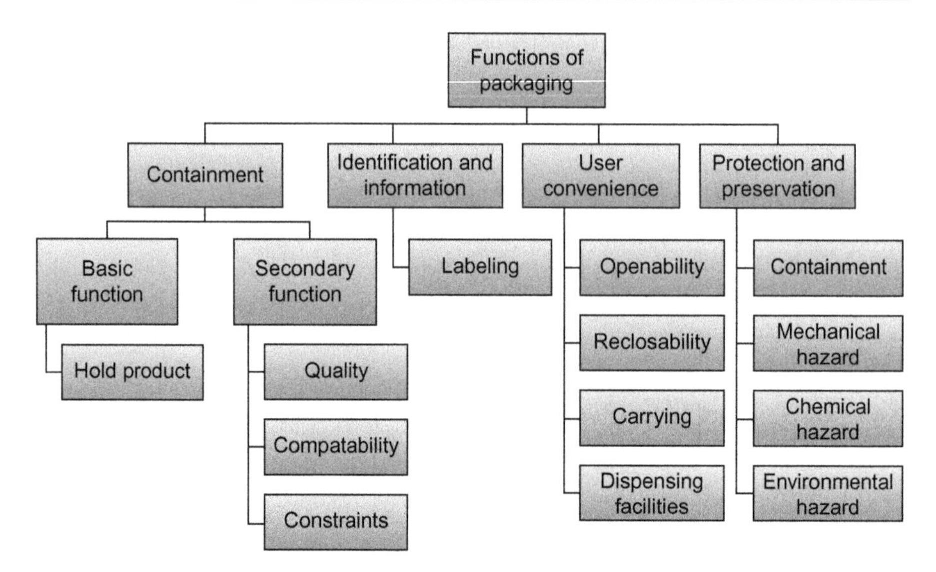

Figure 9.2 Functions of packaging [15].

The logistics perspective focuses on two aspects. First, it provides protection for the product. Second, it is an instrument for improved distribution efficiency Figure 9.2.

More specially, packaging performs the following functions.

1. *Containment*: Products must be contained before they can be moved from one place to another to protect them.
2. *Apportionment*: Make the large output of manufacturing into smaller quantities of greater use to customers. It results in manageable, desirable, and consumer-sized productions.
3. *Convenience*: Packages can have features that add convenience in distribution, handling, display, sale, opening, reclosing, use, and reuse.
4. *Information transmission*: Packages and labels communicate how to use, transport, recycle, or dispose of a package or product. With pharmaceuticals, food, medical, and chemical products, some types of information are required by the governments.
5. *Portion control*: Partitioning large package or bulk commodity into packages makes for more suitable sizes for individual households and aids inventory controls.
6. *Unitization*: To unitize primary packages into secondary packages, the secondary packages are unitized into a container that is loaded with several pallets. This decreases the number of times a product must be handled.
7. *Physical protection*: The package protects the enclosed object from physical items, shock, vibration, compression, and temperature, among other forces.
8. The *barrier protection*: Keeping the contents clean, fresh, and safe for the intended shelf life is a main function. Some packages contain desiccants or oxygen absorbers to help extend shelf life. Modified atmospheres or controlled atmospheres are also maintained in some food packages.
9. *Security*: Packaging can play an important role in reducing the security risks of shipment. Packages can be made with improved tamper resistance to deter tampering and also can have tamper-evident features to help indicate tampering. Packages can be

engineered to help reduce the risks of package pilferage: some package constructions are more resistant to pilferage, and some have pilfered indicating seals. Packages may include authentication seals to help indicate that the package and its contents are not counterfeit. Packages can also include antitheft devices, such as dye packs, radiofrequency identification (RFID) tags, or electronic article surveillance tags, that can be activated or detected by devices at exit points and that require specialized tools to deactivate. Using packaging in this way is a means of loss prevention.

10. *Marketing*: Encouraging potential customers to prepare the product by means of labels and packages is the exact meaning of marketing function. Marketing communications and graphic design are applied to the surface of the package and the point-of-sale display.

9.2.4 Packaging Operations

It is clear that a packaging function includes the basic packaging operations of folding, inserting, wrapping, sealing, and labeling. The procedure is as follows. The packaging is first folded from its collapsed form and then the product is inserted into the folded package, which is then wrapped with a packaging sheet before it is sealed by tape. Finally, identification labels such as bar codes are either stuck or printed on.

9.2.5 Packaging Equipment

Using the proper packaging equipment is so important in accomplishing the packaging functions at lowest possible costs. Equipment for this purpose may be defined as any device or contrivance that assists in the accomplishment of a task [15].

In the broadcast sense, two classes of packaging equipment are employed: first, equipment to fabricate packaging materials and containers; second, equipment employed by the user to utilize packaging materials and containers. Under certain conditions, users may find it economical to incorporate in their operations, complete or partial, packaging material or container fabrication equipment. In these instances, users perform the function of packaging suppliers.

Packaging equipment employed by the user is often designed for a specific material or container; as such, it cannot be isolated in the discussion of a particular packaging practice. Therefore, many of the aids utilized in such operations as closure forming, reinforcing, bundling, and easy opening have been in conjunction with a particular material or container type.

For simplicity, packaging equipment is best classified by function. The following items represent the major functions that can be performed by packaging equipment:

1. Forming and assembly
2. Filling, loading, and overwrapping
3. Weighing and counting
4. Closing and sealing
5. Bundling, unitizing, and reinforcing
6. Identifying
7. Miscellaneous

9.2.6 Labeling

After the material is being packaged, it is placed into a box and enclosed by the cover. At this point, it becomes necessary to identify and label the box. Package labeling is any type of communication such as written, electronic, or graphic on the packaging-associated label. Whether words or code numbers are used depends on the nature of the product and its vulnerability to pilferage. Retroflexed labels that can be ready by optical scanners may also be applied.

Labeling of consumer packages and products such as bottles, cans, and folding cartons is frequently an integral part of the filling, packing, closing, and weighing operations. Labels are affixed by specialized high-speed, fully automatic equipment, and the principal considerations are machine performance, economy, and appearance [15].

The choice of color and typography in labeling often is very important in customer acceptance, use, and response. Indeed, the success or failure of a packaged product can be attributed to the manner and style in which it is identified [17].

Labeling Regulations

Many regulations govern the labeling of customer-size packages, including the labeling of weight, specific contents, and instructions for use. Today, many of these must also be placed outside the larger cartons because some retail outlets sell in carton lots, and buyers do not see the consumer packages until they reach home.

For instance, most countries have some laws governing food labeling. These are spread over many reforms and parliamentary acts, making the subject complex. Nevertheless, the following general laws should be implicit for any food product.[1]

Name: This must also inform the customer the nature of the product. It may also be necessary to attach a description to the product name. However, certain generic names must be used only for their conventional uses—for example, muesli, coffee, and prawns.

Ingredients: All ingredients of the food product must be stated under the heading "Ingredients" and must be stated in descending order of weight. Moreover, certain ingredients such as preservatives must be identified as such by the label "Preservatives," a specific name (e.g., sodium nitrite), and the corresponding European registration number colloquially known as an *E number* (e.g., "E250").

Nutritional information: Although it is not a legal requirement to declare nutritional information on a product, if the manufacturer makes the claim that the product is "low in sugar," then it must be supported with nutritional information (normally in tabulated form). However, it is recommended to declare nutritional information because consumers more than ever are investigating this information before making a purchase.

Medical or nutritional claims: Medical and nutritional claims are tightly regulated; some are allowed only under certain conditions and others are not authorized at all. For example, presenting claims that the food product can treat, prevent or cure diseases, or other "adverse conditions" are prohibited. Claiming that a food is reduced in fat or rich in vitamins requires the food to meet compulsory standards and grades; in addition, the terms must be used in the form specified in regulations.

[1] Refer to http://en.wikipedia.org/wiki/Food_labeling_regulations

Storage conditions: If there are any particular storage conditions for the product to maintain its shelf life, these must be pointed out. However, as a rule it is recommended to always describe the necessary storage conditions for a food product.

Date tagging: There are two types of date tagging:

- A *use by* date must be followed by a day or month (or both) that the product must be consumed by.
- A *best before* date is used to indicate when the product's optimal quality will begin to degrade: this includes when the food becomes stale, begins to taste "off," or decays, rots, or goes moldy.

Business name and address: In addition to the business name and address, it is necessary to indicate the manufacturer or packager if it is independent of the main business and the seller is established within the European Union.

Place of origin: The food is required to specify its place of origin, especially if the name or trademark is misleading—for example, if a product called "English Brie Cheese" is produced in France.

Instruction for use: This is only necessary if it is not obvious how to use or prepare the product, in which case the consumer's own initiative must be used.

Presentation: The label must be legible and easy to read. It must also be written in English although the manufacturer may also include other languages.

Lot mark or batch code: It must be possible to identify individual batches with a lot mark or batch code. The code must be prefixed with the letter L if it cannot be distinguished from other codes although the date mark can be used as a lot mark. Manufacturers must bear in mind that the smaller the size of a batch, the smaller the financial consequences in case of a product recall.

Sectioning: All of the following must be in the same field of vision:

- Product name
- Date mark
- Weight
- Quantity
- Alcohol strength (if applicable)

Standard specification: Indicate the level of the product's standards compliance in manufacturing and packaging.

Bear in mind that there are many other laws and European regulations for different types of food products.

Labeling Techniques

Label materials: Labels are available in both cut and roll-stock forms. An advantage of roll-stock labels is that there is less chance of a wild label occurring in the roll. The rolls can also be easily and automatically inspected offline by a user to double-check the label printer's own quality control. Both forms generally are made from paper, foil, film, or laminate structure. For special applications, they can also be made from paperboard, fabrics, synthetic substrates, and even metals.

Label-application equipment: The equipment selected for the application of labels is influenced by the type of label backing, the type and size of container to which the label is to be affixed, and production or shipping requirements.

Table 9.1 Comparison of Labeling Systems[a]

	Label Type				
Characteristics	**Plain, Cut**	**Heat-Seal, Cut**	**Plain, Roll**	**Heat-Seal, Roll**	**P-S Roll**
Ease of operation	E	D	C	B	A
Changeover time	E	D	C	B	A
Cost of equipment	A	A	B	C	A
Cost of labels	A	B	A	C	C
Quality of printing	A–B	A–B	B	A–B	A–B
Ease of code marking	A	B	A	B	B
Servicing from supplies	E	E	B	A	A
Avoidance of label mix-ups	E	E	A	A	A

[a]A is best; E is poorest.

The methods used for various label backing previously recorded are as follows [15].

1. *Plain-back labels* are used with high-speed labeling machines when economy is of utmost importance. Spot or lap gluing is accomplished on this equipment with minimum glue consumption. Frequently, portions of shipping documents that run through electronic data-processing equipment later become ungummed shipping labels.
2. *Gummed labels* are more expensive than plain labels but require only simple moistening for application. The application of gummed shipping labels is usually a manual operation. Gummed labels for product identification can be applied with semi- or fully automatic equipment.
3. *Pressure-sensitive labels* will adhere to nearly any type of smooth surface and do not require moistening or gluing for application. The labels are adhered to a low-release backing paper; removal of the label from the backing sheet can be accomplished manually or by machine.
4. *Heat-seal-coated labels* require a heating element to activate the thermoplastic adhesive. Pressure of the label on the desired surface can be accomplished manually or by mechanical means.

To compare cut and roll-stock label systems, see Table 9.1.

9.2.7 Protection Packaging

A protective package should perform the following functions [6]:

1. Enclose the materials, both to protect them and to protect other items from their effects.
2. Restrain them from undesired movements within the container when the container is in transit.
3. Separate the contents to prevent undesired contact, such as through the use of corrugated fiberboard partitions used in the shipment of glassware.
4. Protect the contents from outside vibrations and shocks.
5. Support the weight of identical containers that will be stacked above it as part of the building-blocks concept.

6. Position the contents to provide maximum protection.
7. Provide for fairly uniform weight distribution within the package because most equipment for the automatic handling of packages is designed for packages that have evenly distributed weights.
8. Provide enough exterior surface area that identification and shipping labels can be applied along with specific instructions such as *this side up* or *keep refrigerated.*
9. Be tamper-proof to the extent that the evidence of tampering can be noticed.
10. Be safe in the sense that the package itself (both in conjunction with the product carried and after it has been unpacked) presents no hazards to customers or others.

9.2.8 Packaging for Distribution Efficiency

Packaging is an important factor in logistics, and its role and purpose are widely discussed in the literature. The purpose of packaging can be summarized into four main areas: information displays, improved materials handling, improved customer service, and quality security [18].

These are some of the most important concerns to the logistician in packaging design:

- Handling and storage
- Strength, size, and configuration
- Unitization
- Containerization
- Identification

As described, packaging is a significant factor in logistics, and it affects logistics in three important domains: MH, transportation, and customer service.

Packaging Effects on Materials Handling

Packaging has an important impact on the MH function. Good packaging can have positive impact on materials handling and vice versa. Poor packaging such as seen with oversized packages can inhibit the MH operation. If the package is not designed properly, the logistical system's efficiency will decline. The ability to display and provide information is an important advantage of packaging. Information can relieve the MH operations because the information on the package allows warehouse personnel to locate and quantify the products easily. The size and protection of products directly affect MH as well as quality. The size aspect influences the utilization of a warehouse. Size also affects the quantity of products that can be moved at the same time. The level of protection that the package provides enables different transportation alternatives. A package must not only protect the product from physical damage but also be required to support the weight of products stacked above it.

Efficient MH can improve a warehouse's ability to provide customer service in terms of quick and accurate response to customer demands. Efficient MH could also reduce cost by consuming fewer resources such as forklift time, manual labor, and warehouse space [18].

Packaging Effects on Transportation

Another logistical activity that is greatly affected by packaging is transportation because packaging affects the volume of a product. Transportation provides place utility to a product and also time utility in some aspects [19]. *Transportation* in this sense concerns the movement of products from the focal firm to the customer.

From a packaging perspective, the most important factors that affect transportation costs are volume, stowability, and handling because these factors are highly influenced by a product's physical characteristics.

Packaging Effects on Customer Service

Customer service is a logistical activity that is also affected by packaging. Customer service has various meanings throughout an organization, so it is important to consider these aspects.

Customer service can be viewed as something that is provided by a firm to the buyer who is purchasing a product.

Packaging is an important factor in providing customer service. For example, a packaging solution might be good for the firm, but the customer might not be able to handle the package at its premises, and then customer service is lost. This adds to the interrelationship between packaging and logistical activities.

9.2.9 Packaging Costs

Packaging costs account for a significant portion of a product's manufactured cost, so it is important for companies to minimize these costs [20].

The three major components of packaging costs are labor, equipment, and materials. Packaging is generally considered as the process of placing the enclosed item into a shipping container, so the labor costs involved to accomplish either or both of these operations is influenced by the materials, methods, and equipment of the specific packaging system.

About 9% of the cost of any product is likely to be the cost of its packaging; about 90% of this cost of packaging may be attributed to factors other than the packaging material itself. The manufacture, use, and disposal of packaging accounts for about 60% of total production costs or between 15% and 50% of the selling price of a product.

However, the impact of packaging costs is often not considered or measured by logisticians. They often found that the handling of packaging items strongly impacts the overall logistics cost.

In Table 9.2, different packaging consequences are presented that create requirements for trade-offs between the logistic activities: materials handling, customer service, communications, and transportation [19]. From this, we can see that packaging is closely related to logistics activities [7].

As presented by Lambert et al. [19], the issue is improving the advantages of the chosen packaging solution while minimizing the disadvantages. This

Table 9.2 Packaging Consequences [6]

Packaging Consequences	Trade-Offs
Increased package information	Decreases order-filling times Decreases tracking of lost shipments
Increased package protection	Decreases damage in transport Increases weight Decreases cube utilization from larger dimensions Increases product value
Increased standardization	Decreases MH Decreases customer customization

emphasizes the importance of understanding the effects that packaging has on the logistical activities and their internal interrelation in order to find the right balance between them.

9.2.10 Packaging Models

Traditional packaging is usually considered as a cost-driven center rather than a value-added component throughout the manufacturing and distribution processes. If we reconsider the packaging design in a systematic approach, it is easy to develop more cost-effective solutions for manufacturing processing that can support handling and distribution as well as provide product protection [21].

Chan et al. [21] have introduced a systematic approach to packaging logistics. In this approach, they believe that logistics considerations cannot fully dominate the design of packaging. They should be weighted along with marketing and product design to work out the best possible compromise by taking all factors into consideration. Modern packaging needs to compromise between all packaging functions and consider the role of the packaging in a systematic way. As a result, the roles of packaging would be both a cost-driven center and a value-added process in the logistics system.

Chan et al. [21] suggest six important steps in explaining the methodology for systematic approach in packaging.

Step 1: Identify the possible package flow route and the packaging level.
Step 2: Integrate packaging logistics in the product type stage.
Step 3: Create a preliminary package design.
Step 4: Establish information flow among all parties.
Step 5: Redesign the packaging system.
Step 6: Use value-chain model to evaluate the finalized chosen package.

9.3 Case Study

In this section, challenges in the packaging of microelectromechanical systems (MEMS) are explained [22].

MEMS are made of mechanical devices and mechanical components that can be as small as a few microns. They can be mechanical interconnects of microsystems, and they can also receive signals from one physical domain and send them to another, such as mechanical to electrical, electrical to mechanical, and electrical to chemical. These devices are broadly categorized as either sensors or actuators. MEMS sensors are devices such as pressure sensors, accelerometers, and gyrometers that perceive an aspect of their environment and produce a corresponding output signal. Actuators are devices that are given a specific input signal on which

Table 9.3 Current Packaging Parameters, Challenges, and Suggested Possible Solutions for MEMS

Packaging Parameters	Challenges	Possible Solutions
Release etch and dry	Washing away parts during release Must release parts individually after dicing	Freeze drying, coating, or processes that reduce surface tension Use dimples Develop a dicing Laser sawing
Dicing and cleaving	Eliminating contamination caused by cooling fluid and particulates during wafer sawing	Release dice after dicing Cleave wafers Laser sawing Wafer level encapsulation
Die handling	Damages top die's face contact region	Fixtures that hold MEMS dice by sides rather than top face, such as collects that fit existing pick-and-place equipment
Stress	Abating performance degradation and resonant frequency shifts Curling of thin film layers Misalignment of device features	Low modulus Low creep die attach material Annealing Die attach materials with CTE similar to that of silicon
Outgassing	Corrosion Outgassing of organic solvents from polymeric die attach materials	Low outgassing epoxies Low modulus solders New die attach materials Removal of outgassing vapor
Testing	Applying nonelectric stimuli to devices Testing moving device features before release Inability to release parts before dicing	Electrical test structures to mimic nonelectrical functions Modify (where possible) wafer-scale probers to do nonelectrical tests Cost-effective, high-throughput, and parallel-packaged device-test systems

to act and a specific motion or action is produced. Other examples of MEMS actuators are microengines, microlocks, and discriminators.

MEMS packaging is quite different from conventional integrated circuit (IC) packaging. Whereas many MEMS devices must interface with the environment to perform their intended functions, the package must be able to facilitate access with the environment while protecting the enclosed devices. The package must also not interface with or impede the action of the MEMS device. The incomplete attachment material should be low stress and low outgassing while also minimizing stress relaxation over time, which can lead to scale-factor shifts in sensor devices. The fabrication process used in creating the devices must be compatible with each other and not damage the devices. Many devices are specific in application, requiring custom packages that are not commercially available. Devices may also need media compatible packages that can protect the devices from harsh environments in which the MEMS device may operate. Current packaging parameters, challenges, and suggested possible solutions for MEMS are shown in Table 9.3.

References

[1] E.L. Magad, J.M. Amos, Total Materials Management, Chapman & Hall, New York, 1995.

[2] P. Baily, D. Framer, Material Management Handbook, Grower Publishing, England, 1988.

[3] R.H. Ballou, Business Logistics Management, third ed., Prentice Hall, New York, 1992.

[4] R.M. Eastman, Material Handlings, Marcel Dekker, Inc., New York, 1987.

[5] J.A. Tompkins, J.A. White, Y.A. Bozer, E.H. Frazelle, J.M.A. Tanchoco, J. Trevino, Facilities Planning, John Wiley & Sons, New York, 2003.

[6] J.R. Stock, D.M. Lambert, Strategic Logistics Management, fourth ed., McGraw-Hill, Singapore, 2001.

[7] Satich C. Ailawadi, Rakesh Sing, Logistics Management, Prentice Hall of India, New Delhi, 2006.

[8] A. West, Managing Distribution and Change: The Total Distribution Concept, John Wiley & Sons, New York, 1992.

[9] S. Chittratanawat, J.S. Noble, An integrated approach for facility layout, P/D location and material handling system design, Int. J. Prod. Res. 37 (1999) 683–706.

[10] S.G. Lee, S.W. Lye, Design for manual packaging, School of Mechanical and Production Engineering, Nanyang Technological University, Republic of Singapore, Emerald J. 33(2) (2003) 163–189.

[11] J. Paulo, A Mathematical Model of Operation Allocation and Material Handling System Selection Problem Manufacturing System, M.A.Sc. Thesis, Department of Industrial and Manufacturing Systems Engineering, University of Windsor, Canada, 2002.

[12] J. Paulo, R.S. Lashkari, S.P. Dutta, Operation allocation and material handling system selection problem manufacturing system: a sequential modeling approach, Int. J. Prod. Res. 40 (2002) 7–35.

[13] R.S. Lashkari, Towards an integrated model of operation allocation and material handling selection in cellular manufacturing systems, Int. J. Prod. Econ. 87(2) (2004) 115–139.

[14] W. Soroka, Fundamentals of Packaging Technology, Institute of Packaging Professionals, Naperville, IL, 2002. ISBN 1-930268-25-4 http://en.wikipedia.org/wiki/Special:BookSources/1930268254.

[15] F. Albert Paine, The Packaging User's Handbook, Springer, Blackie Academic and Professional, New York, 1991 (published by authority of the Institute of Packaging).

[16] J. García-Arca, J.C. Prado-Prado, Antonio-Garcia-Lorenzo, Logistics improvement through packaging rationalization: a practical experience, J. Packag. Technol. Sci. 19 (2006) 303–308.

[17] W.F. Friedman, J.J. Kipnees, Industrial Packaging, John Wiley & Sons, New York, 1960.

[18] Hanlon J.F., Kelsey R.J., & Forcinio H.E., Handbook of Package Engineering, third ed., CRC Press, London, 1998.

[19] D.M. Lambert, JR Stock, L.M. Ellarm, Fundamental of Logistics Management, McGraw-Hill, Boston, 1998.

[20] J. Hassel, T. Leek, Packaging Effects on Logistics Activities: A Study at ROL International, Jönköping International Business School, Jönköping, 2006.

[21] F.T.S. Chan, H.K. Chan, K.L. Choy, A systematic approach to manufacturing packaging logistics, Int. J. Adv. Manuf. Technol. 29(9–10) (2006) 1088–1101.

[22] A.P. Malshe, C. O'Neal, S.B. Singh, W.D. Brown, W.P. Eaton, W.M. Miller, Challenges in the packaging of MEMS, Int. J. Microcircuits Electron. Packag. 22(3) (1999) Third Quarter (ISSN 1063-1674) 2–4.

10 Storage, Warehousing, and Inventory Management

Maryam Abbasi

Department of Industrial Engineering, Amirkabir University of Technology, Tehran, Iran

10.1 The Reasons for Storage Inventory

The costs of capital, warehousing, protection, deterioration, loss, insurance, package, and administration make stocks expensive [2]. "Inventories can absorb 25–40% of logistics costs and represent a significant proportion of the total assets of an organization" [3], but we keep inventory for the following reasons.

- To protect a firm against unexpected changes in customer demand and lead time and improve customer service [4].
- To take advantage of economics of scale when purchasing, transportation, and manufacturing in large volumes by reducing the cost per unit [4].
- To balance supply and demand when they are not equivalent in a same time. Sometimes demand is higher than the available supply (e.g., products that have a seasonal demand) and vice versa. Keeping inventory thus helps to answer demand as needed [5].
- To hedge against contingencies such as labor strikes, fires, and flood. In such situations, inventory guarantees normal supply for some period of time [3].
- To eliminate manufacturing bottlenecks, in this situation work in process goods should be kept in order to eliminate bottlenecks and increase productivity.
- To hedge against price changes in situations when prices change unexpectedly (most of the time increase) and keeping raw material is economical [1].

10.2 The Role of Distribution Centers and Warehouses in Logistics

The general reasons for installing distribution centers and warehouses are as follows.

Storage of goods: The basic function of warehouses is to store goods for the time they will be needed.

As part of production process: In many cases, a production process needs a period of time (without any operation) to complete a product—e.g., the production of cheese and wine. Thus, warehouses can keep such products as a part of their production process [5].

Logistics Operations and Management. DOI: 10.1016/B978-0-12-385202-1.00010-4

Returned goods center: In reverse logistics, the handling of returned goods becomes important, so warehouses can act as a place to accumulate and make decisions about returned goods [7].

Consolidation: When customers order a number of products from different places and want them to be delivered together, a warehouse can receive the requested products from their separate origins and deliver them altogether to the customer [1].

Breakbulk: Breakbulk warehouses divide large receiving shipments in bulk from manufacturers into small less than truckload (LTL) shipments and send them to the customers [4].

Postponement: Warehouse can also be used as a place to postpone the production process. In these cases, a warehouse is capable of doing light manufacturing activities such as labeling, marking, and packaging. In-process goods are kept in these warehouses until a demand with special characteristics such as mark or package occurs; the requested activities will be done in the warehouse and finished goods will be ready to satisfy the demand [8].

Cross docking: In some cases, a warehouses act as a cross-docking point. Inventory does not stay in more than 12 hours. However, these warehouses receive inventory, transfer it to vehicles, and deliver to retailers. This system leads to reduction of inventory costs and lead times by decreasing storage time [9].

Transshipment: Transshipment is the process of transferring goods from one vehicle to another as necessary [9].

Product-fulfillment center: Fulfillment centers are distribution centers or warehouses that connect directly with final consumers. The following are some of the differences between product-fulfillment centers and other warehouses:

- Higher levels of customer service are available because of direct connects with final customers.
- More orders of smaller size are possible; these are almost always received electronically.
- Fulfillment centers typically must receive customer payments, often by major credit card; some also create customer invoices and handle banking for their clients.
- Returns from customers are more than that in the other warehouses.
- Computerized information systems and task automation are increasingly critical, and the transportation function (especially residential delivery) is more complex [10].

10.3 Warehouse Location

After a need for storage space has been demonstrated, the warehouse location should be determined. Facility location has a long-term impact on supply chains, so it is a strategic decision in supply-chain design [11].

According to Chopra and Meindl [11], there are many factors that help decision makers to choose a place for warehouse; the following factors are some of them:

- Land configuration and developing costs
- Building construction costs
- Community and local government attitude toward the warehouse

- Availability and access to transportation services
- Potential for expansion
- Hazards of the site (fire, theft, flood, etc.)
- Local labor quantity, labor rates, and climate
- Traffic congestion around the site
- Advertising value of the site
- Taxes relative to the site and operation of the warehouse [11].

Facility location decision is typically made at two levels [3]:

1. With respect to the location of all existing warehouses, where should be located a new one to balance transportation costs, inventory costs, order processing costs, etc.
2. After determination of general geographic region, it should be determined whether the warehouse is to be located on this side of town or that, or in this industrial part or that.

10.4 Warehouse Design

10.4.1 Size of Warehouse

When the location of warehouse is specified, the next decision is to determine the size of the building that is needed. Determination of warehouse size is affected by the following factors:

- Customer-service levels
- Size of market(s) served
- Number of products marketed
- Size of the product(s)
- Material-handling system used
- Throughput rate (i.e., inventory turnover)
- Production lead time
- Economics of scale
- Stock layout
- Aisle requirements
- Office area in warehouse
- Types of racks and shelves used
- Level and pattern of demand
- Storage policy [4]

10.4.2 Storage Policies

Three policies are followed to assign products to storage areas: randomized, dedicated, and class based [12]. Under a randomized storage policy, there are no restrictions on where items can be stored in the storage area, so any item can be placed anywhere. Under the dedicated storage policy, items can only be stored in their own special areas. Under class-based storage policy, items are divided into some classes, and each class is assigned to one storage area. Indeed, in the class-based storage policy, when there is only one class for all items, we have a randomized storage policy; when there is one class for each item, we have class-based storage policy. The best selection among different storage policies depends on the costs of order picking and warehouse space. Dedicated or random policy is selected when the order-picking costs or space costs are important. However, if both of costs (order picking and warehouse space costs) simultaneously are important, the best solution is class-based storage policy [13].

The cube-per-order index (COI) is one criterion used to assign product classes to storage locations. The COI of an item is defined as the ratio of the item's storage-space requirement (cube) to its popularity (number of storage and retrieval requests for the items). One procedure used to assign product classes to storage locations in a class-based storage policy is as follows [14]:

1. Calculate the COI for all items: ratio of an item's storage-space requirement (cube) to the number of storage and retrieval (S/R) transactions for that item.
2. Sort the items in a nondecreasing order by their COIs.
3. Allocate the first item in the list to the storage spaces that are nearest to the input–output (I/O) point and so on.

Pareto's law is another way to assign product classes to storage locations. This law divides all items into three classes: A, B, and C. Class A includes the 20% of items that 80% of the S/R activity is directed at. Class B includes the 30% of items with 15% of the S/R activity. Class C includes 50% of items with 5% of the S/R activity. Items of class A must be stored closest to the I/O point, class B next closest, and class C the farthest away [14].

10.5 Types of Warehouses

Warehouses may vary from one to another, but there are many ways to classify them [1].

- *By stage in the supply chain*: Warehouses may hold raw materials, works in progress, or finished goods.
- *By geographic area*: A regional warehouse may cover a number of countries, and a national warehouse will cover just one country.
- *By product type*: Storage may be devoted to individual products of classes of products—e.g., electrical parts, perishable foods, and hazard materials.
- *By function*: Warehouses may hold inventory, postponement materials, or breakbulk.
- *By area*: Size may be important—e.g., warehouses with less than 100 square meters or more than 1000 square meters.
- *By systems*: With respect to the level of automation (equipment and methods used for order picking or storage and retrieval of items), warehouse systems are divided into three types:
  1. In *manual warehousing systems* (picker-to-product systems), order pickers ride in vehicles (like elevators) in a storage area and pick the required items. The vehicles used in manual warehousing vary from completely manual to fully automated, but they are all applied to picking products.
  2. In *automated warehousing systems* (product-to-picker systems), machines and tools are used to facilitate order picking—e.g., a storage carousel that rotates around a

closed loop and delivers requested stock-keeping units (SKUs) to the order picker (horizontal carousel or vertical carousel).
3. In *automatic warehousing systems*, the level of automation is higher than in automated systems. Human order pickers are replaced by robots. Automatic warehouses perform order picking of small- and medium-sized items and shape (e.g., compact disks or pharmaceuticals) [15].

- *By ownership*
  1. Firms sometimes use *private warehouses* that they own or occupy on long-term leases [16]. These warehouses are always used by firms that are stable and have enough stock to fill them. The most important aspect of private warehouses is that they involve large capital investments. The many advantages of private warehouses include the following [5]:
     – They have lower stock-keeping costs than hired warehouses, particularly when their capacity is used most of the time.
     – They have higher service levels and more control over their operations.
     – They have a higher degree of flexibility than rented warehouses.
     – They have specialized personnel and equipment when needed to handle special products.
     – The warehouse space can be used for other applications when the storage space is no longer needed.
  2. *Public warehouses* are rented by firms that cannot justify the initial investment in private warehouses or that prefer to outsource warehouse operations. This type of warehouses has the following advantages.
     – They require no capital investment.
     – Space can be rented as much as needed [16].

There are many types of public warehouses.

– *General merchandise warehouses* are used to keep any kind of product and are used by different firms [6].
– *Temperature-controlled warehouses* are used to keep products and materials that must be kept at controlled temperatures such as perishable foods or some chemical materials [6].
– *Bulk storage warehouses* are used to keep unpacked products and materials in high volume (bulk goods) such as oil [3].
– *Bonded warehouses* are "licensed by the government to store goods prior to payment of taxes or duties. This helps to reduce the inventory value of the firm considerably" [3].
– *Special commodity warehouses* are used to keep agricultural products such as grains, wool, and cotton [4].

1. *Contract warehouses* are "the other type of nonowned warehousing, a partnership arrangement between the user and provider of the warehousing service such as transportation, inventory control, order processing, customer service, and returns" [6].

10.6 Warehouse Components

Space, equipment, and people are the three components of warehousing.

Equipment consists of material-handling devices, racks, conveyors, and all of the hardware and software used to make a warehouse function and that may differ from one warehouse to another (depending on warehouse nature and design). People are another important component in warehousing; it is their performance that makes the difference between high- and low-quality warehousing. Warehouse space is another component that differs in location, size, and design [17].

10.7 Warehouse Tasks and Activities

10.7.1 Material Flow in Warehouse

There are many steps in warehouses from receiving goods until their delivery to customers, including the following.

- *Receiving*: Goods are unloaded from shipment vehicles and delivered to the warehouse personnel.
- *Inspection and quality control*: After receiving, goods should be verified by inspection and quality control (usually randomly checked).
- *Preparation for transportation to the storage area*: Specific tasks such as product labeling may be done here.
- *Put away*: Goods are transported to the storage area.
- *Order picking*: Goods are retrieved when orders are received.
- *Aggregation of SKUs*: Sometime orders with multiple SKUs are combined for shipment.
- *Preparation for transportation to shipping area*: Items are readied for delivery, for example, with packaging.
- *Transportation of goods to the shipping area*: Ready-to-ship orders are made available for shipment to customers [18].

Warehouses also may have the following tasks:

- Inventory tracking
- Cross docking
- Packaging
- Light assembly, blending, filling, and outfitting
- Labeling and shrink wrapping
- Breakbulk and consolidation
- Postponement
- Transportation
- Import and export services
- Tracing, customer service, and billing
- Carrier monitoring
- Site location
- Real estate management
- Network analysis
- Systems development [19]

10.7.2 Order Picking

Order picking is one of the most important activities in warehouses because of its direct effects on customer-service level and warehouse costs [1].

A study in the United Kingdom [20] demonstrated that more than 60% of all operating costs of warehouses are related to order picking (Figure 10.1).

Order-Picking Methods

- In *discreet picking*, a single order is filled.
- In *batch picking*, a group of orders are filled by one order picker.
- In zone picking, each order picker is in charge of a specific zone of the warehouse and selects items that are in that zone. All order pickers do their work until the order is completed [4].
- In *wave picking*, orders may be released in waves (e.g., hourly or each morning or afternoon). This helps to control the flow of goods and replenishment, picking, packing, marshalling, and dispatching. Wave timing is tied to the schedules of outgoing vehicles [1].

Principles for Better Order Picking

The following are among the many ways to improve order picking and make it more efficient [21].

1. *Apply Pareto's law*: According to this law, about 80% of storage and retrieval activities are dedicated to 20% of different SKUs (in cube or value). Some 15% of activities are dedicated to 30% of the SKUs types, and 5% of them are dedicated to 50% of remaining SKUs types. If the more popular products are grouped together, order-picking time will be reduced.
2. *Use a clear and easy-to-read picking document*: A clear document helps order pickers to do their job more effective. Instructions that are written in a clear font and have information about the products that are required and their quantity will help an order picker fill the order in the shortest possible time and with minimal numbers of errors.
3. *Maintain an effective stock-location system*: A location system strongly affects order-picking efficiency, so using the proper system is important.
4. *Eliminate and combine order-picking tasks*: Combining tasks reduces the time required for order picking. For example, pickers can read documents when they travel among areas or put sorted items in their packages simultaneously.
5. *Pick a group of orders to reduce total travel time*: If the number of orders increases, an order picker can pick more items in one tour, decreasing task time.
6. *Dedicate specific locations*: Put items that have the highest I/O rates in the most accessible locations.
7. *Group high-probability items*: Locate similar items or items that are often ordered together in the same place or in places that are close together.

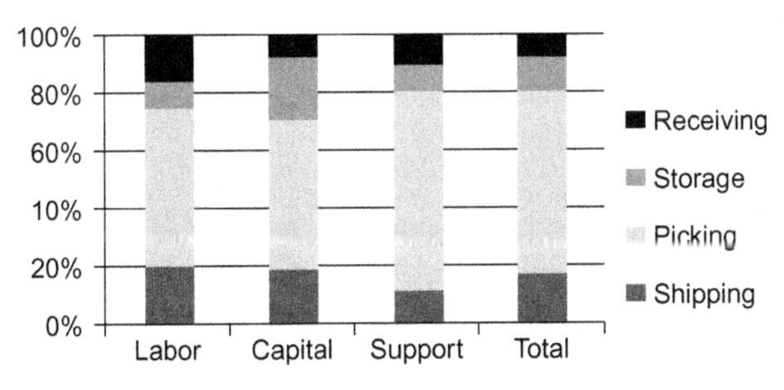

Figure 10.1 Warehousing cost by activity [18].

8. *Balance activities and locations*: Keep picking activities and picking locations balanced to reduce congestion.
9. *Keep order pickers responsible*: By keeping order pickers accountable for their time and effort, extra labor in the form of checkers can be avoided.
10. *Designing proper vehicles*: the time of order picking activities like sorting and packaging or mankind errors can be reduced by designing proper vehicles.
11. *Reduce counting errors*: Electronic weigh scales are accurate and can enhance productivity, especially for very small items.
12. *Eliminate paperwork from the order-picking activities*: Paperwork is one of the major sources of inaccuracies and productivity loss in order picking. New technologies such as radio-frequency data communication and voice I/O can be used instead of paper to reduce errors.

10.8 Inventory Management

Inventories are raw materials, work in process, and finished goods that companies keep for different reasons such as saving time, to meet economic objectives, and as a buffer against uncertainties.

The basic element of customer service for all logistics is inventory availability, and generally the most expensive logistics cost is inventory. Effective inventory management decreases carrying cost and increases customer satisfaction at the same time [22].

10.8.1 Types of Inventory

The following are among the many types of inventory that can be warehoused.

- *Cycle stock* is inventory, i.e., highly predictable in its turnover and need to be replenished [6].
- *Safety stock* is inventory, i.e., concerned with short-range variations in either demand or replenishment. It protects against the uncertainty of demand and lead time.
- *Transit inventory* or *pipeline inventory* is composed of products that are in transit between producer and purchaser locations and are not ready to use or be sold. This stock is equal to the expected demand over the lead time (the time between issuing an order and receiving it) [23].
- *Speculative stock* is inventory kept in case of material shortages, price increases, or unexpected changes in demand rather than to satisfy current demand [4].
- *Seasonal stock* is one form of speculative stock that is held for anticipated demand for a specific time period—e.g., increasing chocolate demand on Valentine's Day.
- *Dead stock* is inventory for which there is no longer demand. These inventories impose tax costs on a firm, so they should be moved out as appropriate.

10.8.2 Costs of Inventory

Inventory costs are a major logistics costs and thus a key factor in decision making about inventory management. The three main categories are holding costs, procurement costs, and shortage costs.

Inventory Holding Costs

This type of cost is incurred when products are stored for some period of time. It includes the following elements:

- An *opportunity (capital) cost* represents "the return on investment the firm would have realized if money had been invested in a more profitable economic activity instead of inventory. This cost is generally estimated on the basis of a standard banking interest rate" [24].
- *Inventory service costs* include costs of tax and insurance.
- *Risk cost* is the result of pilferage, deterioration of stock, damage, and stock obsolescence.
- *Storage-space costs.*

Procurement Costs

The costs of orders from vendors include the following:

- Order forms
- Postage
- Telecommunication
- Authorization
- Purchase-order planning
- Purchase-order entry time
- Purchase-order processing time
- Purchase-order inspection time
- Purchase-order follow-up time
- Purchasing management
- Office space
- Office supplies
- Purchase-order entry systems
- Tracking and expediting
- Setup costs to prepare or change over a machine for a specific item's production run [22].

Shortage Costs

A shortage cost is a penalty that accrues when (1) stock cannot satisfy demand (lost profit because of lost sales) and (2) the desired product is not in stock and the customer must wait to receive it.

10.8.3 Inventory Control

Inventory control is the set of activities that coordinate purchasing, manufacturing, and distribution to maximize the availability of raw materials for manufacturing or the availability of finished goods for customers [25].

There are three basic questions for inventory control [2]:

1. What items should we keep in stock? Warehouse managers should evaluate the advantages and disadvantages of keeping an item in a warehouse before they add it to the inventory. Only add an item if it provides clear benefits.

2. When should we place an order? This question can be answered in three ways:
   First, determine a fixed interval time and then place an order of a product in the needed quantity at each interval. In this approach, order quantity may vary from one time period to another (a fixed time period).
   Second, monitor the stock level continuously. When it falls to a specific level, place an order with a fixed size. In this approach, the time periods between two orders may vary (a fixed-order quantity, FOQ).
   Third, with respect to known demand over a specific time period, enough stock should be ordered. In this approach, both time period and order quantity depend on demand and may vary.
3. How much should we order? Different costs are mentioned above for keeping an inventory. If order size is large and the time period between ordering is long, the cost of ordering and inventory shipment decreases, but inventory holding costs increase. If order size is small and the time period is short, then inventory holding costs decrease but two other costs increase. There should be a trade-off between these costs.

Inventory Control in Certain Conditions

Economic order quantity (EOQ). The classic EOQ introduced by Ford W. Harris in 1915 is a simple model that illustrates the trade-offs between ordering and storage costs.

The model has the following assumptions:

- There is a known, continuous, and constant demand.
- Costs are known and constant.
- Shortages are not permitted.
- The lead time between placing and receiving orders is zero, and replenishment time can be ignored.

To represent a model, we use the following notations:

U = the unit cost of the item
R = reorder cost for the item
H = holding cost of one unit of the item in stock for one period of time
Q = order quantity, which is always a fixed order size
T = cycle time, the time between two consecutive replenishments
D = demand of the item that should be supplied in a given time period

$$\text{Economic order quantity (EOQ)} = \sqrt{\frac{2 \times D \times R}{H}} \tag{10.1}$$

$$\text{Variable cost of EOQ } (\text{VC}_{\text{EOQ}}) = \sqrt{2 \times D \times R \times H} \tag{10.2}$$

$$\text{Total cost of EOQ } (\text{TC}_{\text{EOQ}}) = (U \times S) + (VC_{\text{EOQ}}) \tag{10.3}$$

In this model, optimal order quantity is achieved at the point inventory ordering cost per unit of time (RD/Q) equals inventory holding cost per unit of time ($HQ/2$).

Optimum Order Quantity When Shortages (Back Order) Are Allowed

In this model, we will eliminate the third assumption of the EOQ model. Thus, shortages (back orders) are allowed in the model. For this reason, SC and S represent shortage (back-ordered items) cost and quantity of back orders, respectively.

$$Q_{\text{opt}} = \sqrt{\frac{2 \times R \times D \times (H + SC)}{H \times SC}} \tag{10.4}$$

$$S_{\text{opt}} = \sqrt{\frac{2 \times R \times HD}{SC \times (H + SC)}} \tag{10.5}$$

Optimum Order Quantity When the Rate of Replenishment Is Finite

In the EOQ model, the replenishment is instantaneous, and the replenishment rate will be finite. For this reason, P describes the production rate, so the replenishment rate will be $(P - D)$ (Figure 10.2).

$$Q_{\text{opt}} = \sqrt{\frac{2 \times R \times D}{H}} \times \sqrt{\frac{P}{P - D}} \tag{10.6}$$

$$PT_{\text{opt}} = Q_{\text{opt}}/P \tag{10.7}$$

Optimum Order Quantity with Price Discount

In the previous discussion, the models cited did not consider price discounts. In the real world, however, for many reasons (such as reductions in shipment cost) there is a price discount with respect to order size, so it is economical to order large quantities.

The two main types of discounts are based on (1) *lot size*, which are discounts based on the quantities ordered in a single lot (which is appropriate when there is a

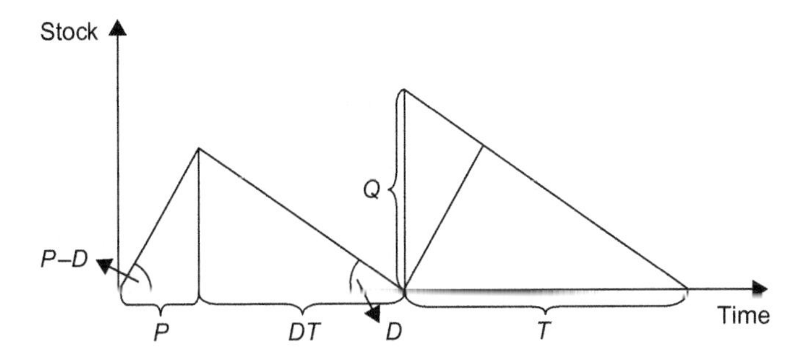

Figure 10.2 Optimal order quantity with finite replenishment rate.

competitive market and price is fixed); and (2) *volume*, or discounts based on the total quantity purchased over a given period (which is appropriate when customer demand increases and price decreases) (Figure 10.3) [11].

In the next model, a supplier offers a reduced price *on all units* for orders above a certain size (Table 10.1).

In this model, the most economical order quantity for each range should be calculated. (If other costs depend on unit costs such as holding costs, there are I Q_i; elsewhere, there is one Q.)

$$Q_i = \sqrt{\frac{2 \times R \times D}{H_i}} \tag{10.8}$$

There are three cases for Q_i:

- $q_{i-1} \leq Q_i < q_i$ (valid range)
- $Q_i < q_{i-1}$
- $Q_i \geq q_i$

After finding the valid range, the total cost of Q_i that is in the valid range and all break points at the left of the valid range should be calculated. The quantity that has lowest cost is the optimal order quantity [24].

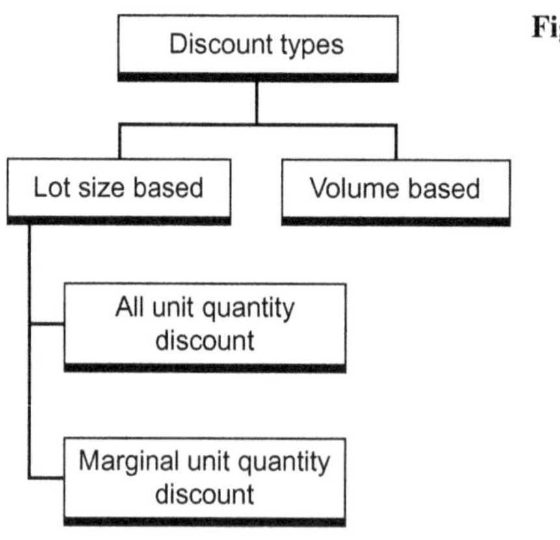

Figure 10.3 Types of discount [11].

Table 10.1 Reduction in Unit Cost by an Increase in Order Quantity [24]

Unit Cost	Order Quantity
U_1	$0 \leq Q < q_1$
U_2	$q_1 \leq Q < q_2$
$\vdots$	$\vdots$
U_I	$q_{I-1} \leq Q < q_I$

Lead Time

Lead time (LT) is the "period between placing a replenishment order and the time it is actually received" [26].

In previous models, lead time was assumed to be zero, but this assumption is not true. Lead time occurs for many reasons such as time needed for order preparation before sending items to a supplier, the time needed for a supplier to process the order and prepare the delivery, the time to get materials delivered from suppliers, and the time to process the delivery—i.e., the total time between receiving a delivery and getting the materials available for use.

If demand does not change with time, then it is not economical to keep inventory from one cycle to the next. For this reason, each order should be timed to arrive as existing stock runs out. The reorder level—the inventory stock level of an item that triggers a reorder—can then be defined [25].

If lead time $<$ cycle time, then

$$\text{Reorder level} = \text{LT} \times D \tag{10.9}$$

If lead time $\geq$ cycle time and between n and $n + 1$ cycle lengths, then

$$\text{Reorder level} = (\text{LT} \times D) - (n \times Q_{\text{opt}}) \tag{10.10}$$

Inventory Management in Uncertain Conditions

The basic assumption of previous models is determinant demand, lead time, and their related variables, but there is always some uncertainty in them. An uncertain parameter is one that does not have an exact quantity but its probability distribution is known [2].

One method used for inventory control under uncertain conditions is the FOQ method. In this model, the EOQ will be ordered whenever demand drops the inventory level to the reorder level. Lead-time demand is a key factor in this model; the variations outside of lead time are not so important because they are compensated for by changing the time period. If demand *outside* the lead time is higher than expected, then we reach the reorder level sooner than expected. If demand *inside* the lead time is higher than expected, then it is too late to make adjustments, and there will be shortages. As mentioned earlier in inventory management, both demand and lead time uncertainty are important (Figure 10.4).

Another method is the fixed-order interval (FOI) method. FOI methods allow for uncertainty by placing orders of varying size at fixed time intervals. If demand is constant, then these two approaches are identical, so differences only appear when uncertainty is entered into the model. The FOI model considers the uncertainty in lead time + order interval [4] (Figure 10.5).

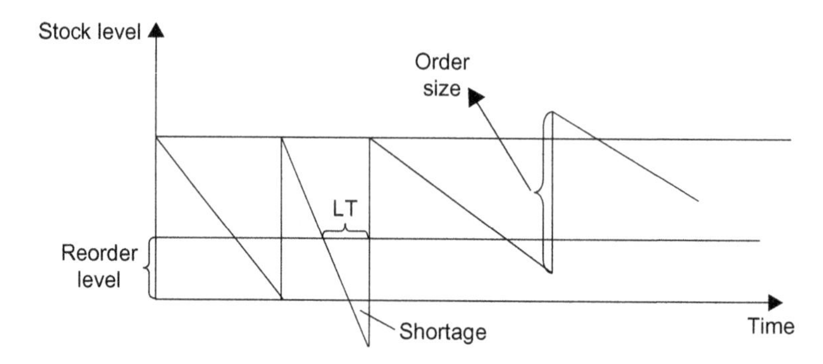

Figure 10.4 FOQ method.

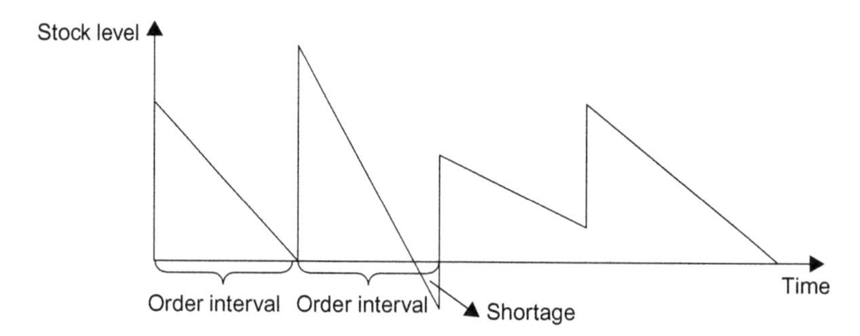

Figure 10.5 FOI method.

Service Level or Product Availability

Product availability is the ability of a firm to fill a customer's order from available inventory. The following are among the several ways to measure product availability:

- The *product fill rate* is defined as the fraction of demand that is satisfied from stock.
- The *order fill rate* is defined as the fraction of orders that are filled from stock.
- The *cycle service level* is the fraction of replenishment cycles that end with no shortage [22].

Safety Stock

Safety stock is stock that is kept because of uncertainties of demand, lead time, or both. The required service level determines the quantity of safety stock kept (a higher service level needs more safety stock). For example, if the demand is higher or lead time is longer than expected, then safety stock will compensate for these variations [16].

Because we know that excess inventory increases holding costs, the important question becomes, how much inventory for safety stock should we hold?

Safety Stock in the FOQ Method

In this method, as previously discussed, the important issue is lead-time demand. If demand has a normal distribution with mean ($\overline{D}$) per unit time and standard deviation (σ_D), and lead time has normal distribution with mean ($\overline{\mathrm{LT}}$) and standard deviation (σ_{LT}), then the lead-time demand has a normal distribution with a mean ($\overline{D} \times \overline{\mathrm{LT}}$) and standard deviation $\left(\sqrt{\overline{\mathrm{LT}} \times \sigma_D^2 \times \overline{D}^2 \times \sigma_{\mathrm{LT}}^2}\right)$.

$$\text{Safety stock} = \mathrm{Z} \times \sqrt{\overline{\mathrm{LT}} \times \sigma_D^2 \times \overline{D}^2 \times \sigma_{\mathrm{LT}}^2} \tag{10.11}$$

Z is the number of standard deviations from the mean that correspond to the specified service level.

Safety Stock in the FOI Method

In this method, safety stock compensates the uncertainty of a LT plus next order interval [25].

If demand over T + LT is normally distributed with mean $\overline{(T+\mathrm{LT}) \times D}$ and standard deviation $\sigma \times \sqrt{(T+\mathrm{LT})}$, then

$$\text{Safety stock} = Z \times \sigma \times \sqrt{(T+\mathrm{LT})} \tag{10.12}$$

10.9 Virtual Warehouses

The virtual warehouse (VW) is a business model that tries to reduce costs and provide high-quality customer service. This concept first appeared in 1995 [27].

According to Fung et al., "The VW is a state of real-time global visibility for logistic assets such as inventory and vehicles" [28]. The key factors to achieve VW are information technologies and real time decision algorithms which provide operating efficiencies and global inventory visibility. Some substantial gains which virtual warehouses provide are: online material visibility for customer [27] service, precise control of transportation and data analysis capabilities for any users capable of accessing the virtual databases. Based on the conventional concept of VW, Figure 10.5 illustrates main elements of the VW approach. The figure shows the basic elements (information technologies and real-time decision) which are needed for efficient operation and inventory visibility [28] (Figure 10.6).

Conceptually, there are two levels in a VW: data level and algorithm level.

Hardware and software constitute the data level of VWs that support real-time algorithms. The following technologies and systems at the data level are essential for real-time data acquisition in mobile applications:

- Standard interfaces and integrated databases
- Wireless communications

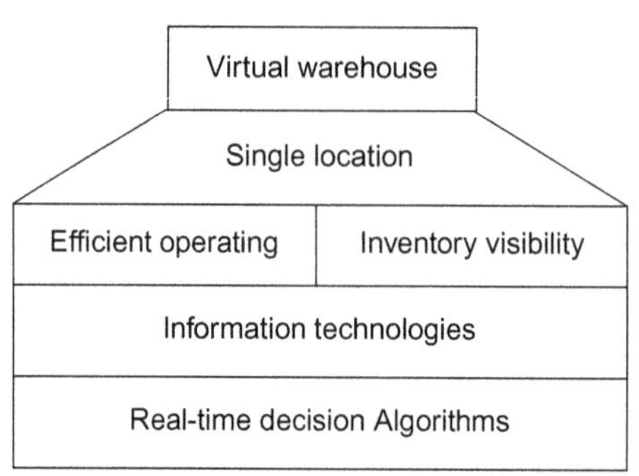

Figure 10.6 Virtual warehousing concept chart [28].

- Global positioning
- Geographic information
- Automatic identification

The algorithm level is the functional level of VWs. At this level, real-time algorithms are used to filter and process data for operational decision making [29].

References

[1] A. Rushton, P.H. Crouche, P. Baker, The Handbook of Logistics and Distribution Management, third ed., Kogan Page Limited, London and Philadelphia, 2006.
[2] D. Water, Inventory Control and Management, second ed., John Wiley & Sons, England, 2003.
[3] R.H. Ballou, Business Logistics Management, third ed., Prentice Hall, London, 1992.
[4] J.R. Stock, D.M. Lambert, Strategic Logistics Management, fourth ed., McGraw-Hill, Singapore, 2001.
[5] R.H. Ballou, Business Logistics/Supply Chain Management: Planning, Organizing, and Controlling the Supply Chain, fifth ed., Pearson/Prentice Hall, Upper Saddle River, N.J., 2004.
[6] S.C. Ailawadi, Logistics Management, Prentice-Hall of India, Delhi, 2006.
[7] P. Baker, M. Canessa, Warehouse design: a structured approach, Eur. J. Oper. Res. 193 (2009) 425–436.
[8] J. Li, T.C.E. Cheng, S.H. Wang, Analysis of postponement strategy for perishable items by EOQ-based models, Int. J. Prod. Econ. 107 (2007) 31–38.
[9] C.F. Daganzo, Logistics Systems Analysis, fourth ed., Springer-Verlag, Heidelberg, 2005.
[10] A. Langevin, D. Riopel (Eds.), Logistics Systems: Design and Optimization, Springer, USA, 2005.
[11] S. Chopra, P. Meindl, Supply Chain Management Strategy, Planning, and Operation, Prentice-Hall, Inc., Upper Saddle River, NJ, 2001.
[12] W.H. Hausman, L.B. Schwarz, S.C. Graves, Optimal storage assignment in automatic warehousing systems, Manag. Sci. 22(6) (1976) 629–638.
[13] V. Reddy Muppani, A.G. Kumar, Efficient formation of storage classes for warehouse storage location assignment: a simulated annealing approach, Int. J. Manage. Sci. 36 (2008) 609–618.
[14] S.S. Heragu, Facilities Design, PWS Publishing Company, Boston, MA, 1997.

[15] J.P. Van den berg, A literature survey on planning and control of warehousing systems, IIE Trans. 31 (1999) 751–762.
[16] J.C. Johnson, D.F. Wood, D. Wardlow, P.R. Murphy, Contemporary Logistics, seventh ed., Prentice Hall, 1999.
[17] K.B. Ackerman, The changing role of warehousing, Warehousing Forum (1993) 8.
[18] J.P. Van den berg, W.H.M. Zijm, Models for warehouse management: classification and examples, Int. J. Prod. Econ. 59 (1999) 519–528.
[19] A. Maltz, The Changing Role of Warehousing, Warehouse Education and Research Council, 1998.
[20] J. Drury, Towards more efficient order picking, The Institute of Materials Management, Cranfield, UK, 1988, IMM Monograph No. 1.
[21] J.A. Tompkins, J.A. White, Y.A. Bozer, E.H. Frazelle, J.M.A. Tanchoco, J. Trevino, Facilities Planning, second ed., Wiley, New York, 2003.
[22] D. Blanchard, Supply Chain Management: Best Practices, John Wiley & Sons, New Jersey, 2007.
[23] J.A. Muckstadt, S. Amar., Principles of Inventory Management When You Are Down to Four Order More, Springer Science + Business Media, New York, 2010.
[24] G. Ghiani, G. Laporte, R. Musmanno, Introduction to Logistics Systems Planning and Control, John Wiley & Sons, Chichester, 2004.
[25] T. Wild, Best Practice in Inventory Management, second ed., Elsevier Science, Oxford, UK, 2002.
[26] C. Mercado (Ed.), Hands-on Inventory Management, Taylor & Francis Group, New York and London, 2008.
[27] D.E. Stuart, J. Owen, T.L. Landers, Establishing the virtual warehouse, Manuf. Sci. Eng. 2(2) (1995).
[28] S.H. Fung, C.F. Cheung, W.B. Lee, S.K. Kwok, A virtual warehouse system for production logistics, Production Planning & Control: the management of operations 16(6) (2005) 597–606.
[29] T.L. Landers, M.H. Cole, B. Walker, R.W. Kirk, The virtual warehousing concept, Trans. Res. 36 (2000) 115–125.

11 Customer Service

Samira Fallah

Department of Industrial Engineering, Amirkabir University of Technology, Tehran, Iran

11.1 Customer-Service Definition

Customer service is a wide concept that varies from organization to organization; any organization has its own policy and view toward customer service [1]. Moreover, it means different things to the parties associated with any business transaction. This broadness of definition makes it difficult to achieve consensus. Alternative explanations of customer service will be discussed in this section.

11.1.1 Customer Service as an Organizational Activity

Customer service can be considered to be a set of functions or activities taking place in the customer-service department of an organization with the aim of dealing with customers and satisfying their demands, including complaints, claims handling, and billing [2,3].

11.1.2 Customer Service as a Process

Another perspective considers service as a process that takes place between a seller, a buyer, and sometimes a third party that creates something of long- or short-term value for the parties. Thus, from this viewpoint, customer service can be defined as a process to effectively provide considerable added value for the whole supply chain [4].

11.1.3 Customer Service from the Customer's Side

Moreover, customer service can be viewed from the customer's side and be defined as providing customers with services that meets and, in some cases, exceeds their needs and expectations [5].

Logistics Operations and Management. DOI: 10.1016/B978-0-12-385202-1.00011-6

11.1.4 How Experts Define Customer Service?

To gain deeper insight into the customer-service concept, we review some definitions from logistics experts.

Christopher [4] defines it as the "transaction of all those factors that affect the process of making products and services available to the buyer." Johnson et al. [6] put forward quite another definition: "The collection of activities performed in filling orders and keeping customers happy, creating in the customer's mind the perception of an organization that is easy to do business with."

11.1.5 Defining Customer Service in Logistics Context

Along with the foregoing definitions, another one is more common and applicable in the context of logistics. This definition, known as the *seven R*'s, views customer service as the output of overall logistics processes and emphasizes seven aspects as prerequisites for delivering satisfactory services to customers. It defines customer service as providing the *right* customer with the *right* product at the *right* place, *right* time, and *right* cost in the *right* condition and *right* quantity. This definition can be applied to all industries [7,8].

11.2 What Is Behind the Growing Importance of Customer Service?

The importance of customer service has grown recently, and everyday more organizations understand the significance of satisfying their customers through delivering good services [9]. Considering how influential customer service can be to any organization's profitability, taking it into account has become a must today. Customer service should not be considered just a tactical method, but a strategic value [10]. This requires changes in an organization's beliefs and culture because organizations and particularly finance departments usually view customer service as a costly function that wastes resources, whereas the strategic view of customer service considers it a competitive tool for gaining market share and differentiation.

What are the reasons for the emerging attention to customer service? What is its magic power? This section discusses some of the main reasons for customer service's growing importance as a strategic and competitive issue.

11.2.1 Customer Service: The Intangible Part of a Product

Today, in different markets around the world, methods of competition between organizations have changed and the "power of brands has declined" [4]. The point is that with impressive advances in technology, each group of products and services has several or many alternatives that are similar in quality. Thus, no organization can be ensured of keeping its current customers just by offering high-quality products; it needs to find more competitive methods.

One suitable approach for any firm to remain competitive is to pay more attention to customer service [11]. Today, the tangible product is not separable from the intangibles of customer service, and not only the product itself should be technically competitive in creating value for customers but also the other service-related issues such as delivery, reliability, ease of ordering, and availability of products must be considered as distinctive differentiators. Hence, an organization's inability to provide satisfactory services to fulfill customers' requirements may result in displeased and lost customers [4,7,12].

Moreover, it is less possible for competitors to imitate the improvement in the area of customer service than in technological or in price reduction aspects [6]. Thus, it can be a strong differentiator for any organization in maintaining competitiveness.

11.2.2 Costs of Attracting New Customers

Attracting new customers is a difficult and costly process. In the first transactions with new customers, firms usually have to offer considerable discounts. The product should have a particular distinctive feature to encourage customers to buy it, and it should have a reasonable quality in order not to be damaged during usage. Providing all these elements is costly [13].

Generally, attracting a new customer is 5 times more costly than keeping existing ones satisfied and pleased. This rate can increase as much as 20 times [10,14]. As a result, it is more logical for organizations to keep their current customers by providing satisfactory services than spending more to find new ones.

11.2.3 Customer Service, Customer Satisfaction, and Loyalty

Customer service can have a considerable impact on the profitability of an organization. Good service increases customer satisfaction, which leads to customer loyalty [15]. A loyal customer is the one who continues to purchase from a specific firm even though the same services or products are available and can easily be provided by other organizations. Regular purchases increase a firm's profitability [16].

11.2.4 Customers as a Means of Marketing

Satisfied and loyal customers are precious assets for any organization, and dissatisfied ones can pose a threat. Besides the direct impact that satisfied customers can have on a firm's revenue, they can also indirectly affect profitability. Satisfied customers can decrease the costs of marketing because they usually can act as sale staff [10]. A pleased customer will tell three people about a firm's services, but a dissatisfied one will dissuade 20 [17]!

With the advances in communication technology and the increasing use of the Internet worldwide, this rate gets worse. Studies have revealed that about 12% of online customers will tell their "buddy list" about their dissatisfaction with a firm's services, which includes an average of almost 60 people [10]. Indeed, satisfied customers can be considered as a means of marketing, and it has been shown that current customers provide as much as 65% of a firm's business [18].

Most organizations use marketing policies to introduce their products, and they mostly focus on traditional features of marketing [4]. Basically, marketing can be considered as the management of the four *P*s: *product*, *price*, *promotion*, and *place* [19]. Most often the major focus and emphasis have been placed on the first three aspects and the fourth one, place, described as physical distribution, has usually been overlooked [4]. However, place can be looked at as a firm's customer-service policy. Hence, customer service is the output of an organization's logistics system and can be considered as an interface between logistics and marketing [8,20].

11.2.5 Customer Service and Organization Excellence

Customer service could indirectly affect a firm's excellence as well. When the services of an organization meet customers' expectations, they are satisfied. Customer satisfaction creates commitment and loyalty that, in higher levels, may lead to the phenomenon known as *owner customers* [13]. Owner customers are interested in being more involved in firm's processes especially in designing new products and improving organizational services. In this respect, an organization manages to achieve the goal of excellence models such as European Foundation for Quality Management (EFQM) in customer involvement and creating innovation chains.

11.2.6 Customer Service and Staff Job Satisfaction

As previously mentioned, being responsive to customers' requirements creates satisfaction and fewer customer arguments and complaints. This brings peace and less stress for a firm's personnel and provides them with the opportunity to save time because they have to spend less time dealing with angry customers.

11.3 Customer-Service Elements

Customer service involves a number of activities referred to as *customer-service elements*. According to a study sponsored by the National Council of Physical Distribution Management, customer-service elements can be categorized into three groups as follows [18]:

1. Pretransaction elements
2. Transaction elements
3. Posttransaction elements

11.3.1 Pretransaction Elements

Pretransaction elements are not directly related to logistics activities and are more concerned with management and organizational issues and policies [4,21]. These elements tend to create a culture and climate for providing customers with a satisfying customer-service system [18]. To have a suitable customer-service system, first it is necessary to prepare *a written customer-service policy statement* that clarifies service standards and the overall mechanism of measuring and controlling the customer-service system [22]. This statement is important from both internal and external perspectives. Internally, it reinforces service commitment within different levels of the organization, especially managers [1]. From the customers' point of view, it provides valuable information about a firm's service levels. Customers should be informed of determined service standards in order to know to what extent they could expect services. This information eliminates complaints and dissatisfactions that might occur because of "unrealistic expectations" [8]. Other pretransaction elements are related to *the structure of an organization*. It is important to have a structure that facilitates cooperation and information flow among departments and personnel in charge of customer-service processes. In addition, the responsibility of each person should be specifically defined, and customers should be able to access responsible individuals easily [21].

Moreover, the system should always be prepared to deal with unexpected events such as natural disasters, labor strikes, and lack of raw materials. It also should have the *flexibility* to respond to specific customer requests [4,8,18,21].

Besides, it is important to involve customers with service processes and develop a good and strong rapport with them [18]. Providing customers with technical information through manuals and seminars may also facilitate the achievement of this goal [8].

11.3.2 Transaction Elements

Transaction elements are directly related to distribution and logistics processes [4]. This category of elements includes dealing with customer orders and filling them promptly and with acceptable accuracy. Some of these elements include:

- The availability of stock [18].
- Providing customers with accurate and on-time, order-related information such as status of inventories, actual shipping dates [4,7,8].
- Providing customers with a convenient and accurate ordering process [8,21].
- Selecting the appropriate transportation mode.
- Focusing on order-cycle elements [8,18,21].

Order cycle is one of the major elements during transaction process, so each stage of it should be focused and managed in an effective way. From customers' point of view, order-cycle time is the interval between when they place orders and when products or services are delivered to them.

- Substituting ordered products [8,21].

During a stock out, an organization can substitute ordered products with others of the same quality but with some differences such as size. In this way, the organization can satisfy customers and increase its service level. Keep in mind, however, that this process needs good communication with customers to ensure it meets their specific needs.

11.3.3 Post transaction Elements

Post transaction elements follow the transaction and delivery of products and are relevant to the "consumption" of products [7]. The main goal of these elements is to provide supporting actions [4]. Among these are the following:

- Installation, warranty, alteration, repairs, and availability of spares [4,8].
- Dealing with customer complaints and claims [21].
- Invoicing accuracy and procedure [7].
- The possibility of temporarily replacing the products [21].
- Product tracking [21].

Firms should have the ability of recalling products that are defective or are recognized to be potentially dangerous to users.

- The policy and procedure for handling returns [6,22]:

A customer may return a delivered product for several reasons: damage sustained during transportation, an order-filling error made by the seller, or an error in ordering made by the customer. Therefore, each seller should have a policy to deal with returns and should determine an appropriate procedure for handling them and responding to customers accurately and rapidly.

- Special Services [13]:

In some cases, the seller can provide customers with special and value-adding services such as diagnostic monitoring and preventive maintenance. It is a differentiation, especially for organizations that provide expensive facilities. By this procedure, some delays in customer organization's processes would be prevented and considerable money and time would be saved.

11.4 Order-Cycle Time

Customers experience a firm's services most frequently when they order a product or service. How a firm handles the customer's order may have a considerable effect on overall customer satisfaction. Having a convenient and accurate ordering process with consistent and reliable order-cycle time is a key factor in this regard.

Order-cycle time can be defined from both the seller's and the buyer's perspectives.[1] From a seller's point of view, order cycle is the interval between receiving a

[1] Order-cycle time from a customer point of view was discussed in Section 11.3.

customer's order and when the product is received. All activities that take place between the order receipt and the time a warehouse is notified to fill it is called *order management* [6].

In fact, managing an order is equivalent to managing each of the following stages of order cycle, which consists of the following steps [6,18]:

- Order preparation and transmittal
- Order processing
- Order picking and packing
- Order transportation and delivery

In the following sections, these stages will be discussed in more detail.

11.4.1 Order Preparation and Transmittal [6,18,21]

The first step for ensuring a suitable response to customer orders is being prepared to receive them in a timely manner and to develop appropriate plans that prevent being overloaded by orders at one time, which usually makes them very difficult to fulfill—for instance, offering discounts to customers who place their orders on specific times.

Being prepared to receive customer orders by applying suitable policies, the next step is to arrange appropriate ordering channels to transmit orders smoothly. Order-transmittal time depends directly on the method used to place and send the orders. There are several solutions for it, from traditional paper-based ones to phone calls, fax machines, ordering via email, websites, and the use of radio. Thanks to technological advances in recent years, new channels such as electronic data interchange (EDI), scanners, and bar codes have been created. These channels can communicate the point-of-sale information directly to a firm, which leads to more accurate and rapid placement of orders.

11.4.2 Order Processing [6,8,18,21,23]

Order processing is a key stage in order cycle, and its speed and accuracy have significant effects on the whole order-cycle time. It involves different operations such as the following.

When an order is placed by a customer and transmitted through the predetermined channel, it should first be checked for completion and clearance, and then related information should be communicated to the warehouse and the availability of the requested items should be checked. If the ordered items are available in the demanded quantities, then they are picked, assembled, and prepared for shipment. In cases in which the requested items are not available, the customer should be notified as soon as possible via a communication channel such as phone, fax, or e-mail. Communications with customers must be kept to know their decisions, whether they accept substitute products, whether they wish to wait until requested items are available, or whether they want to cancel the order.

Other tasks such as checking customer credit, preparing necessary documents for shipping, updating inventory status, and invoicing should also be accomplished.

Order-processing activities are highly dependent on how information flows between related departments. Having an automated, integrated system that connects associated departments such as sales, accounting, marketing, warehouse, production, and transportation and synchronizes the information between them would greatly affect accurate and consistent responses to customers, which helps to guarantee customer satisfaction. Order-processing operations benefit mostly from information and communication technology. Traditionally, order processing used to take as much as 70% of order-cycle time, but today's time has been markedly reduced with the help of advanced technologies such as EDI.

11.4.3 Order Picking and Packing [6,18]

When an order is transmitted to a warehouse, the picking and packing process begins. First, an order-packing list that indicates which items should be picked is prepared and given to a warehouse employee. Computer-based systems are used for order-picking operations, which informs warehouse employees about ordered items, their quantities, and where in the warehouse they can be found. Picked items must be checked to ensure accuracy as well as the necessity of packaging processes before they are passed onto the shipping department. In cases of stock out, warehouse employees should notify the department in charge of handling orders to notify customers.

11.4.4 Order Transportation and Delivery

The final stage of any order cycle begins with loading ordered items on carriers, and the stage finishes when customer receives the desired products [6,18].

11.5 Developing a Policy for Customer Service

11.5.1 Important Points

Customer-service elements have a significant effect on a firm's sales and benefits; therefore, establishing a precise and detailed policy for serving customers is a key to success for any organization.

When setting a policy with respect to customer service, consider the following points in this section.

Customers Should Define the Proper Service Levels

Customer-service policy must be focused on customers. Traditionally, most organizations tend to set service levels based on management judgments and experiences, results of past activities, and some norms of associated industry. This approach, however, usually leads to unreliable service levels because in most cases there are differences between what a firm and its management consider about their services and what the customer actually experiences and perceives. In contrast, an appropriate and applicable customer-service program is based on customers' requirements, viewpoints, needs, and a thorough understanding of the market [6,21].

Customers Are Not the Same

It should be remembered that every single customer has their own needs and expectations, none of which are the same. Moreover, they do not provide equal benefits to the organization. It is exactly the same for a firm's products. Some are more beneficial than others [21]. As a result, in setting policies for customer service, a firm should not assign the same service levels to all customers and products; instead, it should apply some grouping and segmentation based on the differences.

Increasing Service Levels Are Costly

It is important to remember that any service level has its own cost. As the level of services increases, the associated costs also rise [18]. Consequently, in setting customer-service levels, the trade-off between revenue from services and the cost of establishing them should be considered. The purpose of any organization should be to minimize the overall costs of its logistics system while providing a logical and satisfying level of service.

In addition, a firm's customer-service policy should be definite and clear. Following an organized and structured approach will achieve this. The following section introduces the steps for developing an appropriate policy.

11.5.2 Steps for Developing Customer-Service Policy

Determining Major Elements of Customer Service from the Customer's Point of View

The first step in setting any service policy is identifying which service elements are important to a customer when dealing with a firm. For developing such list, besides surveying the current literature and research on major elements, interviews with an appropriate sample of customers may reveal the various concerns of different groups. In this way, relevant elements are derived from customers' viewpoints [4,11,21].

Identifying the Relative Importance of the Major Service Elements

When the list of key service elements is derived, the next step is to determine the elements' relative importance and priority. An appropriate questionnaire can be designed that captures the key elements and asks a sample of customers to weigh each element based on how important they consider that element to be when dealing with the firm. The scale of ranking can be optional. In this step, customers' priorities in service elements are recognized. One important point is asking respondents to ascribe the real weight to each element rather than rating them all as high priority. The designed questionnaire can be sent to the customers in several ways—for example, by e-mail, fax, mail, and through face-to-face interviews [4,7].

Determining Customer Evaluations of Current Service Levels and Firm's Competitiveness [6,7,16]

Identifying the elements thought to be important to customers by itself will not set a proper policy. Managers need a broader view of the firm's current situation in providing services and its position compared to other rivals. To get this information, they require customers' judgments. And for collecting customer's viewpoints, again, a questionnaire is useful. This questionnaire would include the list of main elements that respondents should weigh based on their perception of the firm's current performance; in addition, it should have an extra section for rating each of the firm's rivals in every major element.

Knowing how customers evaluate other competitors in delivering services is important because the results may show that high-level services are provided, but it is not a competitive advantage for that organization if it is not higher than its rivals. On the other hand, in an industry where the level of service provided by most organizations is not very high and customers evaluate the services of the whole industry poorly, then a firm with low-level services may not suffer much and does not need to provide costly high-level services.

Analyzing the results of this survey would provide beneficial information for managers. On the one hand, it shows the areas of improvement in which the firm performs below customers' expectations; on the other hand, managers can recognize areas in which the firm meets or even exceeds customers' demands. In addition, this survey helps the firm evaluate its competitors, identify their strengths and weaknesses in key elements, and appraise its own competitiveness.

Identifying Different Segments of Customer Service

A firm's customers consist of different groups, each with distinctive needs and expectations. They may be from different parts of the country or even from outside national borders and need different packaging methods, distinct modes of transportation, and so on. Thus, they mostly require distinct service levels, so providing the same services for all groups is neither economical nor feasible [24]. It is a fact that customers are not the same; some provide the firm with more profit than others. Some are much more beneficial than others; therefore, they require more attention and higher service levels [25].

As a result, it is necessary for any firm to identify different groups of its customers and set the service levels for each class based on its expectations and profitability. This gives rise to the need for customer segmentation. Each segment represents a group of customers with similar needs, preferences, and behaviors[2] that are different from others[3] [26].

Cluster analysis is the most widely used method for customer segmentation. It involves a collection of statistical methods and techniques for grouping objects based on their similar characteristics. If two respondents weigh the elements similarly, then they would be in the same group [4].

Analyzing Service Requirements for Each Segment and Defining Service Packages

With the identified segments and valuable data from previous steps, a firm can analyze the real requirements of each segment and define proper service packages with the appropriate associated levels.

11.5.3 A Case Study on Customer Segmentation Based on Customer-Service Elements [11]

The importance of market segmentation for a firm's profitability and the effectiveness of its marketing strategy are well discussed in marketing literature. Sharma and Lambert [11] conducted market segmentation for a manufacturer in a high-technology industry based on customer-service elements. One major step in segmentation is selecting an appropriate base to make logical differentiations. In their paper on customer segmentation, Sharma and Lambert [11] declared that a proper base should have certain characteristics: It must be easily translatable to application and be analyzed without complexity. They see customer service as a suitable base for industrial marketing segmentation.

Figure 11.1 gives an overall view of the methodology that Sharma and Lambert [11] incorporated for conducting market segmentation.

The Taken Steps

Identifying Key Service Elements

In addition to studying the current literature, to identify relevant elements of customer service and the marketing mix, the researchers carried out interviews with 30 buyers and asked them to express which elements were important to them when dealing with suppliers. The participants were selected from different sizes, locations, and types of industry to develop a thorough picture of relevant elements that encompassed different requirements and viewpoints.

In this step, 48 major service elements were derived. Among them were the following:

- Order-filling accuracy
- Availability of order-status information

[2] Maximum homogeneity.

[3] Minimum heterogeneity.

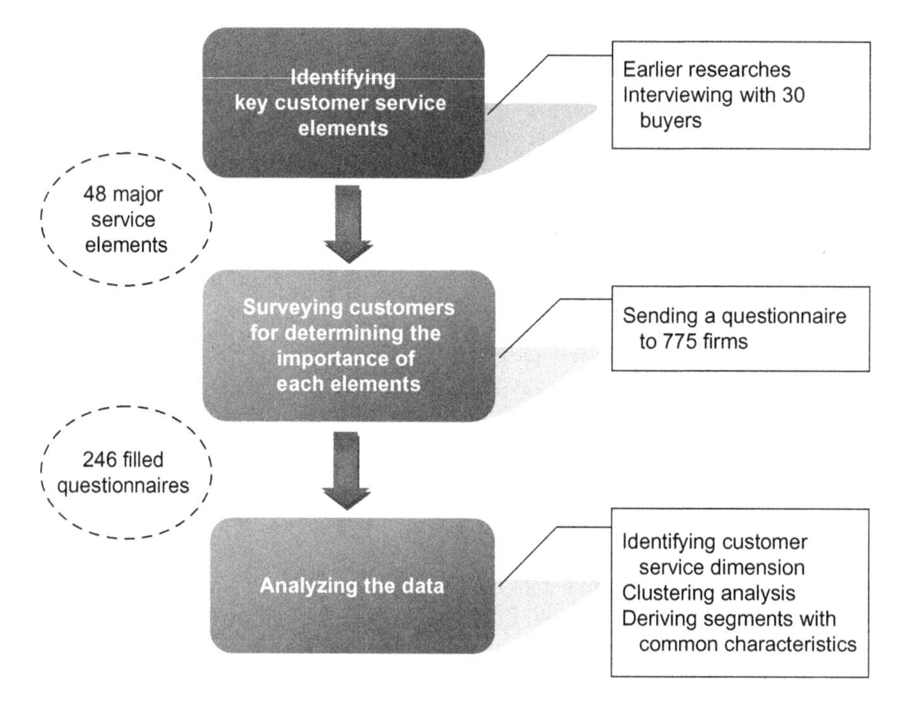

Figure 11.1 The methodology used to conduct market segmentation.

- Consistency in lead times
- Supplier's ability to consolidate orders
- Being able to select delivery carriers
- Location of supplier warehouse
- Locating a supplier's order-processing personnel in the customer's area
- Computerized order entry
- Supplier's ability to meet specific service-level requirements

Determining the Relative Importance of Elements

Having the distinguished elements, the next step was to evaluate their significance. To this end, a questionnaire comprising the key elements was sent to 775 firms, from which 246 filled questionnaires were received.

Data Analysis

At first, for ease of representation, factor analysis was used to group 48 customer elements into more understandable dimensions. The result was five main groups:

1. Information system capabilities
2. Availability of product
3. Various logistics services
4. Lead times
5. Order servicing

Table 11.1 Two Identified Segments

	Group A	Group B
Final cluster centers: Customer-service importance	0.7317	−0.5958
Number of cases	128	118

Table 11.2 Attitudes of Segments Toward Marketing Mix

		Segment A	Segment B
Products	Product quality	0.28	−0.20
	Range of products	0.41	−0.43
	Innovative products	0.31	−0.27
Promotion	Sales support	0.31	−0.36
	Mass media and direct mailing	0.50	−0.44
	General assistance	0.25	−0.23
	Promotional activities	0.14	−0.13
Price	Price sensitivity	0.52	−0.44

Then the scores of respondents to different dimensions were cluster analyzed, and two main segments based on the overall attitude toward customer service were defined. In this step, customers who had the same opinion about the importance of customer service were placed in group A. This group consisted of 128 respondents. Group B, the other 118 respondents, believed that customer service is not a critical factor (Table 11.1).

After recognizing different segments, the researchers continued their studies and interpretations of derived groups to discover other differences between them, in addition to collecting more information about the characteristics of each group. This was necessary in order to better understand the organization's characteristics and set the most appropriate marketing strategy for each group.

Further analysis showed that the segments had a different attitude toward all marketing mixes. The firms in group A were more sensitive than group B toward these elements. Organizations in group A were smaller than in B, but their service requirements were higher. In addition, the proportion of purchases in group A was more than in group B (Table 11.2).

Managerial Implications

In this research, a marketing segmentation was conducted based on customer-service elements. The analyses of gathered data showed two main segments with different viewpoints about the importance of customer service while doing transactions with the organization. In group A, there were small firms with high-rate purchases that expected higher service levels. The firms in group B had fewer transactions with the organization, and their expectations of customer service were lower.

With the gathered information, the management of the firm could set the appropriate marketing policy for different customers. For group A, the organization needed to provide more appropriate service levels. The marketing policy for this group should be intensive marketing in customer service. This requires closer connection with customers to achieve deeper understanding of their needs and requirements and to set service levels.

For customers in segment B, the firm needs to conduct a two-stage marketing policy. In stage one, the customer firm needs to be convinced that it needs to be more sensitive about customer service; this can be achieved by explaining and showing the benefits of good customer service. In stage two, the organization should convince customers that its services are more effective and of higher quality than its rivals.

11.5.4 Setting Customer-Service Level

As mentioned previously, providing customer service is a mixed blessing. On the one hand, high-level services increase customer satisfaction, which leads to high profitability; on the other hand, it is costly because each level of customer service needs to allocate different resources and therefore has a certain amount of cost [18] (Figure 11.2).

Hence, it is important to strike a balance between cost and profitability of customer-service levels [7]. There are different approaches for defining optimal service level from experience-based methods to more scientific and mathematical approaches. In the following section, a sample model for defining optimal customer service will be introduced.

A Sample Model for Defining Cost-Effective Customer-Service Level [27]

Determining a cost-effective customer-service level that simultaneously considers profits and associated costs is a major concern. In most cases, it is unclear to an organization which service level will satisfy customers and what amount of inventory is required to accomplish that.

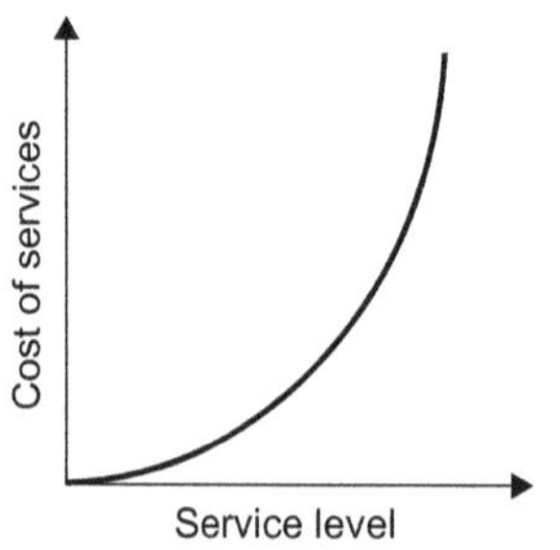

Figure 11.2 The relationship between service levels and their associated cost [4].

According to Jeffery et al. [27] in logistics and supply-chain literature, there are a wide range of models for determining service level and the appropriate investment in inventory, each of which considers different variables and constraints and has its own advantages and disadvantages.

Related models in this area have different approaches for determining the proper service levels, including the following:

- Minimizing inventory or cost while considering service levels as a constraint.
- Maximizing service levels while cost or inventory is a constraint.
- Determining the relative importance of cost and service levels by using the weights chosen by a decision maker.
- Using stochastic programming.

According to Jeffery et al. [27], most of the current models are not suitable for today's complex market and supply networks. In most cases, certain simplifying assumptions are applied that make them invalid in practice. First, most of the models consider demand as stationary, whereas in practice it is not always the case. Second, in most cases, the models determine arbitrary service levels and do not base them on product characteristics.

Jeffery et al. [27] developed a model to determine proper service levels with their associated amount of inventory. The process was carried out in two main steps. First, the researchers used logistic regression to determine how delivery performance, the dependent variable, related to three independent variables: order lead time, errors in forecast, and variation in demand.

Second, to understand the financial effects of these variables, the researchers developed a cost equation. Intel Corporation was used for a case study, and two product families consisting of 18 different products were studied. Over 1 year, 5000 data points were collected, each representing one order.

Step 1: Logistic Regression Modeling

By means of logistic regression, two groups of models were developed: planning models and insight models.

Planning models were developed to model the relationship between delivery performance (a binary dependent variable) and inventory and for use in future model development. The purpose of the insight models was to understand other factors' impacts on the relationship between delivery performance and inventory. The planning models were developed for each individual product in addition to an aggregate model for both groups, whereas insight models were developed for each group of products.

The result of each planning and insight models was a binary variable. The variable is 1 if the order was on time or early, and a value of 0 shows a late order.

The aim of the two modeling groups was determining the weeks of inventory (WOI) value while providing appropriate service levels for customers (Tables 11.3 and 11.4).

Table 11.3 Variable Codes

Code	Value	Formula
X_{n1}	WOI (weeks)	On-hand inventory/(demand for forecast for next 13 weeks/13)
X_{n2}	Forecast error	abs ((Forecasted demand – Actual demand)/Actual demand)
X_{n3}	Order lead time	Requested delivery date – Order placed date
X_{n4}	Coefficient of variation of demand	*s* of demand for product *i*/*m* of demand for product *i*

Table 11.4 The Insight Models

Product Group	Model
1	$Y1 = 0.8068 \times X11$
1	$Y1 = 0.389 \times X11 - 1.195 \times \log(X12)$
1	$Y1 = 0.021 \times X11 - 0.057 \times \log(X13)$
1	$Y1 = 0.288 \times X11 - 1.638 \times \log(X14)$
2	$Y2 = 0.6705 \times X21$
2	$Y2 = 0.639 \times X21 - 1.053 \times \log(X22)$
2	$Y2 = 0.028 \times X21 - 0.279 \times \log(X23)$
2	$Y2 = 0.274 \times X21 - 1.815 \times \log(X24)$

Step 2: Modeling the Cost Equation

After developing equations that relate customer-service level and inventory, a cost equation was developed to derive the cost associated with a given service level:

$$\begin{aligned} \text{C(SL)} = \text{inventory}\frac{\text{units}}{\text{period}} \times \text{inventory holding}\frac{\text{cost}}{\text{unit}} \\ + \text{expected lost sales}\frac{\text{units}}{\text{period}} \times \text{profit margin/unit} \end{aligned} \tag{11.1}$$

Inventory holding costs include warehousing costs, the decrease in the value of products from the time they are manufactured until they are sold, the opportunity cost of investing in inventory, and the scrapping of obsolete products.

The cost of lost orders was determined by asking customers to indicate their reactions to stock outs. Interviews revealed that 80% of customers would wait for the desired product or accept a substitute from the same manufacturer; the other 20% would buy the desired product from a competitor.

The developed cost equation is shown in Table 11.5.

Table 11.5 The Notations of Cost Equation

t	Period index
i	Product index
SL_{it}	Service level (percent of orders delivered when requested) for product i in period t
WOI_{it}	Units of inventory of product i held during period t divided by weekly forecasted demand
F_{it}	Average weekly forecast for product i during period t (units)
D_{it}	Average weekly demand for product i during period t (units)
P_i	Variable production cost for product i
R_i	Revenue for one unit of product i
H_{it}	Inventory holding cost as a percent of variable production cost for product i in period t

The inventory holding cost of customers was determined to be:

$$IHC_{it} = H_{it}\, P_i\, \text{ave}(WOI_{it})F_{it} \tag{11.2}$$

Considering Eqn (11.2), the overall cost equation is:

$$C(SL_{it}) = IHC_{it} = H_{it}\, P_i\, \text{ave}(WOI_{it})F_{it} + 0.2(1 - SL_{it})D_{it}(R_i - P_i) \tag{11.3}$$

Step 3: Determining Proper Service Levels and WOI

After developing the above equations for inventory and cost, some plots were generated that show cost and different service levels at various levels of WOI for each of 18 different products. The plots were used to determine the minimum cost–inventory service level. Figure 11.3 shows the plot for one of the products.

Examining the plot reveals that the proper service level for this product is 93.8%. This level is achieved by holding 3.6 WOI; the associated cost for this level of service is 1%.

Similar evaluation was carried out for all 18 products. The proper service levels derived were between 88.1% and 99%, and the associated week of inventory for achieving this level was about 3.6–6.5. The result of this model was different for all 18 products because of the differences between their profitability and costs. But overall, the derived potential cost savings for the desired service levels ranged from trivial amounts to 42% for some products.

11.6 Measuring Customer-Service Performance

Having an adequate and competitive customer-service policy does not ensure a firm's success in fulfilling customer satisfaction while maintaining profitability. Rather, it is necessary to regularly monitor and control it to assess the extent to

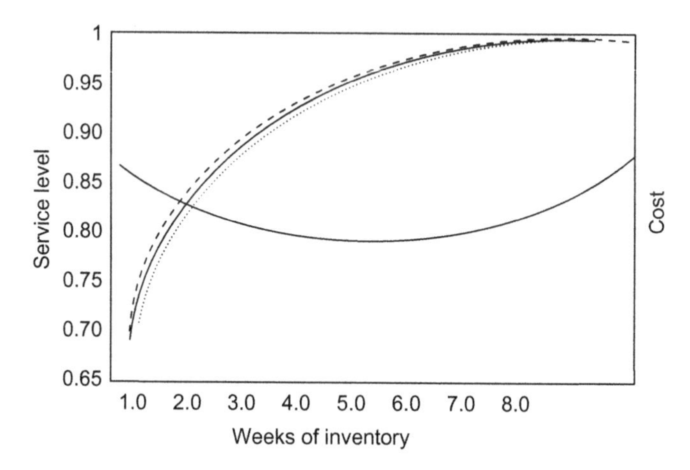

Figure 11.3 Relationship between inventory, customer-service level, and cost for product 1.

which policy objectives are accomplished and which areas need more improvements besides comparing current performance with the past and with other competitors [28].

Put in another way, managers need to conduct regular performance measurement because the ability to measure is the ability to control, and this measurement provides them with valuable information on how well a firm is performing in connection with customer service and to what extent it meets customer requirements [6,28].

Measuring the performance of an action means to quantify its effectiveness and efficiency [29]. This process should be based on predefined metrics and standards [7]. Therefore, determining these measures is a major concern. Most often firms tend to measure the aspects of services that are more accessible and easier to measure, or sometimes they use regular metrics that have been used in previous evaluations or the ones which more support the manager's viewpoints. This inaccurate definition may cause problem because measuring performance based on these metrics does not give managers a real picture of the firm's current situation in providing customer service and the areas of underperformance that may lead to customer displeasure and further profit loss. The appropriate metrics must be defined based on what customers recognize as important factors. For determining them, a close connection and communication with customers is vital [6,28].

Moreover, a proper measure generally must have certain characteristics: it must be objective, measurable, and easily understandable by the relevant people. In addition, it must have a current emphasis and not just focus on previous performance; it also must allow the firm to conduct comparisons over time. Finally, defined metrics should have minimal overlap with each other to ensure they are measuring different aspects of a system [28].

In logistics and supply-chain management literature, different service elements are suggested as appropriate metrics for measuring performance. These can be categorized similar to those service elements in the three groups of pretransaction, transaction, and posttransaction metrics [4,18,21,28]:

1. Pretransaction metrics
   Stock availability
2. Transaction metrics
   Convenience of ordering process
   Consistency of order-cycle time
   Rate of order filling
   Availability of order-status information
   Frequency of delivery
   Reliability of deliveries
   Status of back orders
   Availability of special equipments
   Delays in shipping
   Cost of delivery
   Percentage of on-time deliveries
3. Posttransaction metrics
   Procedure of claims
   Technical support
   Accuracy of invoices
   Replacing products

Besides these elements, order-cycle time is the variable that most often can be referred to as a single and appropriate metric for assessing customer service because its associated steps are among the most important service elements from customers' viewpoints [18].

The important point is that measuring a firm's performance in the area of customer service is not useful unless some analyses and interpretations are conducted on the findings. This allows management to discover the areas of underperformance and plan for corrective action.

References

[1] W.G. Donaldson, Manufacturers need to show greater commitment to customer service, Ind. Market. Manage. 24 (1995) 421–430.

[2] Lim Don, C.P. Prashant, EDI in strategic supply chain: impact on customer service, Int. J. Inform. Manage. 21 (2001) 193–211.

[3] Lee, S.H., Demand Management and Customer Support, IEMS Research Center. http://www.iems.co.kr/CPL/lecture/part4/6.%20Order%20Management%20&%20Customer%20Support.pdf

[4] M. Christopher, Logistics and Supply Chain Management: Creating Value-Adding Networks, third ed., Pearson Education Limited, Great Britain, 2005.

[5] A customer service definition from the customer's point of view. www.customer-servicepoint.com/customer-service-definition.html

[6] C.J. Johnson, F.D. Wood, D. Wardlow, R.P. Murphy, Contemporary Logistics, seventh ed., Prentice-Hall International, Upper Saddle River, NJ, 1998.
[7] A. Rushton, P. Croucher, P. Baker, The Handbook of Logistics and Distribution Management, third ed., Kogan Page Limited, Great Britain, 2006.
[8] D. Lambert, R.J. Stock, M.L. Ellram, Fundamentals of Logistics Management, first ed., Irwin McGraw-Hill, Boston, MA, 1998.
[9] E. Bottani, A. Rizzi, Strategic management of logistics service: a fuzzy QFD approach, Int. J. Prod. Econ. 103 (2006) 585–599.
[10] A.J. Goodman, Strategic Customer Service: Managing the Customer Experience to Increase Positive Word of Mouth, Build Loyalty, and Maximize Profits, first ed., AMACOM, New York, 2009.
[11] A. Sharma, M.D. Lambert, Segmentation of markets based on customer service, Int. J. Phys. Distrib. Logist. Manage. 24 (1994) 50–58.
[12] J.M. Kyj, Customer service as a competitive tool, Ind. Market. Manage. 16 (1987) 225–230.
[13] S. Kaplan, R.P.D. Norton, Strategy Maps: Converting Intangible Assets into Tangible Outcomes, Harvard Business School Publishing, Boston, MA, 2004.
[14] P. Kotler, K. Keller, Marketing Management, thirteenth ed., Prentice Hall, Upper Saddle River, NJ, 2008.
[15] J. Mentzer, B.M. Myers, Cheung Mee-Shew, Global market segmentation for logostics services, Ind. Market. Manage. 33 (2004) 15–20.
[16] L. Fogli, Customer Service Delivery: Research and Best Practices, first ed., John Wiley & Sons, San Francisco, CA, 2006.
[17] S. Chowdhury, Power of Six Sigma, Dearborn Trade, Chicago, IL, 2001.
[18] R.H. Ballou, Business Logistics/Supply Chain Management, fifth ed., Prentice Hall, Upper Saddle River, NJ, 2004.
[19] Gerald J. Tellis, Modelling marketing mix, The Handbook of Marketing Research, SAGE Publications, Thousand Oks, CA, 2006, 506–522.
[20] D. Waters, Global Logistics: New Directions in Supply Chain Management, fifth ed., Kogan Page Limited, London, 2007.
[21] J. Stock, D. Lambert, Strategic Logistics Management, fourth ed., Irwin McGraw-Hill, Boston, MA, 2001.
[22] A.J. Tompkins, D.J. Smith, Warehouse Management Handbook, second ed., Tompkin Press, Raleigh, NC, 1998.
[23] G. Ghiani, G. Laporte, R. Musmanno, Introduction to Logistics System Planning and Control, John Wiley & Sons, Chichester, England, 2004.
[24] William. R. Dillon, Soumen Mukherjee, A guide to the design and execution of segmentation studies, The Handbook of Marketing Research, SAGE Publications, Thousand Oks, California, 2006, 523–545.
[25] J.H. Lee, S.C. Park, Intelligent profitable customers segmentation system based on business intelligence tools, Expert Syst. Appl. 29 (2005) 145–152.
[26] S. Dibb, Market segmentation: strategies for success, Market. Intell. Plann. 16 (1998) 394–406.
[27] M.M. Jeffery, J.R. Butler, C.L. Malone, Determining a cost-effective customer service level, Supply Chain Manage. An Int. J. 13 (2008) 225–232.
[28] D. Waters, Logistics an Introduction to Supplychain Management, first ed., Palgrave Macmillan, Houndmills, Basingstoke, 2003.
[29] A. Gunasekaran, B. Kobu, Performance measures and metrics in logistics and supply chain management: a review of recent literature (1995–2004) for research and applications, Int. J. Prod. Res. 45 (2007) 2819–2840.

Part IV

Special Areas and Philosophies

12 Logistics System: Information and Communication Technology

Shokoofeh Asadi

Industrial Engineering Department, Amirkabir University of Technology, Tehran, Iran

The main purpose of this chapter is to describe how information and communication technology has affected the logistics system in organizations. At the first section, the role of information in a Logistics system is briefly described. Then, in the second part, after reviewing the concept of information flow in logistics system, some important concepts in Logistics Information Systems (LIS) such as its functionality levels, key roles of LIS in company operations and its structure and the main modules are explained. The third section is assigned to some significant information and communication technologies which have been appeared in Logistics systems recently such as RFID, Bar-coding, EDI and etc. Finally, results of a survey which is conducted by "eyefortransport institute" are mentioned to evaluate the real word companies' situation in using state of art technologies in Logistics systems.

12.1 The Importance of Information in Logistics

In recent years, the role and importance of logistics has increased significantly in companies and become a key element of corporate strategies. Nowadays, logistics has been identified as an important element of corporate strategy that potentially can lead to customers' value creation, cost savings, enforced discipline in marketing efforts, and increased flexibility in production [1]. The emergence of new technologies leads to a change in the practice and significance of logistics management and emphasizes its role in strategic function of companies [2].

By involving new technologies in logistics operations, the role of logistics in organizations changed from only a supportive function to a value-added function, and companies can achieve better customer service and higher cost savings so that they can compete in global markets.

IT affects the organization in two ways: it improves the current logistics functions of organization, and it changes the structure of logistics operations (e.g., by eliminating distributors and making contact with customers directly by using IT solutions).

Information and communication systems and technologies can play different roles in supply-chain and logistics management. These systems and technologies

Logistics Operations and Management. DOI: 10.1016/B978-0-12-385202-1.00012-8

can be applied to data gathering and analyses, support decision makers, control and monitor supply-chain operations, use information for forecasting, and facilitate the communication between supply-chain members.

12.2 Logistic Information System

12.2.1 Information Flow

There are three flows in the logistics function: material flow, information flow, and financial flow. According to Hammant [3], logistics is as much about the management and movement of information as it is about the management and movement of physical goods.

Information can be seen as the lifeblood of a logistics and distribution system. Nowadays, the effectiveness and accuracy of distribution systems depend on the transfer of information. To achieve this, developing an appropriate corporate strategy for information requirements in a company is important [4].

Introna [5] demonstrates that the logistical systems and information and communication systems work parallel, as long as the logistics system has converted materials into products, through the creation of value for customers, the information and communication systems convert data into information, in order to facilitate managerial decision making.

The information flow in a logistics system is as important as material flow, and information is considered as a key source in logistics system. When the complexity of logistics systems and channels increases and more members become involve, the role of information flow in the logistics system becomes more significant.

As Chibba and Hörte suggest [6], information flow in logistics system can be divided into two separate categories based on their relation with physical material flow. The first category is the information that is directly connected to the physical material flow such as order, delivery, and replacement information. This type of information is represented by (F) and is necessary for production and other operations of the logistics system [7].

The second category is the information ($-F$) that is indirectly related to the physical material flow. It is not necessary for doing daily operations of logistics system such as processing information about customers' satisfaction and demands and market trends.

In a logistics system, different types of activities and internal or external processes generate data. In the next step, generated data are converted into information. This information can be in different formats such as oral, textual, or paper based [8]. To use this information for a specific purpose such as decision making, it should be converted to knowledge. Converting information to knowledge is done through analysis.

So the major purpose for collecting, retaining, and manipulating data within a firm is to make decisions ranging from the strategic to the operational and to facilitate a business's transactions [9]. Figure 12.1 shows the process of converting data to a decision.

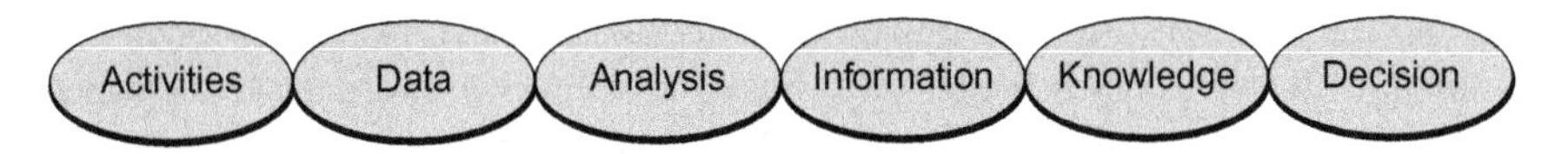

Figure 12.1 Process from activities to decision making [6].

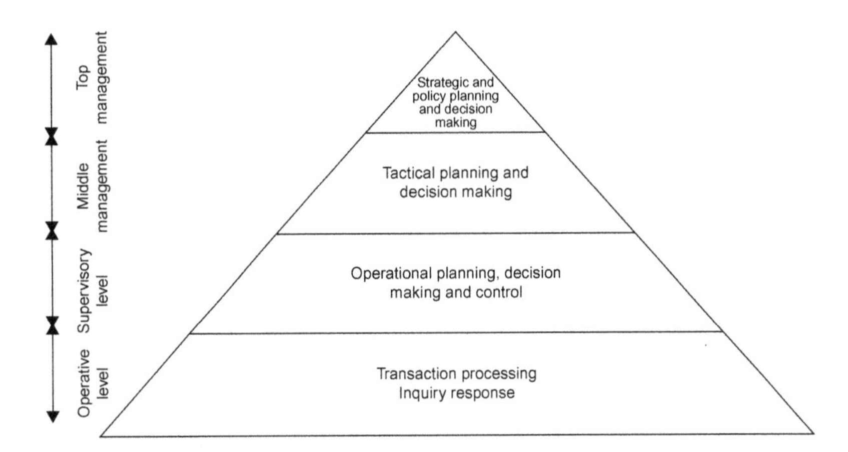

Figure 12.2 Pyramid of LIS functionality levels [12] & [14].

12.2.2 A LISs' Functionality Levels

An information system represents an integration of data, supporting equipment, and personnel and problem-solving methods that are used to assist the logistician in planning and operations [10].

According to Lambert et al. [11], an LIS is a computer-based information system that coordinates and manages all logistics management activities. Such an LIS must be capable of transferring information between the source and demand points [11].

This section reviews three concepts about LIS: functionality levels, information requirements, and structure. As shown in Figure 12.2, however, LIS has four levels of functionality: transaction system, management control, tactical planning and control and decision analyses, and strategic planning [12,13].

Transaction System

This level includes the most frequent activities in logistics management such as order inquiries and receiving, order processing, stock status checks, bill-of-lading preparation, and transportation-rate lookups. Such interactions may be repeated many times each hour, so the speed of the information flow is highly important. Operative personnel such as order-processing and transportation-rate clerks are typical users at this level.

Management Control

The emphasis at this level is on performance measurement and reporting. The main users of this level of the information system are first-line supervisors. For example, warehouse supervisors need information about space, inventory, and labor to control them, or a truck-fleet manager must have the necessary information about people, equipment, and spare parts to schedule deliveries. The interval of generating this information is daily, and the information is presented in report format.

Tactical Planning and Control and Decision Analyses

Tactical planning and control is an extension of management at the supervisory level. Its interval is less than 1 year but not as often as every day. Evaluation of inventory-control limits, planning for suppliers improvement, carrier selection, vehicle routing and scheduling, planning warehouse layout, and planning for seasonal space and transportation needs are examples of tactical planning and control problems. The users of this information system level are middle managers such as warehouse managers and transportation managers. At this level, the planning is a kind of decision making.

Strategic Planning

The strategic planning is the extension of the decision analysis that is more abstract, less structured, and more long-term in focus. At the strategic-planning level, the goals, policies, and objectives of logistics system are established, and the main decisions on the overall logistical structure and resource distribution are taken. At this level, there is no need for high-speed information transfer, and the information system is interrogated infrequently. So off-line systems with manual procedures are enough for this level of planning. Examples of this level of planning include:

- Strategic alliances with other value-chain members
- Identification and development of company capabilities and opportunities
- Improvement of services according to customer opinions

12.2.3 Role of Information in Logistics System Operation and Performance

Information acts as the glue in logistics functions, holding the system together and coordinating all components of logistics operations. As shown in Figure 12.3, logistical information has two major components: (1) planning and coordination and (2) operations [13].

Planning and Coordination

The primary drivers of supply-chain operations are *strategic objectives* derived from marketing and financial goals. These initiatives detail the nature and location of customers that supply-chain operations seek to match to planned products and services. This will include customer bases, breadth of products and services, and promotions. The financial aspect of strategic plans details resources that are required to support inventory, receivables, facilities, equipment, and capacity.

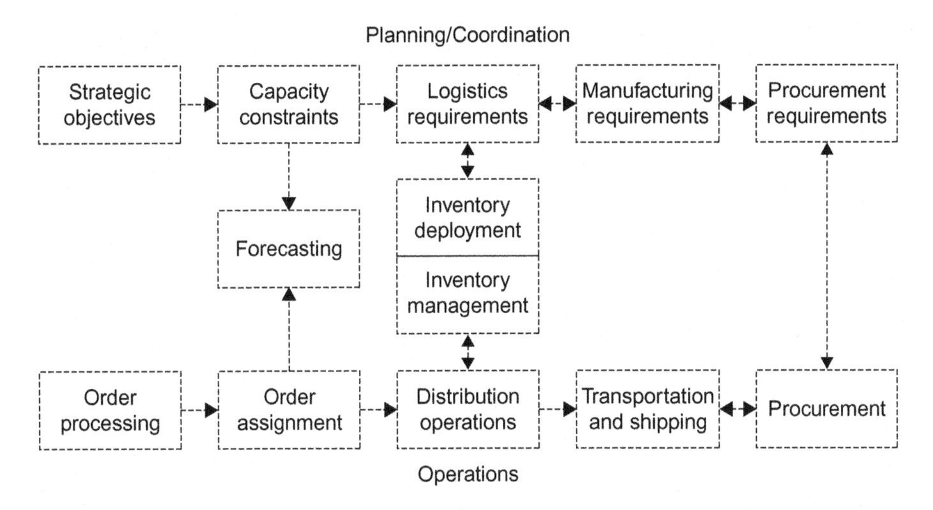

Figure 12.3 Logistics information requirements [14].

Capacity constraints identify manufacturing and market-distribution limitations, barriers, or bottlenecks. It also helps identify when specific manufacturing or distribution work should be outsourced. The output of capacity constraint planning is time-phased objectives that detail and schedule facility utilization, financial resources, and human requirements. For each product, capacity plans determine the where, when, and how much for production, storage, and movement.

Using inputs from forecasting, promotional scheduling, customer orders, and inventory status, *logistic requirements* identify the specific work facilities, equipment, and labor forces required to support the strategic plan. Logistics requirements must be integrated with both capacity constraints and manufacturing requirements to achieve the best performance.

Inventory deployment interfaces with inventory management between planning and coordination and operations, as shown in Figure 12.3. The deployment plan details the timing of where inventory will be positioned to efficiently move inventory through the supply chain. From an information perspective, deployment specifies the what, where, and when for the logistics processes. From an optional viewpoint, inventory management is performed on a day-to-day basis.

In production situations, *manufacturing requirements* determine planned schedules. The traditional deliverable is a statement of time-phased inventory requirements that is used to drive master production scheduling (MPS) and manufacturing requirement planning (MRP). In situations characterized by a high degree of responsiveness, an advanced planning system is more commonly used to time-phase manufacturing. MPS defines weekly or daily production and machine schedules, whereas MRP coordinates the purchase and arrival of material and components to support the manufacturing plan.

Procurement requirements represent a time-sequenced schedule of material and components needed to support manufacturing requirements. In retailing and wholesaling establishment, purchasing determines inbound merchandise. In manufacturing situations, procurement arranges for arrival of material and component parts from suppliers. Regardless of the business situation, purchasing information is used to coordinate decisions concerning supplier qualifications, degree of desired speculation, third-party arrangements, and feasibility to long-term contracting.

Forecasting is the prediction of the future using historical data, current information, and planning goals and assumptions. Logistical forecasting generally has short-term specification; typical forecast horizons are from 30 to 90 days. The forecast challenge is to quantify expected sales for specific products.

Operations

A second purpose of accurate and timely information is to facilitate logistical operations. To satisfy supply-chain requirements, logistics must receive, process, and ship inventory.

Order processing refers to the exchange of requirements information between supply-chain members involved in product distribution. The primary activity of order management is accurate entry and qualification of customer orders. IT such as mail, phone, fax, and electronic data interchange (EDI) has radically changed the traditional process of order management.

Order assignment identifies inventory and organizational responsibility to satisfy customer requirements. Allocation may take place in real time—i.e., immediately or in a batch mode, which means that orders are grouped for periodic processing such as during a day or shift. The traditional approach has been to assign responsibility or planned manufacturing to customers according to predetermined priorities. In technology-rich order-processing systems, two-way communication linkages can be maintained with customers to generate a negotiated order that satisfies customers within the constraints of planned logistical operations.

Distribution operations involve information that facilitates and coordinates work within logistics facilities. LIS functions in distribution operations contain physical activities such as product receiving, material handling, storage, and order selection.

Inventory management is related to information that is necessary for implementing the logistic plan. The main function of inventory management is managing and deploying inventory according to planned requirements. The work of inventory management is to distribute resources in a way that ensures that the overall logistical system performs as planned.

Transportation and shipping information directs inventory movement. The activities in transportation and shipping include shipment planning and scheduling, shipment consolidation, shipment notification, transport documentation generation, and carrier management. The activities ensure efficient transport resource use as well as effective carrier management. In distribution operations, it is important to consolidate orders so as to fully utilize transportation capacity. It is also necessary to ensure that the required transportation equipment is available when needed.

Finally, because ownership transfer often results from transportation, supporting transaction documentation is required.

Procurement is concerned with the information necessary to complete purchase-order preparation, modification, and release while ensuring overall supplier compliance. In many ways, the information related to procurement is similar to that involved in order processing.

The overall purpose of operational information is to facilitate integrated management to market distribution, manufacturing support, and procurement operations. Planning and coordination identify and prioritize required work and identify operational information needed to perform the day-to-day logistics.

12.2.4 LIS Structure

The LISs have three main components [10,12]: input, database, and output.

The structure of the LIS is shown in Figure 12.4.

Input

The input phase is a collection of data sources and data-transfer methods and means for making appropriate data available to the computing portion of the system. The LIS data can be obtained from many sources and in many forms, particularly from the following sources.

Customers: Customer data are captured during their sales activities, order entries, and deliveries. The obtained data are useful for forecasting, planning, and operating decisions. Freight bills, purchase orders, and invoices are typical sources

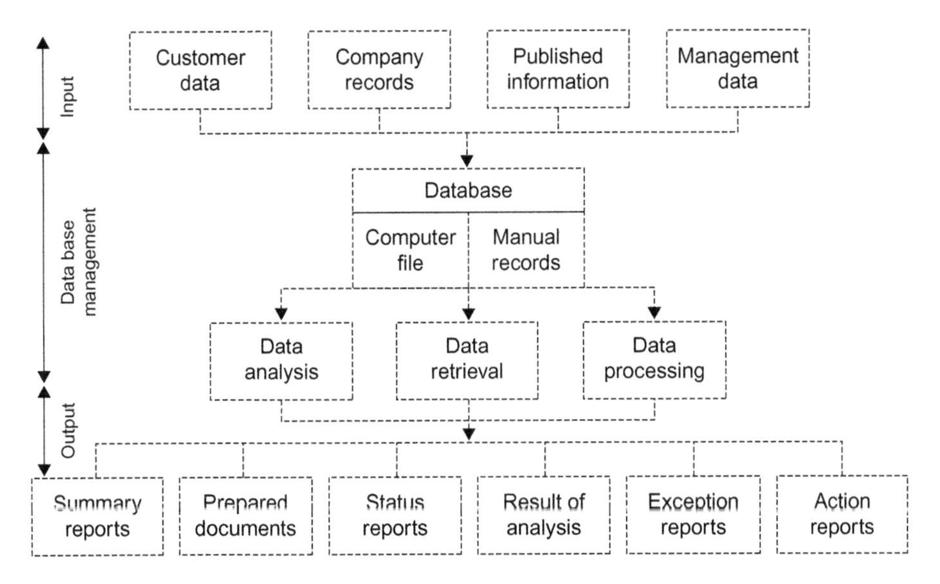

Figure 12.4 Basic structures of the LIS [9,10,12].

for this type of data. The primary source of information in logistics system is sales order because it contains basic data about customer and demanded items. Typical data from the customers are customer locations, their demands, weight and value of demanded items, date of order and date of shipping, shipment size, packaging, transportation mode, and so on.

Company records: Much valuable information can be obtained directly from a company's internal records. Accounting reports, status reports, internal and external study reports, and various operating reports are examples of this type of information source.

Company data are often an untapped source of excellent data. However, such data are neither located at a single point within the company nor are organized in any meaningful way for logistics decision making.

Published data: Professional journals, trade magazine, and government reports are some sources of this type of data. This type of data is more generalized than internally generated data.

Management predictions: Predictions of future sales level, action of competition, availability of purchased materials are just a few of the examples of information that are judgmental.

These types of data are maintained in the minds of company personnel, not in company files, computer records, or libraries. Company personnel such as managers, internal consultants and planners, and activity specialists are close to data sources and become good sources themselves. Also, clerks who receive customer feedback are valuable sources of such data.

Database Management

The most important component of an information system is the converting module in which data are converted to information and information is converted to useful knowledge for decision making. Database management contains three main functions: data selection, analysis method selection, and basic data-processing procedure to implement.

The maintenance of data in a database depends on the answers to these four questions:

1. How critical is the information to the decisions the logistician must make in a particular firm?
2. How rapidly does the information need to be retrieved?
3. How frequently is the information to be accessed?
4. How much effort is required to manipulate the information into the form needed?

Because logistical decisions vary in their frequency and in how rapidly required information for them must be made available, storage and retrieval methods should reflect these needs. Generally, the more accessible the information, the more costly the storage and retrieval. Therefore, computer storage and electronic retrieval and display can be justified for the most frequent planning and storage problems. Demand forecasting, inventory control, freight-bill preparation, shipment scheduling, and cost-report preparation are just a few of the daily, weekly, or monthly logistics management activities.

Activities such as warehouse location, facility layout, and material-handling equipment review require information at approximately yearly intervals. Computer storage is not usually economical. Most of the information is retained in company files as records.

Finally, infrequent planning-and-control activities such as private-warehouse construction and private-transportation equipment review usually do not justify maintaining information in a ready form. Rather, information in its raw form can be generated from primary data sources.

This three-tiered or ABC approach to information-storage requirements is a good approximation method for ranking and identifying how information and data should be stored, if at all. This then becomes a basis for deciding what types of storage capacity are needed and how much.

Data retrieval refers to the capacity of recalling data from a database in essentially its raw form or in only slightly modified form.

Data processing is one of the most popular features of the information system. Data-processing activities are relatively simple and straightforward conversions of data in files to more useful forms such as preparing transport bills of lading. Processing data into information is a very basic function of the information system. Data processing usually contains simple operations on data such as sorting and summarizing, coding, and arithmetic manipulations that convert data to useful information for logistics decision making and reporting.

Data analysis is the most sophisticated and newest use made of the information system. The system may contain any number of mathematical and statistical models. Such models provide information that is useful in dealing with some of the most difficult planning and control problems. These models use the database or the output of data-processing steps to find trends and forecast future level of activities and other information that is useful for planning.

Output

The output of an information system is the interface with the user of system. The outputs in LIS cab are grouped in three types: reports, prepared documents, and results of data analysis from mathematical and statistical models.

- Reports
  - *Summary reports of financial and performance indicators* refer to information on which the logistician may take action. They do not in themselves initiate action. Inventory-level reports are of this type.
  - *Status reports of current activities* are special-purpose reports that help the logistics operation run smoothly. Information reports on the date of order receipt and date shipped are examples.
  - *Exception reports that compare actual performance with goals* are special reports that, for example, report on unplanned events such as when transportation costs as a percent of sales exceed a preplanned ratio.

  - *Reports that initiate actions* are commands sent out by the LIS to perform some activity. Examples are stock-replenishment orders, truck-routing schedules, and order-picking list. These reports are based on management rules that are incorporated into the computer-based information system.
- Prepared documents are common and printable documents such as shipment documents and freight bills.
- Results of data analysis from mathematical and statistical models for instance, demand forecasting is one of the most useful and important outputs can be obtained from data analysis.

12.2.5 System Modules

The LIS should be comprehensive and capable enough to allow for communication not only between functional areas of a firm (marketing, production, finance, logistics, and so on) but also between members of the supply chain (vendors and customers) [9]. According to Frazelle [14], LIS modules can be listed as follows:

- Customer response system (CRS)
- Inventory management system (IMS)
- Supply management system (SMS)
- Transportation management system (TMS)
- Warehouse management system (WMS)

The CRS and SMS can be seen as part of the order management system (OMS), and ordinarily WMS contains the IMS module. So LIS has three main modules: OMS, WMS, and TMS.

The Order Management System

The OMS is the first point of logistics system contact with customers by managing order receiving and placement. It is the front-end system of the LIS. The OMS are closely related to WMS for checking product availability. The customer-ordered items may be available from inventories or may be seen in the production schedules. This provides information about the location of the product in the supply network, quantity available, and possibly the estimated time for delivery. After checking product availability and accepting the delivery time by the customers, the next step is credit checking. In this step, the OMS communicate with the financial information system to check a customer's credit status. Once the order is accepted, the OMS will allocate the product to the customer order, assign it to a production location, decrement inventory, and prepare an invoice when shipping has been confirmed.

The OMS dose not stand in isolation from the firm's other information systems. If the customer is to be served effectively, then information must be shared.

It should be noted that although the discussion has focused on the orders being received by a firm, there is a similar OMS for the purchase orders placed by the company (sometimes called the SMS). Whereas in a customer-based OMS a firm's customer data are important, in a purchase-based OMS the focus is on the company's vendors' data such as their delivery-performance ratings, costs and terms of sale, capabilities, availabilities, and financial strength [9].

The ways customer orders can be placed vary from completely manual to automatic when a customer's computer directly connects to the seller's system without human involvement. There are clear trade-offs in each situation between cost and information quality. In automatic order placement, the speed and accuracy of the process increases. However, initial costs are more than manual orders because of the need for system facilities.

Automating the order-processing function has many advantages for companies. The first one is improving customer services through increases in speed and accuracy. For example, by increasing the speed of the order-placement process, the order-cycle time can be reduced, which means that customers do not need to hold so much safety stock. In this case, when a customer order is received, the system is able to inform customers immediately about the order status, including item availability, shipping dates, and credit availability. If the order is allocated from inventory, the inventory levels are updated automatically; if the item is not in stock, then, according to production planning, the estimated delivery date is provided to customers. Another benefit to a firm is avoiding human interference in order-handling functions because these activities are now largely computerized. Automation also has financial benefits such as generating customer invoices on the same day as shipments, which accelerates cash flow. Finally, there are fewer billing errors and clerical mistakes [15].

The Warehouse Management System

In some LISs, the WMS may contain the OMS or it may be two separate modules within the LIS. The significant point is that WMS and OMS must be related because the sales department should know about product availability. The WMS is an information subsystem that focuses on the management of product flow and storage [9]. The WMS normally gets information such as purchase orders and customer orders from the company's main transaction system such as an enterprise resource planning (ERP) or legacy system. Information such as goods received and dispatched and inventory levels are then fed back. The WMS has itself a wide range of modules that may or may not be applicable for a specific purpose. The key elements can be identified as follows [5]:

- Receiving
- Put away
- Inventory management
- Order processing and retrieving
- Shipment preparation

All of these elements will appear in the WMS of a typical distribution warehouse, but some may not be present in warehouses used primarily for long-term storage or those with very high turnover.

The Transportation Management System

The TMS focuses on a firm's inbound and outbound transportation. Like the WMS, it shares information with other LIS components such as order content, quantity, weight and cube, delivery date, and vendor shipping schedules. The function of TMS as a part of LIS is planning and controlling a firm's inbound and outbound transportation activities. This involves the following:

- Mode selection
- Freight consolidation
- Routing and scheduling shipments
- Fleet management
  - Maintenance scheduling
  - Vehicle parts control
  - Fleet administration
  - Fleet costing
  - Tachograph analysis
- Claims processing
- Tracking shipment
- Freight-bill payment and auditing.

12.2.6 LIS Characteristics

The characteristics that should be concerned in designing and evaluating of LISs are as follows [5].

Availability

The rapid availability of information is absolutely necessary in responding to customers and in improving management decisions. Customers frequently need quick access to inventory and order-status information regardless of managerial, customer, or product-order location. Many times it calls for decentralized logistics operations so that the information system can access information updated from anywhere.

Accuracy

Logistics information must accurately reflect both current status and periodic activity for customer orders and inventory levels. Accuracy is defined as the degree to which LIS reports match actual physical counts or status. Increased information accuracy reduces inventory requirements.

Timelines

Timelines refers to the delay between the occurrence of an activity and the recognition of that activity in the information system. Timely information reduces uncertainty and identifies problems, thus reducing inventory requirements and increasing decision accuracy.

Exception-Based LISs

Sometimes LISs require reviews to be done manually particularly when decisions require judgment on the part of user. The central issue is identifying these exception situations that require management attention and decision making. LISs should be strongly exception oriented and utilized to identify decisions that require management attention, particularly for very large orders, products with little or no inventory, delayed shipments, and declining operating productivity.

Flexibility

An LIS must be flexible in meeting the needs of both system users and customers. It must be able to provide data tailored to meet the requirement of a specific customer.

Appropriate Format

Logistics reports and screens must contain the right information in the right structure and sequence.

12.3 Logistics Information and Communication Technology

As technology costs decline and usage becomes easier, logistics managers are managing information electronically at lower logistics expense with increased coordination results in enhanced services by offering better information and services to customers [13].

Advances in information systems are transforming the way logistics is managed. Automating the order-processing function leads to better customer service and the storage of more information for later analysis. The growing use of decision-support systems (DSSs) in logistics is helping managers to improve both their decision-making and forecasting capabilities. EDI is another technology that is used for transferring information in an efficient, secure, and lower-cost way than manual systems. Finally, technological advances in various types of hardware will continue to enhance the quality of information available to managers, improve customer service, and lower response times [15].

Nowadays, IT is mentioned as an opportunity to change the structure of logistics system in a firm. IT applications in logistics [16] can be listed as follows:

- Data collection: optical scanning, electronic-pen notepads, voice recognition, and robotics
- Identification: bar codes, radio frequency(RF) tags and antennas, smart cards and magnetic strips, and vision systems
- Positional systems (GPS-MPSGIS-Navigator)

- Communication networks and data exchange (EDI-XML-Internet-Satellite-LAN-WAN-EPOS)
- Data storage: data marts and data warehouses
- Software: DSSs, artificial intelligence, general software, and LIS modules

12.3.1 Data-Handling Hardware (Data Collection and Data Identification)

So many different IT technologies exist to make accurate capture, storage, and distribution of information within supply chains as streamlined as the supply chains themselves. Many of these systems are about the initial capture of data. The following describes some of the most popular techniques.

Bar Codes

Bar coding is the best-known technique used in warehousing. A bar code is a readable label that contains bars that represent specific numbers and letters. Bar coding is used widely in warehousing because it is a fast and accurate technology. Bar codes are used to identify goods and their locations. The use of bar codes can speed up operations significantly. Problems can occur if bar codes are defaced or if the labels fall off in transit.

Bar coding facilitates the tracking of goods moving through the logistics system, especially in warehousing operations such as goods receiving, stock checking, finding storage locations, and dispatching. Generally, we can say bar-code systems are used to identify logistics system elements such as the identification of goods, containers, documents, production, location, and equipment. The advantages of using bar-coding technology are its price, speed, and reliability. This technology has disadvantages, too; for example, the labels that are used may be damaged by scuffing, and they may not save a large amount of information, just a few digits of data, such as a product code or a pallet identification code.

Two-dimensional bar codes are available. As the name suggests, these are scanned in two directions simultaneously: horizontally and vertically. They can hold hundreds of numbers or characters, but their use is not widespread because special scanners are required at each stage in the supply chain and common standards are not fully established. They are, however, used in closed-loop situations [5].

Each bar-coding system has three main components: bar-code label, bar-code readers, and bar-code printers [17].

Bar-Code Readers

Although bar codes themselves have changed little since their introduction, the devices used to read them and transmit data onward certainly have. The kind of modern scanning equipment used within warehouses to confirm the movements of pallets, cases, or individual products, for instance, almost invariably uses radio communications to ensure that a central WMS is updated as soon as a scan is carried out.

Bar codes are read by both contact and noncontact scanners. Contact scanners must contact the bar code. They can be portable or stationary and typically come in the form of a wand or light pen. The wand or pen is manually passed across the bar code.

Noncontact readers may be handheld or stationary and include fixed-beam scanners, moving-beam scanners, and charged couple device scanners.

Suppliers have introduced omnidirectional scanners for industrial applications that are capable of reading bar codes passing through a large view field at high speeds, regardless of the orientation of the bar code. These scanners are commonly used in high-speed sorting systems [17].

Optical Character Recognition

Optical character recognition (OCR) technology uses labels that can be read by both machines and humans. It is appropriate in applications such as document handling and interrogation, and text scanning [5].

Radio Frequency Identification

Radio frequency identification (RFID) technology is a new data-capture capability that allows every individual item (each separate can of beans on a pallet, for instance) to be uniquely identified, something bar codes could never practically do.

RFID is being applied increasingly in supply chains to track unit loads (e.g., roll-cage pallets and tote bins), for carton identification (e.g., in trials by food retailers and parcel carriers), and for security and other purposes at the item level (e.g., for high-value goods). As the name RFID suggests, the mechanism uses radio waves. Normally, such a system has the following four components:

1. A *tag* is attached to the goods or container that contains a microchip and an antenna. So-called active tags also contain a battery.
2. An *antenna* transfers data via the tag.
3. A *reader* reads the data received by the antenna.
4. A *host station* contains the application software and relays the data to the server or middleware.

RFID is a rapidly developing technology that allows objects to be tagged with a device that contains a memory chip. The chip has a read-and-write facility that is currently executed using a variety of radio frequencies.

One advantage of RFID over bar codes is that the information contained in the tag can be updated or changed altogether. The tags are less vulnerable to damage, unlike bar-code labels, and not easily defaced. Another advantage is that the tags may be read from a distance and in some cases do not require line-of-sight visibility. RFID tags also can be read through packing materials but not through metal. A mixed pallet of different products may be read simultaneously by one scanner, thus reducing processing time significantly. RFID tags may be used to track many different types of assets, including people and animals.

Active tags tend to be used for high-value units (e.g., car chassis in assembly plants or ISO containers). However, passive tags are used in supply chains more than active tags. These tags do not use batteries, and their power is supplied by incoming signals, so they can be used in ranges between approximately 1 and 4 meters because they need very strong signals to provide the power. Passive tags could not save power. Although they were expensive at first, nowadays their progressively lower costs is an advantage. There remain shortages in the RFID concept and technology in such areas as standards, technical feasibility, operational robustness, financial business cases, and, in some instances, civil liberties [5].

Of all the new technologies, RFID arguably holds the most promise, but the fact that RFID tags cost many times more than bar codes has put off many potential users up to now. Tag prices, however, are only part of the equation, and many firms are now beginning to concentrate more on the overall business case. The time, labor, and potential errors associated with carrying out numerous manual scans of a bar code on its journey through a particular process or part of the supply chain, for example, can all be done away with using RFID because RFID tags are automatically read whenever they are in proximity to a reader. The total number of possible reads for a product through a system makes the cost lower and lower over time, making the technology more economical.

Magnetic Stripes and Optical Cards

Magnetic stripes commonly appear on the back of credit and bank cards. They are used to store a large quantity of information in a small space. The magnetic stripe is readable even through dirt or grease. Data contained in the stripe can be changed. The stripes must be read by physical contact, however, thus eliminating them from high-speed sorting applications. Magnetic stripes systems are generally more expensive than bar-code systems. In warehousing, magnetic stripes are used on smart cards in a variety of paperless applications. Smart cards are now used in logistics to capture information ranging from employee identification to the contents of a trailer load of material to the composition of an order-picking tour [5].

Vision Systems

Vision system cameras take pictures of objects and codes and send the pictures to a computer for interpretation. Vision systems "read" at moderate speeds with excellent accuracy, at least for limited environments. Obviously, these systems do not require contact with the object or code. However, the accuracy of a read is highly dependent on the quality of light. Vision systems are becoming less costly but are still relatively expensive.

A large mail-order operator recently installed a vision system at receiving. The system is located above a telescoping conveyor used to convey inbound cartoons from a trailer into the warehouse. The system recognizes those inbound cartons that do not have bar codes, reads the product and vendor number on the carton, and directs a bar-code printer to print and apply the appropriate bar-code label [17].

Error Rates

Some years ago, the US Department of Defense published the following information about error rates in different data-capturing methods:

Written entry	25,000 in 3 million
Keyboard entry	10,000 in 3 million
OCR	100 in 3 million
Bar code (code 39)	1 in 3 million
Transponders (RFID tags)	1 in 30 million

Although these data are experimental, they illustrate the increasing accuracy of this IT.

12.3.2 Positioning

The geographical information system (GIS) and global positioning system (GPS) are well-known technologies ordinarily used in transportation and distributions. GPS is a satellite-based navigation system, and GIS is used to capture, store, analyze, manage, and present data that are spatially referenced (linked to location). Applications of these technologies include the following:

- In vehicle-tracking systems, a vehicle's geographic position can be monitored using GPS. This can provide a variety of different benefits, from improved vehicle, load, and driver security to better customer service with the provision of accurate delivery times to lower costs through reduced waiting and standing time because exact vehicle arrival times are available.
- Trailer tracking allows vehicles and their loads to be monitored in real time using satellite GPS technology. This can be particularly beneficial for the security of vehicles, drivers, and loads, many of which are of high value. Trailers can be tracked automatically, and so-called red flags can be issued if there is any divergence from set routes. In addition, these systems can be used for consignment tracking to provide service information concerning delivery times and load-temperature tracking for refrigerated vehicles so that crucial temperature changes can be monitored and recorded, an essential requirement for some food chains and some pharmaceutical products.
- In-cab and mobile terminals enable paperless invoicing and proof of delivery. These are used by parcels and home-delivery operators, based on electronic signature recognition, and also by fuel companies for the immediate invoicing of deliveries where quantities may be variable.
- On-board navigation systems are common in many private cars, but they are also used in many commercial vehicles. They can provide driver guidance to postal codes and addresses, which is very beneficial for multidrop delivery operations in which final customer locations may be new or unfamiliar. They can result in significant savings in time, fuel consumption, and redelivery, greatly improving customer service.
- Linked with on-board navigation systems are traffic-information systems. These provide real-time warnings of traffic congestion and road accidents, allowing drivers to avoid these problem areas and considerably reduce delays and their associated additional costs.

In tandem with routing and scheduling software, this information can be used to enable the immediate rescheduling of deliveries and the rerouting of vehicles.

Understandably, these systems can be expensive, but for many large fleets and operations the cost savings and improved service make compelling cases for their adoption.

12.3.3 Communication, Networks, and Data Exchange

Electronic Data Interchange

EDI is the interorganizational computer-to-computer exchange of business documentation in a standard machine-processable format. EDI tries to omit additional human intervention in data processing in the receiving computer.

In supply chains, EDI is used to exchange essential business information, especially between partners that have a long-term trading relationship. For example, some multiple retailers will supply electronic point-of-sale (EPOS) data directly to suppliers, which in turn triggers replenishment of the item sold. As a consequence of this type of strong link, suppliers will be able to build a historical sales pattern that will aid their own demand-forecasting activities. In this context, EDI has many benefits. It is providing timely information about customers' sales, it is highly accurate, and it is very efficient because it does not require staff to manually collate information. EDI is used to send invoices, bills of lading, confirmations of dispatch, shipping details, and any other information that linked organizations choose to exchange [5].

To engage in EDI, business partners must add three components to their existing computer systems: EDI standards, EDI translation software, and some sort of transmission capability. To illustrate the underlying concept, Emmelhainz [18] provides the analogy of an American dealing by mail with a trading partner in Germany. For a successful communication, the partners should be able to write a letter in a generally accepted business format and be able to translate it from English to German and use a transmission method such as a mail service (Figure 12.5).

If we simulate the condition, EDI standards play the role of the format, EDI software provides the translation, and either direct-link or value-added networks are utilized for transmission.

One of the biggest constraints affecting EDI growth is the lack of a single EDI standard or language. The United Nations/EDI for Administration, Commerce, and Transport (UN/EDIFACT) is the international standard of EDI software, whereas the most common format of EDI standard in North America is the American National Standard Institute's ANSI ASC X12. Also, certain industry-specific standards have been developed according to specific needs of certain industries. Two of the most common are Caro Interchange Message Procedure (CARGO-IMP) in aviation, developed by the International Air Transport Association, and Transportation Date Coordinating Committee (TDCC), which was the first developed standard in the transportation era [15].

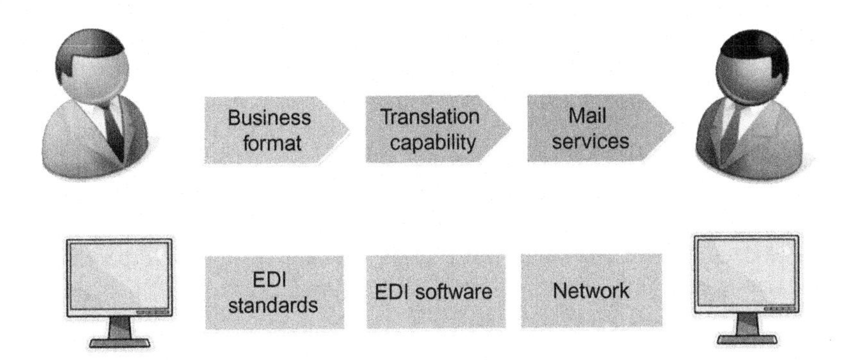

Figure 12.5 EDI components.

EDI transactions can be categorized into the following four main areas:

1. Interactive, query-response transactions are related to customers and contain operations such as ticket reservations by customers in airlines and order-status checking.
2. Trade data interchange covers the interactions between supply-chain members, such as purchase orders, delivery notifications, and invoices.
3. Electronic funds transfer is about the supply chain and banking system interactions such as payment against invoices.
4. Technical data interchange is about technical data transfer such as with engineering and design data.

Using EDI, a totally paperless supply chain can be achieved because EDI establishes connections between manufacturers, suppliers, retailers, customers, and banks, its most important benefit. Using EDI can facilitate coordination between supply chains and improve planning, production, and communications within. The other benefits of using EDI can be improving internal effectiveness and efficiency, saving time and resources, and thus reducing administrative costs. Other advantages of using EDI are as follows:

- Increased internal productivity through faster information transmission as well as reduced information entry redundancy.
- Better accuracy by reducing the number of times and the numbers of individuals involved in data entry.
- Improved channel relationships.
- Increased external productivity.
- Increased ability to compete internationally.

Decreased operation cost through reduced labor, paper-based transactions materials, and tools and telephone and fax communications.

Electronic Point of Sale

Now a common sight in most large retail stores in the developed world, EPOS has revolutionized the process of paying for goods. Equipment includes scanning equipment, electronic scales, and credit card readers. Goods marked with a bar code are scanned by a reader, which in turn recognizes the goods. It notes the item, tallies the price, and records the transaction. In some cases, this system also triggers replenishment of the sold item. One major advantage of an EPOS system is that it provides an instant record of transactions at the point of sale. Thus, replenishment of products can be coordinated in real time to ensure that stock outs in the retail store are minimized. Another advantage of this system is that it has speeded up the process of dealing with customers when large numbers of items are purchased. It reduces errors by being preprogrammed with selling prices, and it avoids staff having to add purchase prices mentally.

Many retailers offer loyalty card systems that reward customers with small discounts for continuing to shop with a given retailer. The advantage to the retailer is that loyalty cards with customers' personal details are linked to their actual purchases; this allows the retailer to obtain vital marketing information about these customers [5].

Radio Frequency Data Communications

RF technology is used for two-way information exchange and is applicable in small areas such as warehouses and distribution centers. These facilities communicate and receive messages on a prescribed RF via strategically located antenna and a host computer interface unit. The applications provide the following:

- Real-time communication with material handlers such as forklift drivers and order selections.
- Updated instructions and priorities for forklift drivers on a real-time basis.
- Two-way communication of warehouse selection instructions, warehouse cycle count verification, and label printing to guide package movement.

Synthesized Voice

The use of a synthesized voice is increasingly popular in warehouse operations. In stationary systems, a synthesized voice is used to direct a stationary warehouse operator. In mobile-based systems, warehouse operators wear a headset with an attached microphone. Via synthesized voice, the WMS talks to the operator through a series of transactions.

The advantages of voice-based systems include hands-free operations; the operator's eyes are free from terminals or displays, and the system functions whether or not the operator is literate. Another advantage is the ease with which the system is programmed. A simple Windows-based software package is used to construct all necessary transaction conversations [17].

Satellite Communications

Many freight carrier companies are using modern technology to provide improved customer service through better shipment tracking. Satellite communication presents the latest technology to be integrated into tracing and tracking systems. In just-in-time systems, where uncertainties in shipment arrivals can cause serious consequences for production operations, navigational satellites are being used to identify the exact location of truckload shipments as they move through the distribution pipeline [10].

Satellite communication is a powerful tool and channel for fast and high-volume information movement around the world. The applications allow the following:

- Communication between drivers and dispatchers by using dishes on the tops of vehicles.
- Dispatchers to use up-to-date information regarding location and delivery for redirecting trucks in response to need or traffic congestion.
- Retail chain headquarters to obtain daily sales information quickly and use that information to activate store replenishment and to gather useful input information for identifying local sales pattern [13].

Networks

A local area network (LAN) is a network of personal computers (PCs) that use phone lines or cables to communicate and share resources such as storage and printers. A LAN is restricted to a relatively small geographical location such as an office or warehouse. Wide area networks operate across a wider geography. Client/server architecture uses the decentralized processing power of PCs to provide LIS operating flexibility. A server is a large computer that allows common data to be shared by a number of users. *Client* implies network to PCs that access and manipulate data in different ways to provide extensive flexibility [13].

E-Commerce

Using the Internet for business-to-business and business-to-consumers relations is common and popular in companies. Many of the transactions and relations between companies and their customers or partners are through these emerging technologies. In logistics system, web-based solutions and the use of the Internet affect relations with both customers and vendors. The following are examples of web-based solutions in logistics [17]:

- Web-based customer response
  - Online ordering
  - Online customer services
  - Online auctions
  - Web-based returns processing
  - Online presentation of product information
  - Proactive order status reporting via web or e-mail
- Web-based inventory management
- Web-based supply
  - Vendor selection from web sites
  - Alternate sources of material
  - E-procurement
  - Online catalog for products
  - Electronic bidding

  - Electronic funds transfer
  - Collaborative planning, forecasting, and replenishment
- Web-based warehouse management
- Web-based transportation management
  - Online bidding, booking, and tracking
  - Online freight payment

Survey

Knowing the status of real-world companies in using IT technologies in supply-chain and logistics management is critical for both managers and researchers. According to this demand, the "eyefortransport institute" [16] conducted a "Technology in Transportation and Logistics" survey at the Seventh Annual North American Technology Forum (April 25 and 26, 2005, in Chicago). Three main discussion questions were posed in this survey:

1. To what degree was a company working with a selection of technologies?
   - To what degree were a company's selection of technology capabilities and services customer driven?
   - What were the most important motivating factors when deciding to upgrade technology capabilities and customer offerings?

In the first question, each company was asked to rate the degree to which it was working with various technology applications and IT solutions: from "Not

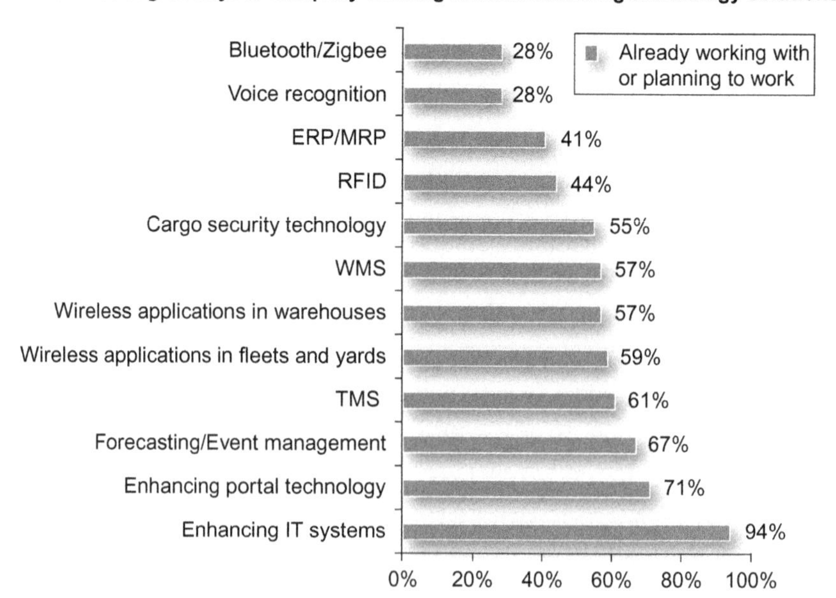

Figure 12.6 Following technology solutions in logistics [17].

interested at this time" to "Interested, but with no plans to invest," "Planning to invest," and "Already investing in." As shown in Figure 12.6, the top three technologies that companies were already investing in were (1) enhancing their IT systems, (2) improving portal technology, and (3) forecasting and event management.

TMS, wireless applications in fleet and yard management are coming next. On the other hand, the top three technologies in which companies were not then interested were Bluetooth and Zigbee, voice recognition, and ERP systems.

The second question the survey posed was, to what degree were various technology investments customer driven? This issue is very important in measuring return on investment (ROI) and determining long- and short-term benefits. The top one, the same as for the first question, was "enhancing IT systems." WMS and TMS came in second and third, respectively.

As shown in Figure 12.7, RFID's prospects are interesting. Although only 17% said their customers were demanding it, more than 50% indicated that their

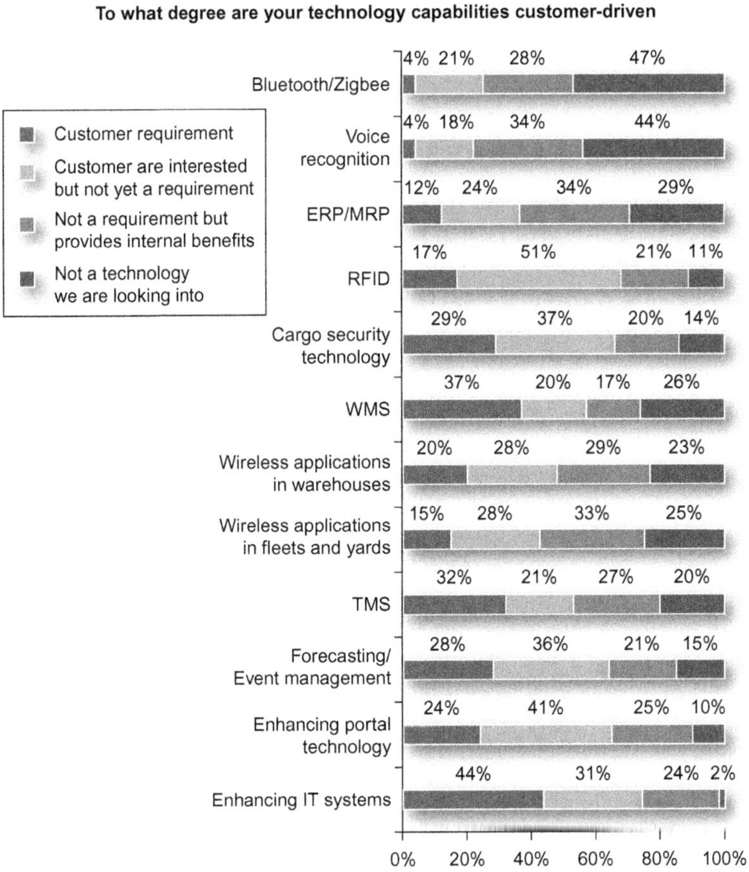

To what degree are your technology capabilities customer-driven

Customer requirement
Customer are interested but not yet a requirement
Not a requirement but provides internal benefits
Not a technology we are looking into

Bluetooth/Zigbee 4% 21% 28% 47%
Voice recognition 4% 18% 34% 44%
ERP/MRP 12% 24% 34% 29%
RFID 17% 51% 21% 11%
Cargo security technology 29% 37% 20% 14%
WMS 37% 20% 17% 26%
Wireless applications in warehouses 20% 28% 29% 23%
Wireless applications in fleets and yards 15% 28% 33% 25%
TMS 32% 21% 27% 20%
Forecasting/ Event management 28% 36% 21% 15%
Enhancing portal technology 24% 41% 25% 10%
Enhancing IT systems 44% 31% 24% 2%

0% 20% 40% 60% 80% 100%

Figure 12.7 Degree of customer-driven in logistics technologies [16].

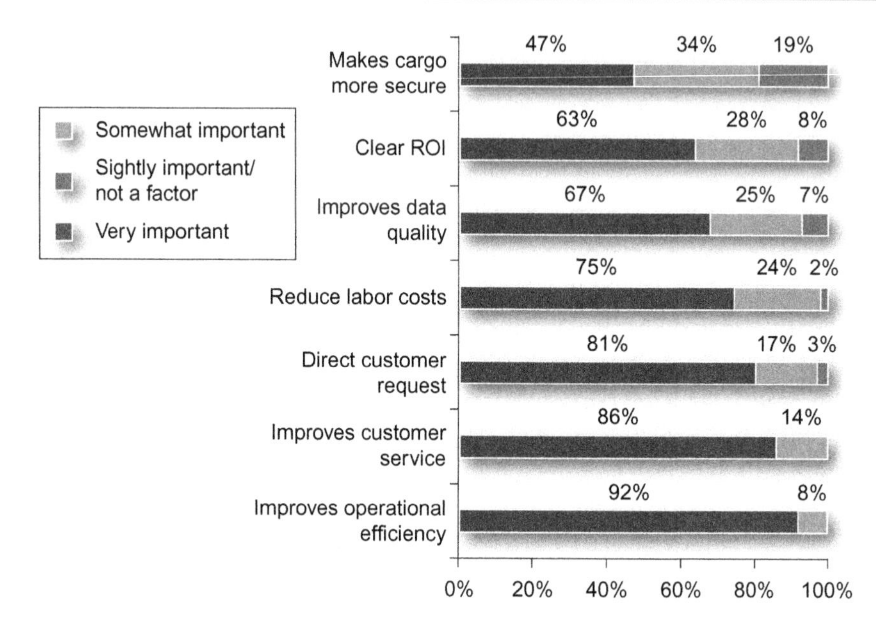

Figure 12.8 Most important motivating factors for upgrading technological capabilities [16].

customers were very interested. This could indicate how RFID might become the most popular technology in the near future, one that many companies aim to invest in.

The third and final question was about the motivating factors for upgrading technology capabilities. As shown in Figure 12.8, the most important driver in this case was improvements in operational efficiencies. The next two were improvements to customer services and direct customer requests. Achieving ROI was surprisingly low on the list, with only 63% claiming it as a significant motivator.

12.4 Conclusion

Changing roles of information in logistics systems and emerging new technologies in this area are transforming the way logistics is being managed or, in some cases, is dramatically changing the structure of logistics systems.

By using new technologies in logistics, competing in global environments is possible, and improvements can be realized in customer services and delivery times, the key elements in logistics systems.

Technological advances in information areas facilitate decision making because the real-time and accurate data and information are available. Also, the emergence of new hardware, supportive software, and networks is helping to improve such logistics functions as warehouse management, transportation management, and order processing.

References

[1] C.M. Gustin, Examination of 10-year trends in logistics information systems. Industrial Engineering 25(12) (1993), 34–39.

[2] Kengpol and Tuominen, 2006. A framework for group decision support systems: an application in the evaluation of information technology for logistics firms. International Journal of Production Economics. v101. 159–171.

[3] Je Hammant, Information technology trends in logistics: the pressure to invest in technology is high and will increase, Logist. Inform. Manag. 8(6) (1995) 32–37.

[4] A. Rushton, P.H. Croucher, P. Baker, Handbook of Logistics and Distribution Management, The Chartered Institute of Logistics and Transport, London, 2006.

[5] Lucas D. Introna, The Impact of Information Technology on Logistics, International Journal of Physical Distribution & Logistics Management, 21(5) (1991), 32–37.

[6] Chibba, A., Hörte, S-Å (2003), Supply chain performance a meta analysis, paper presented at 1st joint EurOMA/POMS conference, Como, Italy.

[7] A. Chibba, Jo. Rundquist, Mapping flows—An analysis of the information flows within the integrated supply chain, Paper presented at 16th annual NOFOMA conference, Linköping, 2004, p.6.

[8] Xian-zhong Xu, G. Roland Kaye, Building market intelligence systems for environment scanning, Logistics Information Management, 8(2) (1995), 22–29.

[9] R. Ballou, Business Logistics/Supply Chain Management Planning, Organizing and Controlling the Supply Chain, Pearson Education, New York, 2004.

[10] R. Ballou, Business Logistics Management, Prentice Hall, 1992.

[11] Lambert, Douglas M, James R Stock, and Lisa M Ellram. Business Logistics/Supply Chain Management. Boston: Irwin/McGraw-Hill, 1998.

[12] R. Ballou, Basic Business Logistics: Transportation, Materials Management, Physical Distribution, Prentice Hall, New York, 1987.

[13] R. Singh, S. Ailawadi, Logistics Management, Prentice Hall of India, New Delhi, 2006.

[14] Frazelle Ed., Supply Chain Strategy: The Logistics of Supply Chain Management. McGraw-Hill, New York, 2001.

[15] eLOGMAR-M Chinese, European Forum on e logistics, Shenzhen, P.R. China, 2006.

[16] Seventh Annual eyefortransport North American Technology Forum, 2005 Technology in Transportation and Logistics Survey Results, Hyatt Regency McCormick Place, Chicago, 2005.

[17] N.K. Gourdin, Logistics Management: A Global Perspective: a Competitive Advantage for the New Millennium, Blackwell Publishing, Oxford, 2005.

[18] M.A. Emmelhainz, Electronic Data Interchange: A Total Management Guide, Van Nostrand Reinhold, New York, 1990.

13 Reverse Logistics

Masoomeh Jamshidi

Industrial Engineering Department, Amirkabir University of Technology, Tehran, Iran

In addition to increased environmental concerns and severe environmental laws, *reverse logistics* (RL) has received increasing attention during this period. There are several explanations for RL in the related literature. For example, the Council of Logistics Management (CLM) defines RL as the "term often used to refer to the role of logistics in recycling, waste disposal, and management of hazardous materials; a broader perspective includes a relating to logistics activities carried out in source reduction, recycling, substitution, reuse of materials, and disposal" [1].

The European Working Group on Reverse Logistics, RevLog (1998), has presented the following definition of RL: "The process of planning, implementing and controlling flows of raw materials, in process inventory, and finished goods, from a manufacturing, distribution, or use point to a point of recovery or point of proper disposal" [1].

Difference Between Waste Management and RL [2]

Waste management is about efficiently collecting and processing waste (products for which there is no new use). RL is concerned with products that have some value to be recovered as a new one.

Difference Between Green Logistics and RL [2]

RL is concerned with recapturing the value of goods at their destination, whereas green logistics (ecological logistics) can be defined as understanding and reducing the ecological impact of logistics. It encompasses measuring the environmental effects of particular transportation modes, decreasing the usage of energy and materials in logistics activities, and following the guidelines of ISO 14000.

In fact, green logistics considers environmental aspects to all logistics manners and especially on forward logistics. Environmentally mindful manufacturing is more than just manufacturing for forward logistics. It is manufacturing that is concerned with the environmental effects of products until the ends of their lives.

Differences Between Forward and Reverse Logistics [3]

In forward logistics:

1. Prediction is relatively straightforward.
2. Transportation is one to many.

Logistics Operations and Management. DOI: 10.1016/B978-0-12-385202-1.00013-X

3. Product quality, packaging, and pricing are relatively uniform.
4. Destination, routing, and disposition options are clear.
5. There is a standardized channel.
6. Accounting systems closely monitor forward distribution expenses.
7. Inventory management is congruous.
8. Product life cycle is controlled.
9. Transactions between parties is straightforward.
10. Real-time data is easily available to track product.
11. There are clear marketing methods.

In RL:

1. Prediction is more difficult.
2. Transportation is many to one.
3. Product quality and packaging are not uniform.
4. Disposition and destinations and routing are unclear.
5. Deviation based in channel.
6. Many items determine pricing.
7. Measuring reverse expenses usually is impossible.
8. Inventory management is not congruous.
9. Estimation of product life cycle is more difficult.
10. Extra discussion brings about complexity in transactions.
11. Process is invisible.
12. Complexity in marketing.

13.1 The Literature on RL

Dowlatshahi [4] defines five categories of RL literature: (1) general summaries and basic RL concepts, (2) research on quantitative approaches, (3) studies of logistical topics, (4) company profiles, and (5) RL applications.

13.1.1 General Summaries and Basic RL Concepts

General summaries and explorations of basic concepts in research logistics include Rogers and Tibben-Lembke (1999), and De Brito and Dekker (2004) [1,2].

Genchev [5] examined returns handling at WCC (wholesale computer company) and described the company's successful RL program turnaround. Zoeteman et al. [6] showed the need and potential for high-level recovery practices on a regional basis. They presented quantitative estimates of current and future e-waste flows between global regions that generate and process waste and analyzed their driving forces. Janse et al. [7] developed a theoretically and empirically grounded diagnostic tool for assessing a consumer electronics company's RL practices and identified potential for RL improvement from a business perspective.

13.1.2 Research on Quantitative Approaches

Studies of various features of and improvements in RL systems can be found in Fleischmann et al. [8,9] and Minner [10].

Krumwiede and Sheu [11] presented a decision-making model for RL to guide the process of assessing the feasibility of RL for third-party providers. Hu et al. [12] presented an RL system with multiple time steps and multitype hazardous wastes with the objective of minimizing total operating costs. Kleber et al. [13] described a recovery problem with continuous time and dynamic framework that multiple demand streams and a single return is considered for different product variants or qualities. A heuristic construction procedure is suggested. Horvath et al. [14] offer a Markov chain approach to modeling the risks, expectations, and potential shocks associated with the cash flows of retail RL activities. Listes and Dekker [15] presented an approach based on stochastic programming in which a deterministic location model for producing recovery network design may be extended to explicitly account for uncertainties. In the same year, Ravi et al. [16] proposed a decision model based on the analytic network process (ANP).

The problem related to RL options for end-of-life (EOL) computers in a hierarchical form and links the dimensions, determinants and enablers of the RL with options available to the decision maker. Min et al. [17] presented a nonlinear mixed-integer programming model for RL problems and suggested a genetic algorithm to solve the problem. The model and its solution explicitly consider trade-offs between inventory cost savings and freight rate discounts because of consolidation and transshipment. Schultmann et al. [18] presented an RL tasks model within closed-loop supply chains. They used a Tabu search (TS) algorithm to solve it. Min and Ko [19] proposed a mixed-integer programming model for RL problems that consider the location and allocation of repair facilities for third-party logistic companies (3PLs). A genetic algorithm is suggested to solve the problem. Sheu [20] proposed a coordinated reverse logistics (CRL) management system for treating multiple hazardous wastes in a given region. A linear multiobjective analytical model is formulated with the objective of simultaneously minimizing the corresponding risks and the total cost of RL.

Pati et al. [21] formulated a framework and model to assist in determining facility locations, routes, and flows of different varieties of recyclable wastepaper. Du and Evans [22] developed a dual-objective MIP (mixed-integer programming) optimization model for the RL network problem; its objective is to minimize total cost and maximize customer satisfaction. To solve the problem, they used a combination of three algorithms: scatter search, constraint method, and dual simplex method. Sheu [23] formulated a linear multiobjective optimization model to optimize the operations of both nuclear power generation and the corresponding induced waste RL. Saen [24] proposed a model for selecting 3PL providers in the presence of multiple dual-role factors.

Fonseca et al. [25] proposed a comprehensive model for RL planning in which they considered many real-world features such as the existence of several facility echelons, multiple commodities, choices of technology, and stochastics associated with transportation costs and waste generation. Moreover, they presented a two-stage stochastic dual-objective mixed-integer programming formulation in which strategic decisions are considered in the first stage and tactical and operational decisions in the second. Pishvaee et al. [26] proposed a mixed-integer linear programming (MILP) model for multistage RL networks that minimize fixed opening and transportation costs. To find the near-optimal solution, they applied a simulated annealing (SA) algorithm with particular neighborhood search mechanisms. Sasikumar et al. [27] designed a multiechelon RL network for product recovery to maximize the profit of remanufacturing operations. The model resented is validated through the example of a tire-retreading business operating in secondary markets.

13.1.3 Studies of Logistical Topics

Particular topics such as distribution, warehousing, and transportation are addressed in Jahre [28] and Pohlen and Farris [29].

Dethloff [30] proposed a vehicle-routing problem (VRP) with simultaneous delivery and pickup and then studied this problem relationship with other VRPs. Dobos [31] presented a minimization model to minimize the sum of the quadratic deviation from described inventory levels in stores and from described operations (e.g., rates of manufacturing, remanufacturing, and disposal). Optimal policies for inventory in an RL system are found with a specific structure. The maximum principle of Pontryagin is applied to solve the problem. Chen et al. [32] addressed the RLRFE (reverse logistic recycling flow equilibrium) problem as a flow equilibrium problem from a systemwide policy-making perspective, focusing particularly on equilibria in situations in which market price and recycling channel flows are coupled interactions and input–output recycled material flows at each agent are not balanced. They propose a three-loop nested diagonalization method in which asymmetric link interactions are gradually relaxed to achieve the equilibrium solution. Kara et al. [33] presented a simulation model of an RL network. The simulation was tested by using an EOL white goods collection process; it is useful also for simulating the collection of other EOL products. Efendigil et al. [34] presented a two-phase model based on artificial neural networks and fuzzy logic for choosing the best 3PLs for RL. Mutha and Pokharel [35] presented a mathematical model for RL network design. In their model, returned products are consolidated in a warehouse and then sent to reprocessing center for dismantling and inspection.

Lee and Chan [36] proposed an RL system based on radio-frequency identification (RFID) technology to demonstrate the benefits of using a computational intelligence technique and RFID to form an integrated model to optimize the coverage of product returns.

13.1.4 Company Profiles

Company profiles that illustrate the critical role of some manufacturing technologies in RL accomplishments can be found, for example, in Thierry et al. [37]. Kim et al. [38] proposed a mathematical model and general framework for a remanufacturing system.

Table13.1 A Summary of RL Applications

Paper	Application	Country
Barros et al. [39]	Logistics network for recycling sand	Netherlands
Fleischman et al. [40]	Copier remanufacturing	European
Blanc et al. [41]	Redesign of a recycling system for liquefied petroleum gas tanks	Netherlands
Listes and Dekker [15]	Recycling sand from demolition waste	Netherlands
Blanc et al. [42]	Collection of containers from EOL vehicle dismantlers	Netherlands
Amini et al. [43]	Designing a repair service for medical diagnostics manufacturers	International
Seitz [44]	Engine remanufacturing	European
Schultmann et al. [18]	EOL vehicle treatment	German
Pati et al. [21]	Paper recycling	India
Asari et al. [45]	Mercury flow for EOL fluorescent lamps	Japan (Kyoto)
Sheu [23]	Nuclear power generation	Taiwan
Kannan et al. [46]	Battery manufacturing industry	India
Geyer and Blass [47]	Detailed economic data on cell-phone collection, reuse, and recycling	United States, United Kingdom
Sasikumar et al. [27]	Truck tire-retreading company	India
Fonseca et al. [25]	RL planning	Spain

13.1.5 RL Applications

Specific applications of RL can be found in, for example, Kroon and Vrijens [48] (Table 13.1).

13.2 Review of Various Aspects of RL

By analyzing the topic and considering three viewpoints, we would like to determine why materials are returned, which ones are returned, and how they are returned. We will discuss the driving forces behind the associations and make them active in RL; what being come back, then we are explaining the products characteristics that make a recovery appealing or obligatory; also, we will say, who involves in this process.

13.2.1 Driving Forces Behind RL

The forces that drive RL were categorized under three headings: economics, legislation, and corporate citizenship.

1. Economics may provide some straight benefits to corporations by their choice of input materials, reduced expenses, and value-added recovery. Indirect benefits to corporations

include anticipating and impeding legislation, protecting markets, fostering a green image, and improving supplier–customer relations.
2. Legislation: in a series of laws, a company is obliged to recover its goods or accept its returning.
3. Corporate citizenship constitutes a set of values that force a corporation to behave responsibly in RL.

13.2.2 Reasons for Return [1]

According to the usual supply-chain hierarchy, returns can be classified as coming from manufacturing, distribution, or customers.

Manufacturing Returns

Manufacturing returns are products that are recovered during production. These include the following:

- Surplus raw materials
- Quality-control returns
- Production leftovers and by-products

Distribution Returns

After goods are produced in a factory and moved into distribution, some products return to production, including the following:

- Product recalls because of safety or health problems.
- Business-to-business commercial returns from buyers because of contractual options that allow the return of products because of damaged deliveries or unsellable products.
- Store adjustments, including outdated products.
- Functional returns or materials used as carriers to move products in distribution such as pallets.

Customer Returns

Customers return products for the following reasons.

Reimbursement guarantees allow customers to change their mind regarding unmet product needs.

Warranty returns allow the return of products because of problems discovered during usage.

End-of-use returns include products such as bottles that cannot be used again but which are returnable.

EOL returns are products that are at the end of their economic or physical life.

13.2.3 Types and Characteristics of Returned Products [49]

The characteristics of returned products are very important. Three relevant aspects are composition, use pattern, and deterioration.

Composition

Elements of composition that are of concern to the manufacturer of a returned product include the following:

- *Ease of disassembly*: like removing some small piece of old electronic devices, which can be reused.
- *Homogeneity of constituent elements* is concerned with products that contain dangerous material that must be removed before they can be recycled (e.g., batteries).
- *Ease of transportation* is concerned with the specific transport of certain products. These products need separate distribution systems to avoid contamination from old products.

Use Pattern

Patterns of use can be divided into two issues: place of use and length and intensity of use.

- For *place of use*, if a product is applied in various usages and different places, its correction operation will be more difficult. As a result, the collection difficulty is directly related to the number of use place.
- If the *length and intensity of use* is short, then the item can be reused without the recovery.

Product Types

We can categorize product types as follows:

1. Food
2. Civil objects
3. Consumer products
4. Industrial facilities
5. Transportation facilities
6. Items for packaging and distribution
7. Oil and chemical products
8. Pharmaceuticals
9. Army equipment

All of these products take different place and have different processes for their recovery following their different aspect.

Deterioration

At the end of a product's use, how much functionality remains with the product in whole or in part strongly determines the recovery option. These aspects have several important roles.

- Through *inherent deterioration*, some products are consumed completely during their use (e.g., gasoline), or they lose their useful life quickly (e.g., batteries).
- *Fixability* is how easily a product can be improved or refurbished to a better condition (e.g., rechargeable batteries are easily restored).

- *Homogeneity of deterioration* is a measure of whether all parts of a product age equally or not. This option directly affects the method of recovery.
- *Economic deterioration* is a measure of a product's economic functionality—that is, as it expires will it become obsolete?

13.2.4 RL Processes

The main reverse logistic processes are (1) collection; (2) inspection, selection, and sorting; (3) reprocessing (including repair, refurbishing, remanufacturing, retrieval, recycling, and incineration) or direct recovery; and (4) redistribution. Further definitions follow (Figure 13.1):

In *collection*, used products are moved to a place for some specific treatment [49].

In *inspection and separation*, products are inspected and separated by both their reusability and how they can be reused. Inspection and separation include activities such as disassembly, shredding, sorting, testing, and storage.

Reuse determines whether products still have enough quality and are in good enough condition that they can be used again. Examples are reusable bottles, containers, and most rented facilities [50].

In *reprocessing*, a used product is converted into a usable product. This can happen at different levels: material (recycling), component (remanufacturing), product (repair), selective part (retrieval), module (refurbishing), and energy (incineration).

In *recycling*, product forms are changed into more basic forms such as scrap metal, glass, plastic, and paper [8].

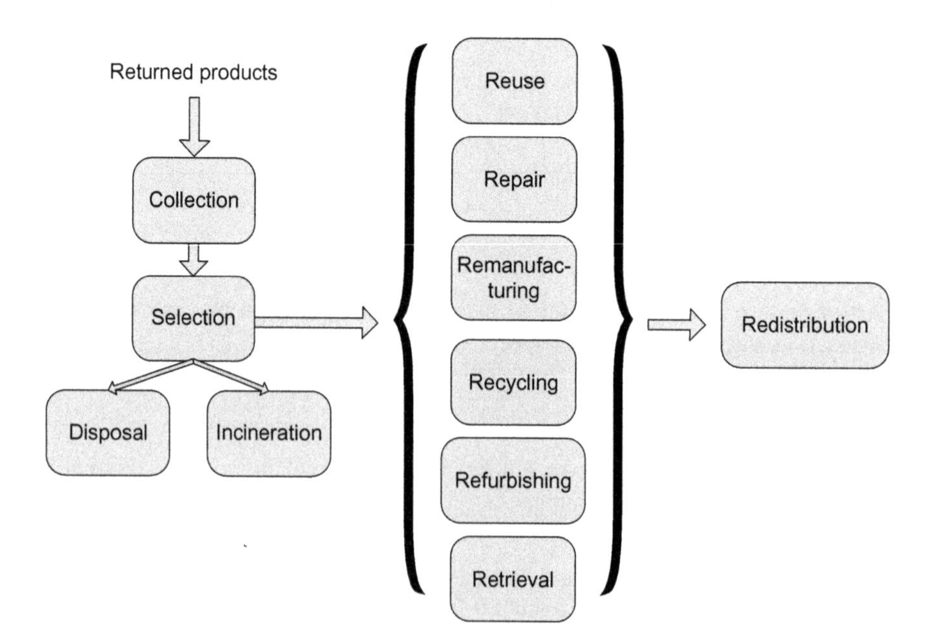

Figure 13.1 Product flow in reverse logistics.

In *remanufacturing*, a product in whole or in part is used to create a new and usable product. Some of these activities include cleaning, disassembly, replacement, and reassembly [50].

In *repairing*, broken products have some aspect of their life cycles restored, possibly with a loss of quality [8].

Refurbishing refers to upgrading a product.

In *incineration*, products are burned and the released energy is captured.

In *disposal*, useless products that cannot be reused because of technical or economical reasons are discarded.

In *recovery*, used material is captured, repaired, and remanufactured, a process that adds value [8].

In *redistribution*, products are distributed to different markets. This step consists of storage, sales, and transportation [50].

13.2.5 RL Actors [1]

Many participants have different roles in RL, such as the supply-chain players (e.g., manufacturers, suppliers, wholesalers, and retailers), specialized reverse-chain actors (e.g., jobbers and recycling specialists), and players who are opportunistic (e.g., charity organizations) (Figure 13.2). There are many actors with different objectives that may compete. For example, a manufacturer that wants to prevent jobbers from reselling its products at lower prices may do recycling itself.

Third-Party Providers and RL [34]

Many companies are outsourcing most or all of their logistics activities to 3PL service providers, including RL activities such as transportation, warehousing, and material disposal. These service providers execute RL activities better and at a

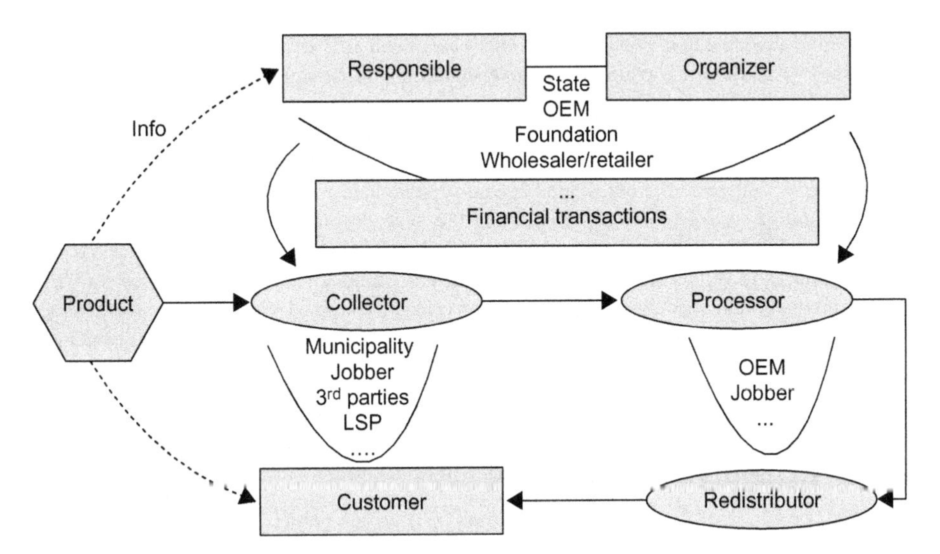

Figure 13.2 Actors in reverse logistics [34].

lower cost. These outsourced suppliers control reverse flows and perform key value-added services such as remanufacturing, repackaging, and refurbishing. Choosing the best specialist who offers the most benefits in the market for the lowest price is a critical time-consuming and complex decision that requires assessing multiple decision-making criteria. In this case, the new literature offers many methods for choosing suppliers (partners):

- Matrix or weight approaches
- Mathematical programming methods
- Probabilistic approaches
- Artificial intelligent techniques
- Integrated approaches

In RL management, a few studies have used such decision-making techniques as fuzzy methods to consider clear and unclear criteria, human judgments, priorities, and trade-offs between criteria and goals.

13.3 Information Technology for RL [3]

One of the biggest duties in the proposing of RL activities is controlling the natural uncertainty in the systems included in products recovery and reuse, where the used products compared to the raw materials, are less standardized and homogeneous. The new technology has an important role with this uncertainty, for example when some estimates for returning merchandise guess that online-driven products realize return rates in excess of 30%, the historical rate is about 5%. By suggesting the efficiencies of web technology and e-commerce, we have the following four directions:

1. *Proactive minimization of returns*: In this case, databases help decrease the number of returned products just because of misunderstanding without any fault.
2. *Minimization of returns' uncertainty*: In this case, a system registers the number of returning products and determines when, where, and why they are being returned.
3. *Returns and 3PL operators*: Increasingly, 3PL offers Web-enabled applications with real-time access to data across customers reverse-supply chains.
4. *Consolidating returns channels*: Exploiting the Web, in order to make a central stream, the original equipment manufacturers (OEMs) mix their channels. Today, the electronic marketplaces have the most popular model for e-commerce for RL that are used for both new products and second-hand ones. So for offering the used pieces or remanufactured equipment, there are some sites which used the Web. There is also a Web-based paradigm that includes collection, selection, reuse, and redistribution.

13.4 RL and Vehicle Routing [30]

At the strategic level, decisions about which activity, where the activity is done, and who does the activity must be made for the RL system. On the medium-term level, the relationship between forward and reverse channels and the redistribution system's operator have to be determined. On the operational level, depending on the forward and reverse relationships, different variants of the popular and extensively studied VRP occur with other planning problems. A separate basic VRP should be solved for each channel if independent forward and reverse channels are selected.

In many distribution or redistribution systems, operating the forward and reverse channels separately may result in use of inessential vehicle. By combining pickups at customer locations and deliveries within the same vehicle routes, this problem can be avoided. In many practical applications, the customers would like simultaneous pickup and delivery because of environmentally motivated distribution or redistribution systems. They may not accept pickup and delivery separately, to do the handling effort just once. This latter situation may be called a *VRP with simultaneous delivery and pickup* (VRPSDP). As law forces corporations to take responsibility for the age of their products, they become increasingly interested in gaining control over entire life cycles in order to enhance the quality of recovered products. Consequently, it is attractive to lease items and periodically exchange them. Also planning for such exchanges is a VRPSDP. For the effective collection of used products and the delivery of reusable products, effective vehicle routes have to be determined.

13.5 Quantitative Models for RL [8]

Quantitative models are divided into distribution planning, inventory control, and production planning. In this section, the basic models in these related areas are presented after an introduction to reverse distribution. For more information, refer to the related references.

13.5.1 Reverse Distribution

The packaging and collecting and following the transportation of used products are reverse distribution. Reverse distribution can occur in a separate reverse channel, through the original forward channel, or through a combination of the reverse and forward approaches. A key topic in reverse distribution systems is whether and how forward and reverse channels can be combined. Considerable system uncertainty and a many-to-few network structure are special aspects of reverse distribution. To set up an effective reverse distribution channel, the following must be considered.

- Who are the actors? An actor can be a member of the forward channel or a specialized party. This is a very important constraint on the potential integration of reverse and forward distribution.
- What is the relationship between the forward and the reverse distribution channel? Recycling is often defined the open-loop systems which in this case the product do not return to their main producer and are used in other industries. The combination

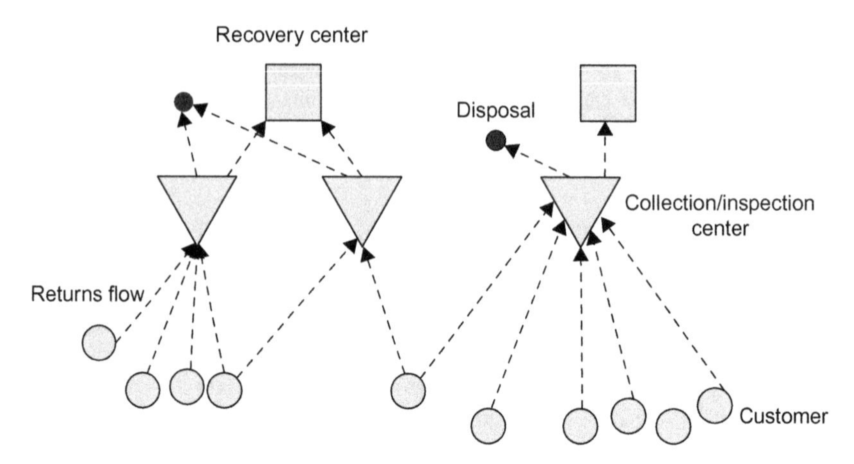

Figure 13.3 Recovery network design [26].

possibilities of forward and reverse distribution are scant in both channels as the different actors. Remanufacturing and reuse are often described as closed-loop systems into which packed products are returned to their original producers.

Reverse distribution can either take place directly through the main network or through specialized logistical suppliers. Because delivery and pickup can be handled differently, the combination of forward and reverse distributions at the routing level can be more difficult, even if the same actors are included.

Mixed-Integer Location Models for the Design of RL Networks [26]

The most widespread modeling approach to logistics network design problems in various contexts is concerned with facility location models based on MILP.

Multistage network is considered for this problem which includes customer, collection and inspection centers, factories for recovering or new production, and disposal centers. Two proclivity for the collected goods is considered: recovery (for certain parts of the collected goods) and disposal. Figure 13.3 shows the general structure of this network.

Sets

I = potential plant locations for collection/inspection
J = fixed recovery locations
K = fixed disposal locations
L = fixed customer locations

Variables

$$Y_i = \begin{cases} 1 & \textit{if a collection/inspection center is open at location } i \\ 0 & \textit{otherwise} \end{cases}$$

X_{li} = number of returns from customer l to collection or inspection center i
Z_{ij} = number of recordable items transferred from collection or inspection center i to recovery center j
W_{ik} = number of scrapped items transferred from collection or inspection center i to disposal center k

Costs

cf_{li} = transportation cost for a unit of returned items from customer center l to collection or inspection center i
cs_{ij} = transportation cost for a unit of recordable items from collection or inspection center i to recovery centers j
ct_{ik} = transportation cost for a unit of scrapped item from collection or inspection center i to disposal centers k
caf_i = capacity of the collection or inspection centers i
cas_j = capacity of the recovery centers j
cat_k = capacity of the disposal centers k

Parameters

d = average percentage of disposed items
r_l = number of returned items from customer l
r_i = fixed cost to set up collection or inspection centers i

The objective function and its related constraints are as follows:

$$\min \sum_{i \in I} f_i \ Y_i + \sum_{l \in L} \sum_{i \in I} \mathrm{cf}_{li} X_{li} + \sum_{i \in I} \sum_{j \in J} \mathrm{cs}_{ij} Z_{ij} + \sum_{i \in I} \sum_{j \in J} \mathrm{ct}_{ik} W_{ik} \tag{13.1}$$

subject to

$$\sum_{j \in J} X_{li} = r_l \quad \forall \ l \in L \tag{13.2}$$

$$\sum_{j \in J} Z_{ij} = (1-d) \sum_{l \in L} X_{li} \quad \forall \ i \in I \tag{13.3}$$

$$\sum_{k \in K} W_{ik} = d \sum_{l \in L} X_{li} \quad \forall \ i \in I \tag{13.4}$$

$$\sum_{l \in L} X_{li} \leq Y_i \mathrm{caf}_i \quad \forall \ i \in I \tag{13.5}$$

$$\sum_{l \in L} Z_{ij} \leq \mathrm{cas}_j \quad \forall \ j \in J \tag{13.6}$$

$$\sum_{l \in L} W_{ik} \leq \text{cat}_k \quad \forall \ k \in K \tag{13.7}$$

$$Y_i \in \{0,1\} \quad \forall \ i \in I \tag{13.8}$$

$$0 \leq X_{li}, Z_{ij}, W_{ik} \quad \forall \ i \in I, j \in J, k \in K, \ l \in L \tag{13.9}$$

In this formulation, inequality (13.2) shows logical constraints that ensure that total returns and customer demand will be considered. Inequalities (13.3) and (13.4) ensure there are balanced flows at a collection or inspection center. Inequality (13.5) ensures that the returned items are transferred to a collection or inspection centers provide that the centers are built up and is assured the capacity constraint. Inequalities (13.6) and (13.7) ensure the balance flow and capacity constraint. Finally, inequalities (13.8) and (13.9) are the usual facility opening conditions.

13.5.2 Inventory Control Systems with Return Flows

The inventory management objective is controlling the external ingredient orders and the internal ingredient recovery process, for supporting a needed service level and minimizing the variable and fixed cost. There are two folds following the returns flow's effects. One side is that overhauling an old product is cheaper than producing the new one. The other side increasing the uncertainty, which can lead to higher safety stock levels, cause to reliable planning becomes more difficult. The process of recovery, in reuse system might vanish with the returned products which enter to the usable inventory directly.

The other system's input parameters which require to be described externally for system are the predicting of future returns and an appropriate economic valuation of the returned items. Three essential aspects differs from those in traditional inventory control systems. First, as the return flow results the level of inventory between new item replenishments, is no longer necessarily reducing but can rise also. Second, for satisfying demands impose, external orders and recovery have to be coordinated. Third, by differentiate between products yet to be repaired and usable the situation explained at the above naturally causes to a two-echelon inventory system. Therefore, in this context, surveys on adequate strategies for echelon stock control, such as PULL against PUSH policies are relevant.

Inventory models can be classified as deterministic or stochastic (Figure 13.4).

In a deterministic model, quantity of returns and the time of the demands and returns are known.

In stochastic models, the demands and returns are not known and included (i) repair systems and (ii) product recovery systems that is classified into periodic and continuous review models.

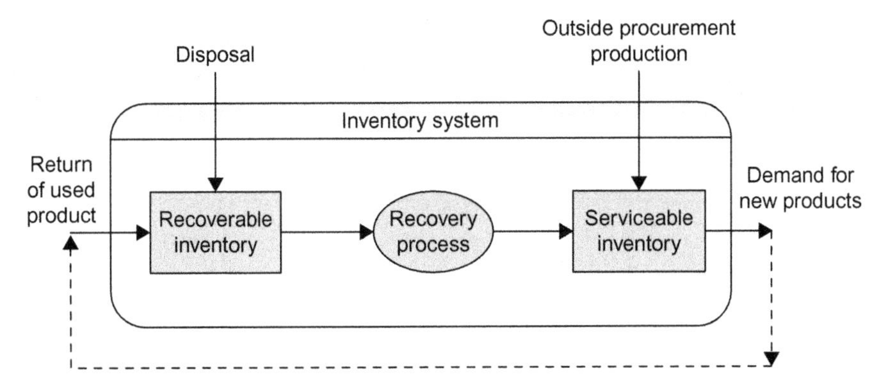

Figure 13.4 Framework of inventory management with returns [8].

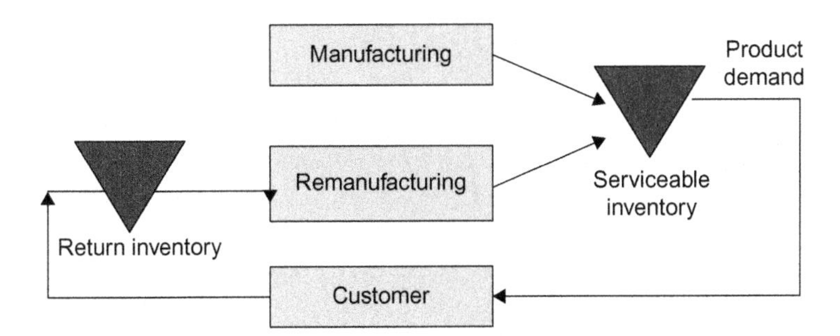

Figure 13.5 An inventory system with remanufacturing [51].

An Inventory System with Remanufacturing [51]

Using the remanufacturing operation, a usable stock could be improved into a newly manufactured stock. Figure 13.5 demonstrates an inventory system with remanufacturing. The model has the following assumptions:

- There is no disposal option for returned products.
- The holding cost for serviceables is larger than that for returns.
- Variable costs of manufacturing and remanufacturing are not included.

The objective is to minimize total setup and holding costs. Two alternatives are considered: (1) a joint setup for manufacturing and remanufacturing when the same production line is used for both processes, and (2) separate setups for manufacturing and remanufacturing.

Joint Setup Cost Model

This model is appropriate for a system that has manufacturing and remanufacturing activities on the same production line and that use the same resources.

This model is formulated as follows:
Variables

I_t^r = number of returned items
h^r = holding cost per unit time for a returned item
R_t = number of items returned in period t
x_t^r = amount remanufactured in time t

Minimize

$$\sum\nolimits_{t=1}^{T} \{K\delta_t + h^r I_t^r + h^s I_t^s\} \tag{13.10}$$

subject to

$$I_{t-1}^r + R_t - x_t^r = I_t^r \quad \forall\ t \in \{1, \ldots, T\} \tag{13.11}$$

$$I_{t-1}^s + x_t^m + x_t^r - D_t = I_t^s \quad \forall\ t \in \{1, \ldots, T\} \tag{13.12}$$

$$x_m^t + x_t^r \leq M_t \delta_t \quad \forall\ t \in \{1, \ldots, T\} \tag{13.13}$$

$$M_t = \sum\nolimits_{i=t}^{T} D_i \quad \forall\ t \in \{1, \ldots, T\} \tag{13.14}$$

$$x_t^m, x_t^r, I_t^s, I_t^r \geq 0, \quad \delta_{\mathrm{t}}, \in \{0, 1\} \quad \forall\ t \in \{1, \ldots, T\} \tag{13.15}$$

Equations (13.11) and (13.12) implicate the balance of return inventory and usable stocks. Equation (13.13) keeps track of the setups.

Separate Setup Cost Model

This model is appropriate for a system that has manufacturing and remanufacturing activities on separate production lines. The formulation of the model is as follows:

K^r = setup costs for remanufacturing
K^m = setup costs for manufacturing

Minimize

$$\sum\nolimits_{t=1}^{T} \{K^r \delta_t^r + K^m \delta_t^m + h^r I_t^r + h^s I_t^s\} \tag{13.16}$$

$$I_{t-1}^r + R_t - x_t^r = I_t^r \quad \forall\ t \in \{1, \ldots, T\} \tag{13.17}$$

$$I_{t-1}^r + x_t^m + x_t^r - D_t = I_t^s \quad \forall\ t \in \{1, \ldots, T\} \tag{13.18}$$

$$x_t^r \leq M_t \delta_t^r \quad \forall\ t \in \{1, \ldots, T\} \tag{13.19}$$

$$x_m^t < M_t \delta_t^m \quad \forall\ t \in \{1, \ldots, T\} \tag{13.20}$$

$$M_t = \sum_{i=t}^{T} D_i \quad \forall\ t \in \{1, \ldots, T\} \tag{13.21}$$

$$x_t^m, x_t^r, I_t^s, I_t^r \geq 0, \quad \delta_t^r, \delta_t^m \in \{0, 1\} \quad \forall\ t \in \{1, \ldots, T\} \tag{13.22}$$

13.5.3 Production Planning with Reuse

Returned items and parts are converted into raw materials by processes such as melting and grinding. These activities are not separate from other production processes. The products must be disassembled before recycling.

Selection of Recovery Options

For returned items, we need to know whether or not they are reusable. For example, for products with complex structures, the appropriate level of disassembly and processing must be selected for the component released with the technical and economical considerations taken into account.

Scheduling in a Product Recovery Environment

Two aspects must be considered in handling corresponding production activities: (1) MRP for product recovery and (2) shop floor control in remanufacturing.

13.6 Classification of Product Recovery Networks [9]

The main differences between recovery networks are the following:

- The degree of centralization
- Number of levels
- Connections with other existing networks
- The supply-chain structure (open vs. closed loop)
- The degree of branch cooperation

Centralization is a measure of the horizontal integration or width of a network by the number of locations that carry out similar activities.

Number of levels is a measure of the vertical integration or depth of a network by the number of facilities and how good visits sequentially.

Connections with other existing networks indicate how well a new network integrates with previously existing networks.

Open loop versus closed-loop network identifies the relationship between incoming and outgoing network flows.

Degree of branch cooperation refers to the parties who are responsible for setting up the network.

Recovery Network Characteristics

Recovery situations can be categorized by characteristics into products, supply chain, and resources.

The physical and economical features of discarded products are concerned to choose the recovery options.

Supply chain characteristics that are listed follow determine the recovery network characteristics: investigates the actor's behavior and the connections between them in the supply chain, driving force exist for recovery and reuse of products, obligations in the supply chain, re-user's category and behavior of disposer.

Resources include both human resources and those of recovery facilities.

Product Recovery Network Types

Product recovery networks can be assigned to three major categories: (1) bulk recycling networks, (2) assembled-product remanufacturing networks, and (3) reusable items network.

References

[1] M.P. De Brito, R. Dekker, Reverse Logistics—A Framework. Econometric Institute Report, Erasmus University Rotterdam, Rotterdam, 2002.

[2] D.S. Rogers, R.S. Tibben-Lembke, Going Backwards: Reverse Logistics Trends and Practices, Reverse Logistics Executive Council, Pittsburg, PA, 1999.

[3] F. McLeod, A. Hickford, S. Maynard, T. Cherrett, J. Alen, Developing innovative and more sustainable approaches to reverse logistics for the collection, recycling and disposal of waste products from urban, 2007. www.greenlogistics.org.

[4] S. Dowlatshahi, Developing a theory of reverse logistics, Interfaces 30(3) (2000) 143–155.

[5] S.E. Genchev, Reverse logistics program design: a company study, Business Horizons 52 (2009) 139–148.

[6] B.C.J. Zoeteman, H.R. Krikke, J. Venselaar, Handling WEEE waste flows: on the effectiveness of producer responsibility in a globalizing world, Int. J. Adv. Manuf. Technol. 47 (2010) 415–436.

[7] B. Janse, P. Schuur, M.P. de Brito, A reverse logistics diagnostic tool: the case of the consumer electronics industry, Int. J. Adv. Manuf. Technol. 47 (2010) 495–513.

[8] M. Fleischmann, J.M. Bloemhof-Ruwaard, R. Dekker, Quantitative models for reverse logistics: a review, Eur. J. Oper. Res. 103 (1997) 1–17.

[9] M. Fleischmann, H.R. Krikke, R. Dekker, P.F. Simme Douwe, A characterization of logistics networks for product recovery, Omega 28 (2000) 653–666.

[10] S. Minner, Strategic safety stocks in reverse logistics supply chains, Int. J. Prod. Econ. 71 (2001) 417–428.

[11] D.W. Krumwiede, C. Sheu, A model for reverse logistics entry by third-party providers, Omega 30 (2002) 325–333.

[12] T. Hu, J.B. Sheu, K.H. Huang, A reverse logistics cost minimization model for the treatment of hazardous wastes, Trans. Res. Part E, 38 (2002) 457–473.

[13] R. Kleber, S. Minnera, G. Kiesmuller, A continuous time inventory model for a product recovery system with multiple options, Int. J. Prod. Econ. 79 (2002) 121–141.

[14] P.A. Horvath, C.W. Autry, W.E. Wilcox, Liquidity implications of reverse logistics for retailers: a Markov chain approach, J. Retailing 81 (2004) 191–203.

[15] O. Listes, R. Dekker, A stochastic approach to a case study for product recovery network design, Eur. J. Oper. Res. 160 (2005) 268–287.

[16] V. Ravi, R. Shankar, M.K. Tiwari, Analyzing alternatives in reverse logistics for end-of-life computers: ANP and balanced scorecard approach, Comput. Ind. Eng. 48 (2005) 327–356.

[17] H. Min, H.J. Ko, C.S. Ko, A genetic algorithm approach to developing the multi-echelon reverse logistics network for product returns, Int. J. Manage. Sci. 34 (2006) 56–69.

[18] F. Schultmann, M. Zumkeller, O. Rentz, Modeling reverse logistic tasks within closed-loop supply chains: an example from the automotive industry, Eur. J. Oper. Res. 171 (2006) 1033–1050.

[19] H. Min, H.J. Ko, The dynamic design of a reverse logistics network from the perspective of third-party logistics service providers, Int. J. Prod. Econ. 113 (2007) 176–192.

[20] J.B. Sheu, A coordinated reverse logistics system for regional management of multi-source hazardous wastes, Comput. Oper. Res. 34 (2007) 1442–1462.

[21] R.K. Pati, P. Vrat, P. Kumar, A goal programming model for paper recycling system, Int. J. Manage. Sci. 36 (2008) 405–417.

[22] F. Du, G.W. Evans, A bi-objective reverse logistics network analysis for post-sale service, Comput. Oper. Res. 35 (2008) 2617–2634.

[23] J.B. Sheu, Green supply chain management, reverses logistics and nuclear power generation, Transport. Res. Part E, 44 (2008) 19–46.

[24] R.F. Saen, A new model for selecting third-party reverses logistics providers in the presence of multiple dual-role factors, Int. J. Adv. Manuf. Technol. 46 (2010) 405–410.

[25] M.C. Fonseca, Á. García-Sánchez, M. Ortega-Mier, F. Saldanha-da-Gama, A stochastic bi-objective location model for strategic reverse logistics, Top 18(1) (2010) 158–184.

[26] M.S. Pishvaee, K. Kianfar, B. Karimi, Reverse logistics network design using simulated annealing, Int. J. Adv Manuf. Technol. 47 (2010) 269–281.

[27] P. Sasikumar, G. Kannan, A.N. Haq, A multi-echelon reverse logistics network design for product recovery—a case of truck tire remanufacturing, Int. J. Adv. Manuf. Technol. 49(9–12) (2010) 1223–1234.

[28] M. Jahre, Household waste collection as a reverse channel: a theoretical perspective, Int. J. Phy. Dist. Logistic Manag. 25(2) (1995) 39–55.

[29] L. Pohlen, M. Farris, Reverse logistics in plastics recycling, Int. J. Phy. Dist. Logistic Manag. 22(7) (1992) 35–47.

[30] J. Dethloff, Vehicle routing and reverse logistics: the vehicle routing problem with simultaneous delivery and pick-up, OR Spektrum 23 (2001) 79–96.

[31] I. Dobos, Optimal production–inventory strategies for a HMMS-type reverse logistics system, Int. J. Prod. Econ. 81–82 (2003) 351–360.

[32] H.K. Chen, H.W. Chou, Y.C. Chiu, On the modeling and solution algorithm for the reverse logistics recycling flow equilibrium problem, Trans. Res. Part C, 15 (2007) 218–234.

[33] S. Kara, F. Rugrungruang, H. Kaebernick, Simulation modeling of reverse logistics networks, Int. J. Prod. Econ. 106 (2007) 61–69.

[34] T. Efendigil, S. Onut, E. Kongar, A holistic approach for selecting a third-party reverse logistics provider in the presence of vagueness, Comput. Ind. Eng. 54 (2008) 269–287.

[35] A. Mutha, S. Pokharel, Strategic network design for reverse logistics and remanufacturing using new and old product modules, Comput. Ind. Eng. 56 (2009) 334–346.

[36] C.K.M. Lee, T.M. Chan, Development of RFID-based reverse logistics system, Expert Syst. Appl. 36 (2009) 9299–9307.

[37] M. Thierry, M. Salomon, J. Van Nunen, L. Van Wassenhove, Strategic issues in product recovery management, California Management Review 37(2) (1995) 114–135.

[38] K. Kim, I. Song, J. Kim, B. Jeong, Supply planning model for remanufacturing system in reverse logistics environment, Comput. Ind. Eng. 51 (2006) 279–287.

[39] AI. Barros, R. Dekker, V. Scholten, A two-level network for recycling sand: a case study, Eur. J. Oper. Res. 110 (1998) 199–214.

[40] M. Fleischmann, J.M. Beullens, L.N. Bloemhof-Ruwaard, Van Wassenhove, The impact of product recovery on logistics network design, Prod. Oper. Manag. 10(2) (2001) 156–173.

[41] H.M.L. Blanc, H.A. Fleuren, H.R. Krikke, Redesign of a recycling system for LPG-tanks, OR Spectrum 26 (2004) 283–304.

[42] I.L. Blanc, M.V. Krieken, H. Krikke, H. Fleuren, Vehicle routing concepts in the closed-loop container network of ARN – a case study, OR Spectrum 28 (2006) 53–71.

[43] M.M. Amini, D. Retzlaff-Roberts, C.C. Bienstock, Designing a reverse logistics operation for short cycle time repair services, Int. J. Prod. Econ. 96 (2005) 367–380.

[44] M.A. Seitz, A critical assessment of motives for product recovery: the case of engine remanufacturing, J. Cleaner Prod. 15 (2007) 1147–1157.

[45] M. Asari, K. Fukui, S.I. Sakai, Life-cycle flow of mercury and recycling scenario of fluorescent lamps in Japan, Sci. Total Environ. 393 (2008) 1–10.

[46] G. Kannan, P. Sasikumar, K. Devika, A genetic algorithm approach for solving a closed loop supply chain model: a case of battery recycling, Appl. Math. Model. 34 (2010) 655–670.

[47] R. Geyer, V.D. Blass, The economics of cell phone reuse and recycling, Int. J. Adv. Manuf. Technol. 47 (2010) 515–525.

[48] L. Kroon, G. Vrijens, Returnable containers: An example of reverse logistics, Int. J. Phy. Dist. Logistic Manag. 25(2) (1995) 56–68.

[49] R. Dekker, M. Fleischmann, L. Wassenhove, Reverse Logistics, Springer-Verlag, Berlin, Heidelberg, New York, 2003.

[50] A.I. Kokkinaki, R. Dekker, J.V. Nunen, C. Pappis, An Exploratory Study on Electronic Commerce for Reverse Logistics Econometric Institute, Report EI-9950/A, 1999.

[51] A. Sbihi, R.W. Eglese, Combinatorial optimization and Green Logistics, 4OR (5) (2007) 99–116.

14 Retail Logistics

Hamid Afshari and Fatemeh Hajipouran Benam

Iran Khodro Industrial Group (IKCO), Iran

14.1 Overview

14.1.1 Introduction

The distribution process finishes with retailing, where each transaction sells a product or service that has personal, family, or domestic use whether in the form of food, clothes, or cars.

As a science, logistics also has affected retailing, and today's mix of retailing and logistics provides great benefits. Logistics is concerning with producing, executing, transporting, sorting, providing services, and managing inventories in ways that interact with each other. The big cycle of logistics starts with planning the physical movements of products from wholesaler to retailer to customer, and then implementing and controlling them. The plan must be effective in time and costs. There are three common and simultaneous advantages for companies if their logistics system is working and useful: (1) reducing stock outs, (2) decreasing inventories, and (3) improving customer services.

An optimum Retail logistics is the one which leads us to a 100% satisfaction in accessibility to on shelves goods for customer when they are needed [41]. Retail logistics is concerned with product availability. It means we must know what the customer wants, how to produce it, and where and when to deliver it [1].

Today is a high point in retail sales history. Many companies are now leaders in terms of sales, including WalMart, General Motors, and other manufacturing giants. Because of more opportunities, it is easy to start a new retail business or become a franchisee. However, some consumers are bored with shopping or do not have time for it, and many retailers sell goods at low profit and try to satisfy customers' expectations, so there are many challenges that retailers face. A retail decision maker must be able to answer such questions as, how can we serve customers while earning a fair profit? How can we remain competitive in an environment in which consumers have so many choices? How can we improve our business with loyal customers? These questions must be addressed in any well-structured retail strategy.

Logistics Operations and Management. DOI: 10.1016/B978-0-12-385202-1.00014-1

14.1.2 Retail Strategy

A retail strategy is a plan that guides the retail firm. It affects the firm's business activities and its reactions to competitors and markets. Every retailer must follow six steps in strategic planning.

1. Define the category of goods or services and the firm's specific mission (such as full service or no frills).
2. Define long- and short-term targets for sales and profit, market share, and so on.
3. Determine customer types and characteristics (such as gender and salary level) and needs (such as brand preferences).
4. Plan a long-term target to define direction for the company and its employees.
5. Develop a complete strategy that includes factors such as store locations, classified products, costs, and advertising in order to achieve targets.
6. Assess and execute the plan and solve ongoing problem.

As mentioned already, retailing is the final part of the distribution chain and includes businesses and people transporting products and services from producers to consumers. A typical distribution channel is shown in Figure 14.1. What consumers expect from a retailer is to have large variety of products from which to choose and buy a limited quantity, because a retailer collects goods in large quantities from various producers but sells in small amounts. Some producers choose a basic system of distribution and sell their products to a few retailers. Consequently, retailers play an important role between manufacturers, wholesalers, and customers. The main role of retailers is a sorting process.

Retailers play a critical role between customers and wholesalers, so they can provide considerable valuable information to wholesalers in anticipating sales, such as consumers' needs. Manufacturers can improve goods and services according to retailers' feedback. Small suppliers and their retailers can maintain close relationships that may help each other in transporting, storing, advertising, and prepaying for products, and the relationship also can affect costs and profits when retailers accomplish their goals. Retailers also keep close relations with customers via accessible locations, prompt responses to customer's demand and accurately and being able in credit purchases processing. Some retailers also offer customers special activities such as wrapping, delivery, and installation. In addition, most large retailers use multichannel retailing in selling goods and services to customer, and both face-to-face selling and selling by websites to make shopping easier for customers.

As mentioned, there are many ways for manufacturers to sell their products such as through retailers, mail-order catalogs, websites, and toll-free phone numbers. This means manufacturers can have more customers, reduce costs, increase sales,

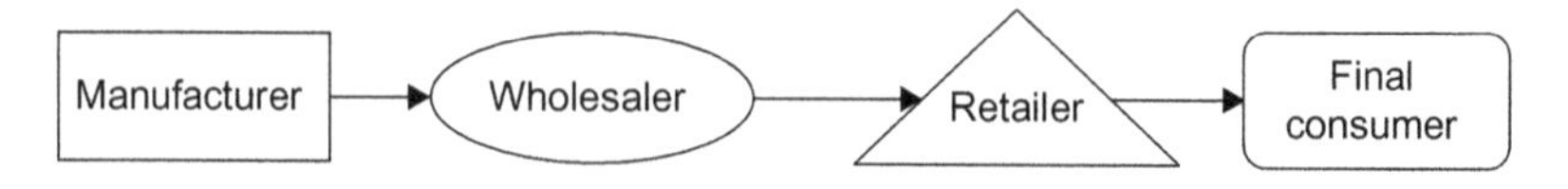

Figure 14.1 A typical channel of distribution [2].

and improve profits. For example, Sherwin-Williams and Polo Ralph Lauren both have full ranges of retailing functions as well as selling through traditional retailers [2].

14.1.3 Goods and Service Retailing

One of the most important retailing items to be considered is the difference between retailers. Two types of firms offer goods and services: store-based and nonstore-based. There are also differences between retailers that sell services and those that sell goods. Customers want to know the differences in the services that retailers offer [2].

Goods retailing includes selling tangible (physical) products. Service retailing involves intangible issues; customers do not buy or gain physical products. Some retailers just focus on goods, some just provide service retailing, and others work on both.

Service retailing includes many diversified business areas from one-to-one services (personal), accommodations (hotel and motel), automobile repair services, and so on. Furthermore, although some services are not exactly retailing activities, they should be assumed so when they are involved in customer sale package.

Service retailing is categorized into three main groups: renting goods and services, repair and improvement services, and nongoods services.

1. *Renting goods and services*: In this type of retailing, services are acquired for specific times or goods are leased for limited time periods. After the specified time is up, the goods must be returned or the service must end. Examples are car-rental services and video rentals.
2. *Repair and improvement services*: In this category, retailers do not own any goods. Customers give their goods or property to retailers for specific purposes such as maintenance, servicing, and so on. Home-appliance repair services and seasonal apartment repair services are examples.
3. *Nongoods services*: By using these services, consumers profit by intangible personal services. Sellers have to offer personal expertise for a specified time because they are not offering tangible goods. Travel agents, real-estate brokers, stockbrokers, and personal trainers are examples.

14.1.4 Factors That Affect International Retailing

Nowadays retailers do not limit themselves to national borders, and international retailers have focused on new markets. The following elements affect the level of productivity of a retailing strategy [2].

Timing: Being first in a market is less important than being in the market before there is serious competition.

A balanced international program: Selecting the suitable market is critical.

A growing middle class: According to current trends, middle class markets are growing which will lead to more sales and more income.

Matching concept to market: Improving quality and mixing fashion in a market will make a business more successful. If a market has developed, then retailers that offer discounts are more successful because consumers are more interested in price, variety, and convenience [2].

14.1.5 Information Flow in a Retail Distribution Channel

Technologies now are changing the roles of business players more than ever. Information technology is one of the largest influences. In an effective retail distribution channel, efficient information passes through three main players: providers (manufactures or wholesalers), retailers, and consumers. The outcome is close relationship among these parties with the purpose of forecasting the needs of each party.

A supplier needs the following kinds of information.

1. From the retailer, he/she needs sales prediction for each group, rates of inventory turnover, information on rivals, the amount of customer returns, and so on.
2. From the consumer, he/she needs to know attitudes about styles and models, how loyal customers are to brands, how willing customers are to pay more for better quality, and so on.

The information a retailer needs consists of the following:

1. From a wholesaler, he/she needs advance notice of new styles and model changes, instructions for complex products, sales forecasts, price changes, and so on.
2. From consumers, he/she needs to know why people shop at a particular retailer, what satisfies and dissatisfies them about retailers, where else they shop, and so on.

Also, consumers want the following:

1. From wholesalers, they want instructions on how to assemble and operate a product, how long warranty coverage will last, what they can expect in after-sale service and support, and so on.
2. From retailers, they want to know in which stores they can find specific products and how they can pay for them.

The main role in gathering data for wholesalers, suppliers, and customers belongs to retailers. They can assist other channel members by performing the following functions:

- Permit data gathering according to their principles. Many research firms like to conduct surveys at shopping centers because of the large and broad base of shoppers.
- Collect needed information for suppliers such as how shoppers react to displays.
- Pass along information on the attributes of consumers buying particular brands and models. Because credit transactions account for a major portion of sales, many retailers link purchases with consumer age, income, occupation, and other factors.

For the best information flows, collaboration and cooperation are necessary; especially between suppliers and retailers. This is not always easy. According to one senior retail executive, the traditional supply chain has one important problem: retailers and suppliers do not like to share information. This is the main reason for disorganized supply chains.

Table 14.1 The World's Top 10 Retailers [3]

Rank	Retailer	Base	Type	Regionalization
1	WalMart Stores, Inc.	United States	Discount store	Global
2	Carrefour Group	France	Hypermarkets	Global
3	The Kroger Company	United States	Supermarkets	Single country
4	Metro AG	Germany	Diversified	Global
5	The Home Depot, Inc.	United States	Hardwares	Global
6	Albertson's, Inc.	United States	Supermarkets	Single country
7	ITM Enterprises SA	France	Supermarkets	Regional
8	Sears, Roebuck and Co.	United States	Department stores	Regional
9	Kmart Corporation	United States	Discount stores	Single country
10	Target Corporation	United States	Discount stores	Single country

14.1.6 The World's Top Retailers

WalMart is number one in world retailing, its sales totaling more than the second, third, and fourth largest retailers in the world combined.

Several US retailers are the biggest retailers in the world, but Europeans engage in more international retail business and have more branches in many countries. Japan retailers have weak remaking among world retailers, according to a report by Ira Kalish (Table 14.1) [3].

As noted, WalMart is the world's largest retailer. It owes this position to its strategy of expanding the number of its US stores and improving the productivity of its old stores. It also joined with British supermarket chain in 2000.

Second place in the world's top 100 retailers is France's Carrefour. Previously, second place belonged to Germany's Metro, which is now in fourth place. Seven out of the ten largest retailers are based in the United States, and only one of those has a presence outside of North America. The others are focused on their home market. Except for US and Japanese retailers, a rule is suggested that retailers cannot gain a comprehensive share of single markets. This concept can also be followed in Europe, especially in France, that due to regulatory restrictions on large store development, the growth of retailers in their home markets is limited. That is why they have strong incentives to invest in new markets.

14.2 Typology

14.2.1 Introduction

The main structure of a business is its retail establishments. In the United States, there are 2.3 million retail firms (including those with no payroll in which only owners or family members work), and they operate 3 million establishments. An institutional discussion shows the proportional sizes and differences between some kinds of retailing. It also shows the influence of external environments on different retailers. Institutional analysis is important in strategic planning when selecting and setting missions, selecting possible ownership strategies, organizing goods and services, and, in particular, defining objectives.

We examine retail institutions from these perspectives: ownership sets of strategies related to stores and nonstore-based electronic and nontraditional retailing. An institution might wisely allocate in more than one group.

14.2.2 Ownership Institution

From the ownership point of view, stores are categorized as independently owned; members of chains; franchised, leased, managed, and owned by manufacturers; wholesalers; or customers.

Retailers at first were small, but they are very large now. (Most of the stores are run by small firms that have one outlet, and over one-half of all firms have two or fewer paid employees). Besides, there are also very large retailers. The top five US retailers have more than 2.5 million employees and more than $550 billion income. Ownership opportunities abound. For example, according to the US Census Bureau [4], women own about 1 million retail firms, African Americans (men and women) about 100,000 retail firms, and Asian Americans (men and women) about 200,000 retail firms.

Each ownership format serves a marketplace niche, if its strategy is executed well.

Independent retailers stress on specific customers and attempt to satisfy customers friendly verbal communications that are vital. It is suggested that independent retailers do not focus on great number of customers and do not debate about prices.

Chain retailers benefit during their widely known images, economies of Scale and mass promotions. They should widely maintain their image chain and should not be inflexible in adapting themselves to changes that take place in the marketplace.

Franchisors have strong geographic coverage because of franchisee investments and the motivation of franchisees as owner-operators. They should not get bogged down in policy disputes with franchisees or charge excessive royalty fees.

Leased departments enable store operators and outside parties to join forces and enhance the shopping experience while sharing expertise and expenses. They should not hurt the image of the store or place too much pressure on the leasees to bring in store traffic.

A firm can have more control over sources of supply by a vertically integrated channel, but it should not provide consumers with too little choice of products or too few outlets.

Cooperatives provide members with price savings. They should not expect too much involvement by members or add facilities that raise costs too much.

14.2.3 Store-Based Strategy Mix Institution

According to store-based strategy mix point of view, there are 14 store-based retail institutions, among which, some of them sell general merchandise products and others sell food-oriented goods. These groups are discussed in this section. Although not all are included, the strategy mixes do provide a good overview of store-based strategies. It must be clarified that more different product lines that retailers sell will be lead to more assortments in retailing strategies.

Food-oriented strategic retail formats are described and are as follows: convenience store, conventional supermarkets, food-based superstores, combinational stores, box (limited-line) stores, and warehouse stores.

A *convenience store* is properly located, usually is open for a long hour, and undertakes fair amount of services. The store facility is small—considerably smaller than conventional supermarkets—and presents moderate prices and customer services. The ease of shopping and the impersonal nature of many large supermarkets make these convenience stores particularly appealing to their customers, many of whom are men.

A *supermarket* is a self-service food shop that has sales of more than $2 million per year.

A *food-based superstore* exceeds a conventional supermarket in size and goods but is reversely smaller than a combination store. This format originated in the 1970s as supermarkets sought to stem sales declines by expanding store sizes and the numbers of nonfood items carried. Some supermarkets merged with drugstores or general merchandise stores but more of them grew into food-based superstores.

A combination store unites supermarket and general merchandise in one facility, with general merchandise accounting for 25–40% of sales. The format began in the late 1960s and early 1970s, as common checkout areas were set up for separately owned supermarkets and drugstores or supermarkets and general merchandise store. The natural offshoot was integrating operations under one management. If an economy supermarket merges with a discount department store, a supercenter will be formed. It is the US version of the even larger *hypermarket* (the European institution pioneered by firms such as Carrefour that did not succeed in United States).

The *box* (limited-line) store is a discounter concentrating on fewer issues, moderate working hours, fewer services, and fewer brands, and it is usually food based. Prices are on shelves or overhead signs. Items are displayed in cut cases.

A *warehouse store* is a food-based discounter offering a moderate number of food items in a no-frills setting. It appeals to one-stop food shoppers, concentrates on special purchases of popular brands, uses cut-case displays, offers little service, posts prices on shelves, and locates in secondary sites.

We now examine the following *general merchandise* strategy retail formats: specialty stores, traditional department stores, full-line discount stores, variety stores, off-price chains, factory outlets, membership clubs, and flea markets.

A *specialty store* provides only particular services or sells only particular goods. These stores include fewer goods but more types of each. This will help the store to advise customers on products better than its rivals.

Department stores are large retailing units that include many types and assortments of goods and services. Each type of good or service is located in separate department and helps customers select the appropriate goods and services.

A full-line discount store is a type of department store with these features:

- It conveys the image of a high-volume, low-cost outlet selling a broad product assortment for less than conventional prices.
- It is more apt to carry the range of product lines once expected at department stores, including electronics, furniture, and appliances, as well as auto accessories, gardening tools, and housewares.
- Shopping carts and centralized checkout service provided.
- Customer service is not usually provided within store departments but in a centralized area. Customers are free to select goods and services with minimal assistance.
- Nondurable (soft) goods feature private brands, whereas durable (hard) goods emphasize well-known manufacturer brands.
- Infrastructures and equipments are not expensive and decrease operational costs more than those in traditional department stores.

A *variety store* offers assortments of reasonably priced services and goods such as apparel and accessories, costume jewelry, notions and small wares, candy, toys, and other items in its general price range. It has fewer staff members in sales, and shelves are open for customers to select. These stores do not have many product lines, may not be largely departmentalized, and may even do not have home-delivery services.

An *off-price chain* features brand-name (sometimes designer) apparel and accessories, footwear (primarily women's and family), fabrics, cosmetics, and housewares, and sells them at everyday-low prices in an efficient, limited-service environment. It frequently has community dressing rooms, centralized checkout counters, no gift wrapping, and extra charges for alterations. The chains buy merchandise opportunistically as special deals occur.

A *factory outlet* is a store managed by a manufacturer. More factory stores now operate in clusters or in outlet malls to expand customer traffic, and they use cooperative ads.

A *membership (warehouse) club* tends to customers with price information who are shop members. It straddles the line between wholesaling and retailing. Some members are small business owners and employees who pay a membership fee to buy merchandise at wholesale prices. They make purchases for use in operating their firms or for personal use and yield 60% of club sales.

At a *flea market*, many retail vendors sell a range of products at discount prices in plain surroundings. It is rooted in the centuries-old tradition of street selling in which shoppers can touch and sample items and haggle over prices. Vendors used to sell only antiques, bric-a-brac, and assorted used merchandise. Today, they also frequently sell new goods, such as clothing, cosmetics, watches, consumer electronics, housewares, and gift items. Usually, flea markets are found in nontraditional areas such as racetracks, stadiums, and arenas.

14.2.4 Nonstore-Based Institution

We now examine retailing channels, whether single or multiple, and then nonstore-based retailing.

Initially, retailers undertake single-channel retailing, and one retail format may be selected that can be store-based or nonstore-based (catalog retailing, direct selling, etc.). Multichannel retailing may be used when corporations reach a substantial growth. This means that multiple formats of retailing will help retailer and customer via sharing costs and grouping suppliers. Retail leader WalMart sells through stores (including WalMart stores, Sam's Club, and Neighborhood Market) and a website [5].

Nontraditional retailing also comprises video kiosk and airport retailing, two key formats that do not fit neatly into store-based or non-store-based retailing. Sometimes they are store based, other times they are not. What they have in common is their departure from traditional retailing strategies.

14.2.5 Types of Locations

To explain the *types of locations* is a matter of importance in typology before deciding the desirable location. There are three types of location: isolated, unplanned business district, and planned shopping center. Each has its own attributes: makeup of competitors, parking, proximity to nonretail institutions (such as office buildings), and other factors.

An isolated store is located somewhere other than a road or street. The important point is that there are no competitive retailers beside and the stores should afford to supply all of the needs. The advantages of this type of retail location are many:

- There is no competition in close proximity.
- Rental costs are relatively low.
- There is flexibility; no group rules must be followed in operations, and larger space may be obtained.
- Isolation is good especially for stores that are located in midways or proper shopping areas.
- Better visibility from road traffic is possible.
- Facilities can be adapted to individual specifications.
- Easy parking can be arranged.
- Cost reductions are possible, leading to lower prices.

There are also various disadvantages to this retail location type:

- Initially, it may be difficult to attract customers.
- Many people will not travel very far to get to one store on a repeating basis.
- Most people like variety in shopping.
- Advertising expenses may be high.
- Costs such as outside lighting, security, grounds maintenance, and trash collection are not shared.
- Other retailers and community zoning laws may restrict access to desirable locations.
- A store must often be built rather than rented.

Now we describe two popular retail stores among customers which are names as Unplanned district and Planned shopping centers.

First of two is Unplanned district which is not planned for a long range, before establishment, commences with two or more stores and usually they are very near each other.

Stores locate based on what is best for them, not a district or neighborhood. Four shoe stores may exist in an area that lacks a pharmacy.

In contrast, planned shopping centers are structured and established according to unified architecture and central ownership and management. These shopping centers usually include parking areas. Some parameters such as location and store size are influenced by the trading areas where they are located.

This is a benefit of planned shopping center: there is enough variety of stores based on population, and the composition of stores and products meets customers' requirements for quality and diversity. In these business areas, managers allocate particular stores for each type of product, and each retailer is informed about the quantity and quality of product that can be sold. Therefore, in properly managed centers, well-considered strategies of coordination and cooperation ensure long lives for the stores and retailers located in these centers.

14.3 Techniques

14.3.1 Location and Site Evaluation

Location and site evaluation are essential techniques in retailing. Many analyses have been conducted of stores' general and specific positions. The "100% location" is a common expression among experts who recognize that a specified location for each store is optimal. Because different retailers need different kinds of locations, a fully optimized place for one store might be malfunctional for others. An upscale ladies' apparel shop would seek a location unlike that sought by a convenience store. The apparel shop would benefit from heavy pedestrian traffic and close proximity to a major department store and other specialty stores. The convenience store would rather be in an area with ample parking and heavy vehicular traffic. It does not need to be close to other stores. In this situation, retailers should use decision-making strategies including defining proper criteria and proper mechanisms of decision making. Two firms may rate the same site differently [2].

Another issue in deciding a location in supply-chain management is where to establish warehouses, distribution points, and even administrative offices to coordinate all activities regarding distribution-system management. To further explain, new researcher in facility location is presented. As described by Gebennini et al. [6], the generic facility location problem in logistic systems is defined by taking simultaneous decisions regarding design, management, and control of a generic distribution network [7–11]. Melo et al. [12] presented a review of the literature on facility location and supply-chain management. According to this review, the following are the most important decisions in facility location:

1. *Location of new supply facilities in a given set of demand points.* The demand points correspond to existing customer locations.
2. *Demand flows to be allocated to available or new suppliers*—i.e., production or distribution facilities.
3. *Configuration of a transportation network*—i.e., design of paths from suppliers to customers and the management of routes and vehicles in order to supply demand needs simultaneously.

In fact, in much of the research, the facility location problem is strongly associated with effective management of multistage production and distribution networks. Many papers in recent years have studied the facility location [8,13–17], and all of them have brought contributions to the main concept.

In traditional supply-chain management, the focus of the integration of Supply Chain Network (SCN) is usually on single objective such as minimum cost or maximum profit. For example, the total cost of the supply chain as an objective function was considered by more studies [18–25]. Real cases are usually accompanied with multiple objectives and designers should consider the issue. The design/planning/scheduling projects are usually involving trade-offs among different incompatible goals. Recently, multiobjective optimization of SCNs has been considered by different researchers in literature. Sabri and Beamon [26] developed an integrated multi-objective supply-chain model for strategic and operational supply-chain planning under uncertainties of product, delivery, and demand.

Although cost, fill rates, and flexibility were considered as objectives, the e-constraint method had been used as a solution methodology. Chan et al. [27] proposed a multi-objective genetic optimization procedure for the order-distribution problem in a demand-driven SCN. They considered minimization of the total cost of the system, total delivery days, and the equity of the capacity utilization ratio for manufacturers as objectives. Chen and Lee [7] developed a multiproduct, multistage, and multiperiod scheduling model for a multistage SCN with uncertain demands and product prices. As objectives, fair profit distribution among all participants, safe inventory levels, maximum customer service levels, and robustness of decision to uncertain demands had been considered, and a two-phased fuzzy decision-making method was proposed to solve the problem. Erol and Ferrell [28] proposed a model that assigns suppliers to warehouses and warehouses to customers. They used a multi-objective optimization modeling framework for minimizing cost and maximizing customer satisfaction. Guillen et al. [29] formulated the SCN design problem as a multi-objective stochastic mixed-integer linear-programming model, which was solved by the e-constraint method and branch-and-bound techniques. Objectives were Supply Chain (SC) profit over the time horizon and customer satisfaction level. Chan et al. [27] developed a hybrid approach based on genetic algorithm and the analytic hierarchy process (AHP) for production and distribution problems in multifactory supply-chain models. Operating cost, service level, and resource utilization had been considered as objectives in their study.

Some authors in recent years have developed multiperiod models. Freling et al. [30] presented a model for simultaneously optimizing inventory and designing a distribution network. They explored the single-sourcing version of the problem by using a branch-and-price optimal-solution procedure. Ambrosino and Scutellà [31] proposed linear models to solve the simultaneous warehouse location-inventory management-routing problem. Snyder et al. [9] formulated the stochastic version of the joint location-inventory management problem by introducing the likelihood of occurrence of each cost factor into the objective function. Thanh et al. [32] proposed a mixed-integer linear model for the design and planning of a production-distribution system. Gebennini et al. [6] presented a nonlinear model supporting strategic, tactical, and operational choices of decision makers in the field of facility location, inventory, and production management that are formulated in a multiperiod perspective. Afshari et al. [33] introduced a multi-objective model for optimizing the multicommodity distribution facility location problem. This model improved inventory decisions in distribution network design.

The last attempt to accost proposed models with real cases is how to make facility location decisions where multiple objectives are considered. In real cases, corporations include many goods and commodities, and solutions for single commodities will not be satisfactory. A model that illustrates the above-mentioned cases is presented. This model includes a multi-objective mixed-integer programming formulation for location within a network distribution problem. Objectives are to minimize total cost, including establishment and transportation costs, and to maximize customer satisfaction. The problem describes two location layers in multiple periods.

Components of a supply chain and their parts are described as follows:

Central warehouses: The main stocks of supply-chain demands are supplied here. There are L potential locations for central warehouses.

Regional warehouses: Stocks between central warehouses that customers demand are distributed here. There are M potential locations for regional warehouses and they are located in the capital of provinces.

Customers: There are N customers located in the cities of the provinces.

Goods: O types of commodities can be supplied for the customers demanding O families of cars.

Assumptions of the problem are as follows:

- There are limited capacities for both central and regional warehouses.
- Transportation cost per unit is a coefficient of the distances between central and regional warehouses and between regional warehouses and customers.
- There is a minimum level of customer satisfaction.

There are two objectives for a supply chain: (1) minimizing total cost, including establishment, transportation, and inventory management costs and (2) maximizing customer satisfaction.

Sets and Indices

L = sets of central warehouses ($|L| = l,\ k \in L$)
M = sets of regional warehouses ($|M| = m,\ j \in M$)
N = sets of customers ($|N| = n,\ i \in N$)
O = sets of good types ($|O| = o,\ t \in O$)
F = sets of periods ($|F| = f,\ p \in F$)

Variables

$$V_k = \begin{cases} 1 & \text{If the potential point of } k \text{ for central warehouses is located} \\ 0 & \text{Otherwise} \end{cases}$$

$$u_j = \begin{cases} 1 & \text{If the potential point of } j \text{ for regional warehouses is located} \\ 0 & \text{Otherwise} \end{cases}$$

x_{pijt} = percentage of demand customer i for commodity t that is supplied by regional warehouse j in period p
y_{pjkt} = percentage of demand regional warehouse j for commodity t that is supplied by central warehouse k in period p.

Parameters

a_{pit} = demand of customer i for commodity t in period p
b_{pjt} = capacity of regional warehouse j for commodity t in period p
c = cost of transportation per unit
d_{ij} = distance between regional warehouse j and customer i
d'_{jk} = distance between regional warehouse j and central warehouse k
e_{pkt} = capacity of central warehouse k for commodity t in period p
s_{it} = minimum level of customer satisfaction i for commodity t
q_k = cost of installation central warehouse k
w_j = cost of installation regional warehouse j
h_w = warehousing cost per unit goods in warehouses
h_s = warehousing cost per unit goods in stocks
π = back-ordered cost per unit goods
dws_{pjkt} = demand of regional warehouse j from commodity t to central warehouse k in period p.

Mathematical Model

$$\begin{aligned}
\text{Min } z_1 = & \sum_p \sum_j \sum_i \sum_t c \cdot d_{ij} a_{pit} x_{pijt} \\
& + \sum_p \sum_k \sum_j \sum_t c \cdot d'_{jk} b_{pjkt} + \sum_j w_j u_j + \sum_k q_k v_k \\
& + h_w \sum_p \sum_j \left(\sum_k \sum_t b_{pjt} y_{pjkt} - \sum_t \sum_i a_{pit} x_{pijt} \right) \\
& + \sum_p \pi \sum_t \sum_i a_{pit} \left(1 - \sum_j x_{pijt} \right) + h_s \sum_p \sum_t \sum_k \left(e_{pkt} - \sum_j y_{pjkt} \right)
\end{aligned}$$

$$\text{Min } Z_2 = \frac{1}{n \cdot o \cdot f} \sum_{p} \sum_{t=1}^{o} \sum_{i=1}^{n} \sum_{j=1}^{m} x_{pijt}$$

$$\sum_{t=1}^{o} \sum_{i=1}^{n} x_{pijt} \leq n \cdot o \cdot u_j \quad \forall j, p \tag{14.1}$$

$$\sum_{i=1}^{n} a_{pit} x_{pijt} \leq b_{pjt} \quad \forall j, p = 1 \tag{14.2}$$

$$\sum_{i} \sum_{t} a_{pit} x_{pijt}$$

$$- \sum_{t} \left(\sum_{k} b_{(p-1)jt} y_{(p-1)jkt} - \sum_{k} a_{(p-1)it} x_{(p-1)ijt} \right)$$

$$+ \sum_{t} \sum_{i} a_{(p-1)it} \left(1 - \sum_{j} x_{(p-1)ijt} \right) \leq \sum_{t} b_{pjt} \quad \forall j, p > 1 \tag{14.3}$$

$$\sum_{j} x_{pijt} \leq 1 \quad \forall i, p, t \tag{14.4}$$

$$\sum_{j} x_{pijt} \geq s_{it} \quad \forall i, p, t \tag{14.5}$$

$$\sum_{k} y_{pjkt} \leq 1 \quad \forall j, p, t \tag{14.6}$$

$$\sum_{t} \sum_{j} y_{pjkt} \geq m \cdot o \cdot v_k \quad \forall k, p \tag{14.7}$$

$$\sum_{j} \sum_{t} b_{pjt} y_{pjkt} \leq \sum_{t} e_{pkt} \quad \forall k, \ p = 1 \tag{14.8}$$

$$\sum_{j} \sum_{t} b_{pjt} y_{pjkt} - \left(\sum_{t} e_{(p-1)kt} - \sum_{j} b_{(p-1)jt} y_{(p-1)jkt} \right)$$

$$\leq \sum_{t} e_{pkt} \quad \forall k, \ p > 1 \tag{14.9}$$

$$\sum_{k} b_{pjt} y_{pjkt} \geq \sum_{i} a_{pit} x_{pijt} \quad \forall j, p = 1, t \tag{14.10}$$

$$b_{pjt}u_j \geq \sum_k \mathrm{dws}_{pjkt}$$

$$+\sum_j \sum_t \left(\sum_k b_{(p-1)jt}y_{(p-1)jkt} - \sum_i a_{(p-1)it}x_{(p-1)ijt}\right) \quad \forall j, p > 1, t \quad (14.11)$$

$$\mathrm{dws}_{pjkt} \leq b_{pjt}u_j$$

$$-\sum_j \sum_t \left(\sum_k b_{(p-1)jt}y_{(p-1)jkt} - \sum_k a_{(p-1)it}x_{(p-1)ijt}\right) \quad \forall j, p > 1, t \quad (14.12)$$

$$\mathrm{dws}_{pjkt} \leq \sum_i \sum_t a_{pit}x_{pijt} + a_{(p-1)it}(1 - x_{(p-1)ijt})$$

$$-\sum_j \sum_t \left(\sum_k b_{(p-1)jt}y_{(p-1)jkt} - \sum_i a_{(p-1)it}x_{(p-1)ijt}\right) \quad \forall j, p > 1, t, k \quad (14.13)$$

$$e_{pkt}v_k \geq \sum_j \mathrm{dws}_{pjkt} \quad \forall p > 1, t, k \quad (14.14)$$

$$u_j \in \{0, 1\} \quad \forall j \quad (14.15)$$

$$v_k \in \{0, 1\} \quad \forall k \quad (14.16)$$

The first objective function, Z_1, is multiplied by the weighting coefficient of P and is a summation of the following:

- Transportation cost between central and regional warehouses, $\sum_{t=1}^{o}\sum_{j=1}^{m}\sum_{i=1}^{n} c \cdot d_{ij}a_{pit}x_{pijt}$
- Transportation cost between regional warehouses and customer, $\sum_{t=1}^{o}\sum_{k=1}^{l}\sum_{j=1}^{m} cd'_{jk}a_{pit}x_{pijt}$
- Installation cost for central warehouses, $\sum_{j=1}^{m} w_j u_j$
- Installation cost for regional warehouses, $\sum_{k=1}^{l} q_k v_k$
- Warehousing costs for commodities in all periods in regional and central warehouses.

The second objective, Z_2 is the summation of the level of the customer satisfaction that is multiplied by $(1 - P)$.

Constraints (14.1), (14.4), (14.6), and (14.7) state that if regional warehouse j or central warehouse k satisfies the demand in period p, then it has been installed. Constraints (14.2), (14.3), (14.8), and (14.9) show capacity restriction for each regional warehouse for each commodity in all periods. Constraint (14.5) implies that there is a minimum level of customer satisfaction i for commodity t. Constraints (14.10–14.14) consider that amount of supply should be greater than amount of demand. Constraints (14.15) and (14.16) are related to integer programming.

14.3.2 Human Resource Management

According to Ed Sweeney from the United Kingdom's Advisory, Conciliation, and Arbitration Service (ACAS) [34], workers in the retail, hospitality, and construction sectors are more likely to file employment claims because these industries often do not have good human resources practices, according to an interview in *Personnel Today* magazine.

ACAS chair Sweeney noted that many of the employers in these sectors had little experience and knowledge in human resource issues. In addition, the lack of effective unions and high staff turnover in these industries led to more employment disputes than in any other.

As shown, human resource management (HRM) is a critical activity in retailing. Also some researches show that frontline service employees (FLSES) actions will lead to significant results in successful service operation. In contrast, much customer dissatisfaction is reported by the firms that did not implement suggested idea.

HRM includes five major and related steps: recruitment, selecting, training, compensating, and supervising. The outcomes of this process are qualified employees who had the chance to enter the firm, then to have the opportunity to develop and progress the people who work in organization.

In the case of labor diversity, two rules should be notified: (1) the labor should be hired and empowered fairly, without regard to gender, ethnic background, and other related factors and (2) in a diverse society, the workplace should be representative of such diversity.

Retailers must be careful not to violate employees' privacy rights. Only necessary data about workers should be gathered. Also these data must be secured, and employers should be sensitive about disclosing such data about employees.

We will now explain each activity in HRM that is used to fill sales and middle-management positions.

In *recruiting retail personnel*, a retailer generates a list of job applicants. The sources for recruitment might be inside or outside a company. In addition to these sources, the Web is playing a bigger role in recruitment.

Selecting retail personnel is the next step. This is done by matching the traits of potential employees with specific job recruitments. Job analysis and description, the application blank, interviewing testing (optional), references, and a physical examination (optional) are tools in the process. They should be integrated.

The third step is *training retail personnel*. Every new employee should receive pre-training, organization's structure and policies, working hours, and his/her duties.

New employees should also be introduced to co-workers. After proper recruitment and suitable selection, training starts. Training programs are useful for both new and existing personnel and help them to perform better. Training can range from one-day session that covers topics such as operating a computerized cash register, selling techniques, or registering a complaint regarding affirmative action programs to long-term education that covers many operational issues in a retailer's main business.

Compensating retail personnel, whether through direct payment or indirect and nonmonetary means, should be satisfying enough for employees to ensure long-time cooperation. To better motivate employees, some firms also have profit-sharing programs.

As an element of HRM in retailing, supervision is chiefly a role of retailer to prepare environmental activities that lead to better performance for both employee and retailer. The retailer, by communicating personally or through meetings, tries to maintain the culture of the organization which finally benefits those who work in the firm. Retailers can succeed by highlighting morale, by explaining the relationship among employees, and by motivating and innovating, which finally benefits those who work in the firm.

14.3.3 Pricing in Retailing

Pricing is the result of a trade-off between retailer and customer. A customer seeks satisfaction from using services and goods, while a retailer looks for profitability. Some believe that the pricing strategy should be compatible with the main business parameters such as the position of the retailer among competitors, total profit, and rate of return on investments [2].

Each retailer may benefit by applying one of three pricing options. First, low prices can offer a competitive advantage, although it primarily obliges a retailer to be content with low profit per unit margin. However, it simultaneously leads to lower operational costs, more inventory turnover, and more sales. Many stores such as off-price shops and full-line discount stores use this strategy. Second, a retailer can move in the middle, which means that goods and services are sold at moderate prices. The problem with this strategy is that it faces competition from both low-cost retailers and prestige stores. The quality of products is from average to above average. Conventional department stores and some of drugstores apply this strategy in pricing. Third, a prestigious image can be adopted. This is the aim of some retailers. Smaller target markets, higher prices, less turnover, and higher profit margins per unit are characteristics of stores that use this strategy. Customers who use these stores are usually more satisfied and more loyal. Specialty stores and upscale department stores are in this category.

On the whole, for all types of pricing strategy, the success factor is to create a valuable position in customer's mind if he or she selects a specific type of shop with specific pricing strategy. That is why Sports Authority has shifted from its longtime three-tier pricing strategy—good, better, and best—to one that emphasizes only better and best products in order to improve gross margins and achieve greater customer loyalty: "We don't want to be in the $199 treadmill business because we will never win that war."

Whether buying an inexpensive $4 ream of paper or a $40 ream of embossed, personalized stationery, every customer wants to feel the purchase represents a good value. The consumer is not necessarily looking only for the best price. He or she is often interested in the best value, which may be reflected in a superior shopping experience.

Today, customers can easily find and compare prices on the Internet, and this fact should be considered by retailers. In the past, when consumers could only compare prices by visiting individual stores, the process was time consuming, which dissuaded most people from shopping around. Now, with a few clicks of a computer mouse, a shopper can quickly gain online price information from several retailers in just minutes without leaving home.

14.3.4 Customer Satisfaction in Retailing

Customer satisfaction is a key concept in all businesses, including retailing. Many researchers have tried to highlight key factors and the benefits of customer satisfaction. Surely, loyalty is the best result of customer satisfaction, one that ensures business results.

The research of Luo and Humborg [35] showed two achievements in customer satisfaction with no previously discussed background. This research showed that besides increasing future advertisement and promotion efficiency of the firm, customer satisfaction can increase human resource performance accordingly.

A conceptual model was presented for understanding shopper satisfaction with entertainment consumption. This research was investigated on satisfaction structure of those who were looking for shopping-mall entertainments [36]. Although entertainment consumption is a usual activity in shopping centers, only some researchers have worked on this issue. In their research, they examined entertainment consumption in shopping center context. They presented a model comparing five key constructs. Therefore, by introducing and discussing about the relationship among these five constructs, the researchers presented new models and concepts about retailing and satisfaction. As new concepts are developed in retailing, it is important for retailers to adopt them.

14.3.5 World Retail Congress [37]

To bring leading international players in retailing such as TESCO, Carrefour, and IKEA together, the World Retail Congress was established to conduct essential debates with international policy makers. Usually, discussions lead to better growth policies and decisions for the global retail economy and sustains the earning and competitive issues of both national and international retailing markets.

Some of the speakers and lecturers in this congress have presented new achievements and experiences. For example, Lan McGarrigle, Director of the World Retail Award in 2008, highlighted the importance of excellence across the entire spectrum of retail activity and emphasized that it is obligation in today's competitive market. He also referred to the role of the congress, which encourages worldwide retailers to accept and implement higher standards in innovation, responsibility, management, and operation.

14.4 Future Trends

The Economist Intelligence Unit has reported on global economic, industrial, and corporate trends in the world with what it calls "2020 foresight" [38]. This section denotes consumer goods and retailing industry in 2020.

Studying future trends is an essential activity to ensure that strategies are sustainable. It is also useful to forecast possible failures and problems and take actions to avoid them.

Looking at the future, two significant growth areas have been diagnosed. First, opportunity is increasing sales in current markets; second, this opportunity includes entering emerging markets. Each opportunity has its own challenges and differences. Many researchers predict that the greatest growth will happen in countries that are not members of the Organization for Economic Cooperation and Development. At currently trending market exchange rates, the United States is expected to be the biggest consumer market in 2020, and the country's share of world consumer spending will be approximately constant. The European Union's share will decrease step by step. In 2020, the leading emerging markets, especially China and India, will see the largest increase in the share of world consumer spending. It even has been predicted that China will close the gap with the United States in many key indexes like those mentioned in this chapter.

Another challenge is maturity increasing sales and seems to be harder in developed markets. Matured markets and high debt level is a problem in most Western countries. The competition in low-price markets will increase considerably. Besides all the earlier mentioned issues, new growth is predicted. Changing demographic patterns will create new markets. Another source of growth through either immigration or high birthrates in subgroups of the population will establish new markets that encourage competition and innovation.

Even though many retailers think of population changes as an opportunity, some believe them to be a threat, mainly because aging changes popular trends and habits such as fast-food-chain stores and mortgage broking. So it will be hard for those who establish their business targeting younger aged groups. In developing markets, more investments according to current trend will diminish the returns in future. In all conditions, competition is available. In spite of strict competitiveness among industries, cost efficiency is still vital. Multichannel marketing may be applied to ensure return on investments, but still there are some barriers that limit the benefits. Some firms use resources from low-cost countries, but this is not a sustainable solution. Cost-control strategies may be a useful solution for some years but cannot be relied upon in 2020. Some experts suggest that firms may use multichannel marketing via the Internet and also may use other techniques such as customer-relationship management technologies. Threats are predictable even for these new techniques and technologies (e.g., market saturation, decreasing margins and conflict between channels), but firms should find their own way for long-term solution and maintain their competitive advantages.

14.5 Case Study

14.5.1 History of Russian Retail Chains [39]

The strategies and policies of the former Soviet Union prevented the development of a national retail system. In 1990, the stocks in state store were greatly diminished, and a retail system was found only in small markets (20–50 square meters) operating in open-air areas and general areas such as stadiums and railway stations. Still, some of these markets attracted people from as far away as 200 kilometers. In Russia, big cities such as Moscow and Saint Petersburg were developed into retail centers, whereas some markets in smaller cities were still developing. The claim that emphasizes in 1992 Moscow constituted 12% of whole Russian consumer goods turnover which shifted to 35% in 2000 has statistical proof. After economic infrastructures were improved in smaller cities, Moscow's share was decreased and this share was divided among those cities. During this period of time, the purchasing process gradually returned from open markets to store and simultaneously store chains were established. New stores were open 24 hours a day. Each brand was clearly separated, and distinctive services were offered. After these developments, the growth of chain stores was rapid, and the turnover of leading Russian retailers almost doubled each year. Retailing became the second-leading sector for investment after the excavation of natural resources, a trend that continued to 2000. On the eve of the twenty-first century, the Russian retail market saw the entry of new retail chains from Western European countries. METRO cash and carry in 2001, IKEA from 2002, and Auchan in 2002 were examples. Now the purchasing process shifted again, but this time from supermarkets to hypermarkets. It means that the format and size of markets were changed. The first markets in the 1990s were just from 70 to 400 square meters; in the middle of the 1990s, the average stores were between 1000 and 2500 square meters. The first hypermarkets were 10,000 square meters.

When Russian retail chains were faced with strong competition from Western retail chains, they established their own Western-style hypermarkets. Supermarket chain Seventh Continent, discounter chain Piaterochka, and supermarket chain Perekriostok were pioneers. Step by step, Russian retail chains progressed, and in January 2008, Russian retailers (along with Chinese retailers) figured in the list of the world's 250 largest retailers for the first time as reported by Deloitte Touche Tohmatsu in conjunction with *Stores* magazine.

14.5.2 Conventional Food Retailing with a Spotlight on Differentiation [40]

Another case study is about chasing the third larger conventional grocer in the United States, Safeway Inc., which appointed Orangetwice. This is an example of shifting from old and conventional way to modern food retailing.

Shortly, the founders believed that if a change is necessary, then do it and make it fast. Orangetwice believes in differentiation. It means that a buying experience must differ from day to day and that a customer should feel it. Therefore, adding value to customers is a new concept that is emphasized by this firm. According to changed situation, customers are not loyal to brands and seek added value without respect to brands. One choice is price but time is another parameter. It means that customers prefer to enjoy the time they spend in shops. By this view, shopping is a leisure activity. Retailers also undertake to decorate and design the environment to seem fascinating. They even offer customers meal solutions instead of just selling goods and foods.

Orangetwice understood this fact and they transferred from white box design to fresh and organic design with natured environment.

Consultant is a new role for grocery retailers. They suggest customers what to buy and how to use the products. Presentation skills thus are critical, and staff members of these stores are friendly and knowledgeable experts in their fields. Safeway Inc. started new strategy and initialized new relationship with customers. These actions encouraged customer to consider Safeway Inc. instead of grocery retailer for a lunch or dinner option. The strategic thought led revolution and extraordinary gaining. After implementation, sales were increased up to 50%.

References

[1] J. Fernie, L. Sparks, Logistics and Retail Management, Kogan, London, 2004.

[2] B. Berman, J.R. Evans, Retail Management: A Strategic Approach, Prentice Hall, Upper Saddle River, New Jersey, USA, 2006.

[3] Available from: http://www.siamfuture.com/retailbuscenter/retailindus/top100.asp.

[4] Available from: www.census.gov.

[5] Available from: www.walmart.com.

[6] E. Gebennini, R. Gamberini, R. Manzini, An integrated production–distribution model for the dynamic location and allocation problem with safety stock optimization, Int. J. Prod. Econ. 122 (2009) 286–304.

[7] C. Chen, W. Lee, Multi-objective optimization of multi-echelon supply chain networks with uncertain product demands and prices, Comput. Chem. Eng. 28 (2004) 1131–1144.

[8] R. Manzini, M. Gamberi, A. Regattieri, Applying mixed integer programming to the design of a distribution logistic network, Int. J. Ind. Eng. 13 (2006) 207–218.

[9] L.V. Snyder, M. Daskin, C.P. Teo, The stochastic location model with risk pooling, Eur. J. Oper. Res. 179 (2007) 1221–1238.

[10] N.A.H. Agatz, M. Fleischmann, A.E.E. van NunenJo, E-fulfillment and multi-channel distribution—a review, Eur. J. Oper. Res. 187 (2008) 339–356.

[11] C.S. ReVelle, H.A. Eiselt, M.S. Daskin, A bibliography for some fundamental problem categories in discrete location science, Eur. J. Oper. Res. 184 (2008) 817–848.

[12] M.T. Melo, S. Nickel, F. Saldanha-da-Gama, Facility location and supply chain management—a review, Eur. J. Oper. Res. 196 (2009) 401–412.

[13] D.R. Sule, Logistics of Facility Location and Allocation, Marcel Dekker Inc., New York, 2001.

[14] L. Zhang, G. Rushton, Optimizing the size and locations of facilities in competitive multi-site service systems, Comput. Oper. Res. 35 (2008) 327–338.

[15] S. Wadhwa, A. Saxena, F.T.S. Chan, Framework for flexibility in dynamic supply chain management, Int. J. Prod. Res. 46 (2008) 1373–1404.

[16] Y. Hinojosa, J. Kalcsics, S. Nickel, J. Puerto, S. Velten, Dynamic supply chain design with inventory, Comput. Oper. Res. 35 (2008) 373–391.

[17] P. Baker, The design and operation of distribution centers within agile supply chains, Int. J. Prod. Econ. 111 (2008) 27–41.

[18] V. Jayaraman, H. Pirkul, Planning and coordination of production and distribution facilities for multiple commodities, Eur. J. Oper. Res. 133 (2001) 394–408.

[19] S.S. Syam, A model and methodologies for the location problem with logistical components, Comput. Oper. Res. 29 (2002) 1173–1193.

[20] A. Syarif, Y. Yun, M. Gen, Study on multi-stage logistics chain network: a spanning tree-based genetic algorithm approach, Comput. Ind. Eng. 43 (2002) 299–314.

[21] V. Jayaraman, A. Ross, A simulated annealing methodology to distribution network design and management, Eur. J. Oper. Res. 144 (2003) 629–645.

[22] H. Yan, Z. Yu, T.C.E. Cheng, A strategic model for supply chain design with logical constraints: formulation and solution, Comput. Oper. Res. 30(14) (2003) 2135–2155.

[23] M. Gen, A. Syarif, Hybrid genetic algorithm for multi-time period production/distribution planning, Comput. Ind. Eng. 48(4) (2005) 799–809.

[24] T.H. Truong, F. Azadivar, Optimal design methodologies for configuration of supply chains, Int. J. Prod. Res. 43(11) (2005) 2217–2236.

[25] A. Amiri, Designing a distribution network in a supply chain system: formulation and efficient solution procedure, Eur. J. Oper. Res. 171(2) (2006) 567–576.

[26] E.H. Sabri, B.M. Beamon, A multi-objective approach to simultaneous strategic and operational planning in supply chain design, Omega 28 (2000) 581–598.

[27] F.T.S. Chan, S.H. Chung, S. Wadhwa, A hybrid genetic algorithm for production and distribution, Omega 33 (2004) 345–355.

[28] I. Erol, W.G. Ferrell Jr., A methodology to support decision making across the supply chain of an industrial distributor, Int. J. Prod. Econ. 89 (2004) 119–129.

[29] G. Guillen, F.D. Mele, M.J. Bagajewicz, A. Espuna, L. Puigjaner, Multi objective supply chain design under uncertainty, Chem. Eng. Sci. 60 (2005) 1535–1553.

[30] R. Freling, H.E. Romeijn, D.R. Morales, A.P.M. Wagelmans, A branch-and-price algorithm for multi period single-sourcing problem, Operat. Res. 51 (2003) 922–939.

[31] D. Ambrosino, M.G. Scutellà, Distribution network design: new problems and related models, Eur. J. Oper. Res. 165 (2005) 610–624.

[32] P.N. Thanh, N. Bostel, O. Peton, A dynamic model for facility location in the design of complex supply chains, Int. J. Prod. Econ. 113 (2008) 678–693.

[33] H. Afshari, M. Amin Nayeri, A. Ardestanijaafari, Optimizing Inventory Decisions in Facility Location within Distribution Network Design, Proceedings of the International Multi Conference of -Engineers and Computer Scientists, Vol. III, IMECS 2010, March 17–19, 2010, Hong Kong.

[34] Available from: *www.clickajob.co.uk/news/professional-men-planning-office-escape-7789.html*

[35] X. Luo, C. Humborg, Neglected outcomes of customer satisfaction, J. Market. 71 (2007) 133–149.

[36] J. Sit, B. Merrilees, 2005. Understanding satisfaction formation of shopping mall entertainment seekers: a conceptual model, ANZMAC, 2005 Conference.

[37] Available from: www.worldretailcongress.com.

[38] Available from: www.eiu.com.

[39] V. Kolossov, O. Vendina, J. O'Loughlin, Moscow as an emergent world city: international links, business developments, and the entrepreneurial city, Eurasian Geography and Economics, 43(3) (2002) 170–196.
[40] Available from: http://www.tatatinplate.com/differ_1.shtm.
[41] A. Mckinnon, The Fundamental Drivers of Transport Demand, Heriot-Watt University, the Netherlands, 1997.

15 Humanitarian Logistics Planning in Disaster Relief Operations

Ehsan Nikbakhsh[1] and Reza Zanjirani Farahani[2]

[1]Department of Industrial Engineering, Faculty of Engineering, Tarbiat Modares University, Tehran, Iran
[2]Department of Informatics and Operations Management, Kingston Business School, Kingston University, Kingston Hill, Kingston Upon Thames, Surrey KT2 7LB

15.1 Introduction

Although human technological advancements have cured many diseases and solved many problems, it is still widely believed that the capacities of many societies are not enough to cope with the massively destructive effects of natural and human-made disasters. Hence, disasters have always been identified with their huge negative impacts on the humans' condition, nations' critical infrastructures, and societies' planning systems. To be able to cope with effects of a disaster, humans try to prepare themselves by creating and enhancing necessary infrastructures and planning for various relief operations in advance.

As much as creating and enhancing infrastructures can mitigate the effects of disasters, humans are still required to devise better proactive plans and improve the implementation of relief operations. One main aspect of such planning and implementation is the logistics of relief operations. *Humanitarian logistics* can be simply defined as a branch of logistics dealing with logistical aspects of a disaster management system, including various activities such as procuring, storing, and transporting food, water, medicine, and other supplies as well as human resources, necessary machinery and equipments, and the injured before and after disasters have struck.

As a result of recent natural disasters such as the Asian tsunami in 2004 and hurricane Katrina in 2005, the field of logistics in the context of humanitarian operations has gained considerable attention from academics as well as practitioners [1]. There are several reasons for this attraction, including the need for agile and capable logistics systems that can deal with different kinds of disasters [2], specialized large-scale risk and disruption management, and the effects of disasters on human lives and economies as well [3]. It also has been estimated that the number

Logistics Operations and Management. DOI: 10.1016/B978-0-12-385202-1.00015-3

of both natural and human-made disasters will increase fivefold over the next 50 years [4]. Therefore, humanitarian logistics is one of the most important aspects of disaster management systems.

The main goal of this chapter is to introduce humanitarian logistics fundamentals, including various topics such as disasters, disaster management systems, humanitarian logistics, mathematical modeling and solving logistical decisions, coordination, and performance measurement in humanitarian responses. To achieve this goal, first a classification of different types of disasters and their effects on human lives are given in Section 15.2. After introducing the concept of the disaster management system cycle and important activities in each phase in Section 15.3, humanitarian logistics and its particular characteristics and main stages are discussed in Section 15.4. Then mathematical modeling of some humanitarian logistics decisions and their optimization solution techniques are discussed in Section 15.5. Next, concepts of coordination and performance measurement in the context of humanitarian logistics are discussed in Sections 15.6 and 15.7, respectively. Case studies regarding the success factors of humanitarian logistics system are presented in Section 15.8. Concluding remarks and possible research directions are given in Section 15.9.

15.2 Disasters

A *hazard* can be defined as a rare severe event that has negative effects on human life, properties, and the environment, and may lead to a disaster [5]. Hence, any event that endangers or devastates these can be considered a disaster. In other words, disasters can create extensive pain and discomfort for human beings and disrupt a society's normal day-to-day activities. Therefore, to be able to cope with the massively destructive effects of a disaster, humans should prepare themselves by creating and enhancing necessary infrastructures and planning many relief operations in advance.

Some researchers such as Cuny and Russell believe that the only true disasters are economic disasters [6,7]. This is because a better disaster management system requires long-term investments in infrastructure and public awareness. Similarly, Akkihal [8] explains that disasters happen when fluctuations in ecological and geological systems exceed a civilization's capacity to absorb such fluctuations. Intuitively, the success of humanitarian logistics practices in every society can thus be considered highly related to society's socioeconomic conditions. In the remaining parts of this section, we first review some of disaster classification systems and then discuss the effects of disasters on human beings.

15.2.1 Classification of Disasters

Broadly speaking, disasters can be divided into two main classes: natural and human made [5]. Natural disasters are direct or indirect consequences of natural

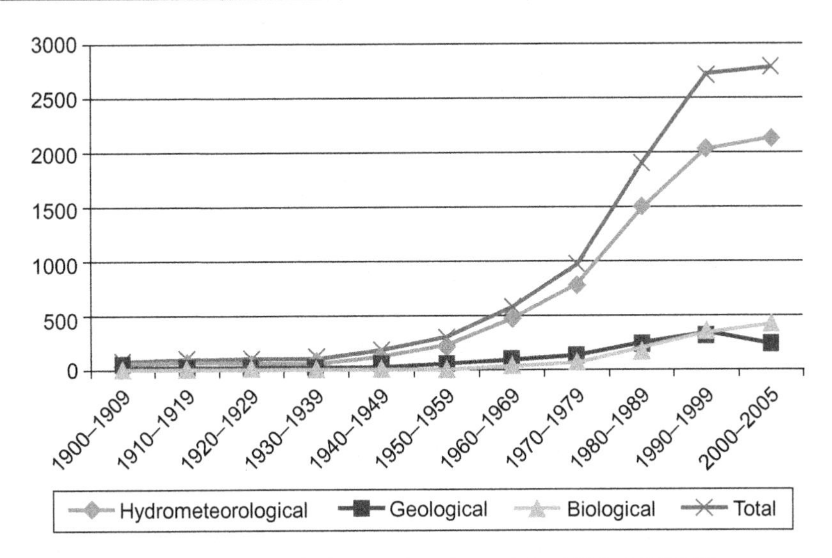

Figure 15.1 Occurrence of natural disasters based on origin, 1900–2005 [10].

phenomena. Their origin can be from three main sources: hydrometeorological,[1] geological,[2] or biological[3] [9]. Figure 15.1 compares the number of occurrences for each class of natural disasters based on their origins for the last century. This figure clearly shows a distinct growth pattern in hydrometeorological disasters in comparison with other types of natural disasters that can be attributed to global warming and climate change. Examples of natural disasters are storms, earthquakes, floods, droughts, epidemics, and volcanic activities (see Figure 15.2 for a more comprehensive classification of natural disasters and their occurrence rate for 1990–2005). On the other hand, human-made disasters or technological disasters are the direct consequences of human activities, whether deliberate (e.g., wars and terrorist attacks) or not (e.g., industrial accidents and infrastructure failures).

Disasters, whether natural or human-made, have various consequences, including loss of human lives, destruction of infrastructures, and ruptured socioeconomic conditions. It is noteworthy that hazardous events that happen outside the boundaries of human habitats are not considered to be disasters. For example, a severe earthquake happening in a remote and uninhabited desert has little impact on human life and surroundings, so it is not a disaster. The consequences of such an earthquake in a populated region, however, would be considered catastrophic.

[1] Natural processes or phenomena of atmospheric, hydrological, or oceanographic nature.

[2] Natural earth processes or phenomena.

[3] Processes of organic origin or those conveyed by biological vectors, including exposure to pathogenic microorganisms, toxins, and bioactive substances.

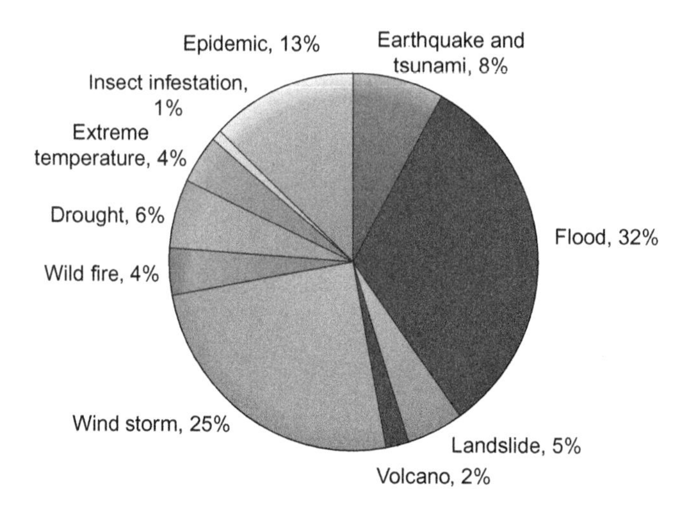

Figure 15.2 Distribution of natural disasters by type, 1990–2005 [11].

15.2.2 Effects of Disasters on Nations

Disasters have been identified with massive casualties and destruction for a very long time. In the ancient world, because of a lack of preplanning and limited capacities of societies to respond to disastrous events of large magnitude, disaster could even lead to the destruction of a complete civilization (e.g., the catastrophic eruption of the volcanic Mount Vesuvius in AD 79 buried the city of Pompeii). Today, disasters still cause many casualties and considerable destruction mainly because of ineffective preventive measures, incomplete preparedness, and weak relief logistics systems.

Quantification of the effects of disasters would enhance our understanding of the sometimes irrecoverable effects of disasters (Tables 15.1–15.3). The first factors to be considered in analyzing each disaster are usually the number of human casualties and people displaced. Table 15.1 shows disaster statistics for four countries: China, India, Iran, and the United States. During the 28-year-period considered, more than 376,000 people had been killed in these four countries alone. Also, more than 4 billion people have been affected by disasters in a way. Based on the statistical data in the twentieth century alone, more than 3.5 million people were killed by natural disasters such as floods, earthquakes, and volcanoes (drought and famine are not included in this total) [15]. Also, 15 million people are estimated to have been killed by disasters during the second millennium. See [16] for a bibliography of research on deaths caused by natural disasters.

Other important factors in disaster analysis are the cost of relief operations and the economic damages of disasters. Over the last decade, huge sums of money have been either spent on or lost because of disasters. One main cost after a disaster is that for relief operations. In 2003, about $6 billion was spent on humanitarian relief operations around the world. Also, the tsunami of March 22, 2005 required

Table 15.1 Statistics of Disaster Damages in Four Countries, 1980–2008 [12]

	China	India	Iran	United States
Number of events	533	395	138	601
Number of people killed	147,204	139,393	77,984	12,030
Average number of people killed per year	5076	4807	2689	415
Number of people affected	2,500,735,703	1,506,794,740	42,657,823	24,482,933
Average number of people affected per year	86,232,266	51,958,439	1,470,959	844,239
Economic damage (US$1000)	230,947,214	45,184,830	21,374,696	483,481,510
Economic damage per year (US$1000)	7,963,697	1,558,098	737,058	16,671,776

Table 15.2 Total Amount of Economic Damage (US$ Billion) from Natural Disasters, 1991–2005 [13]

	Hydrometeorological	Geological	Biological
Africa	3.93	6.14	0.01
America	400.82	29.98	0.13
Asia	357.70	219.74	0.00
Europe	142.83	16.17	0.00
Oceania	14.51	0.87	0.14

Table 15.3 World's Five Most Important Industrial Accidents Based on Economic Damage, 1900–2010 [14]

Place	Date	Damage (US$1000)
Spain	November 17, 2002	9,960,407
Soviet Union	April 26, 1986	2,800,000
Russia	August 17, 2009	1,320,000
United Kingdom	July 7, 1988	1,200,000
United States	October 23, 1989	1,100,000

allocation of about $6.4 billion for the response alone [17]. Besides relief operations costs, economic damages on societies are substantial (see Table 15.2 for natural disasters and Table 15.3 for human-made disasters). Natural disasters imposed nearly $1200 billion in damages and economic losses during 1991–2005 [13]. For example, many economic sectors of Asian countries, including fisheries, agriculture, livestock, tourism, and microenterprises, were severely affected by the 2004 Asian tsunami [18]. Also, in the United States alone, economic damage of more than $483 billion from 601 disasters during 1980–2008 has been reported.

Table 15.4 Consequences of Some Common Natural Disasters [19]

Consequences	Natural Disasters				
	Earthquakes	**Cyclones**	**Floods**	**Fires**	**Drought or Famine**
Casualties	†	†	†	†	†
Injured and diseased	†	†	†	†	†
Epidemic diseases		†	†		
Destruction of agricultural crops		†	†	†	†
Destruction of houses	†	†	†	†	
Damaged infrastructures	†	†	†	†	
Communication disruption	†	†	†	†	
Transportation disruption	†	†	†	†	
Public panic	†	†	†	†	
Pillage and insecurity	†	†	†	†	
Public-order disturbance	†	†	†		
Temporary migration			†		†
Permanent emigration or immigration					Δ
Disabling or halting industrial sector	†	†	†	†	Δ
Disabling service sector	†	†	†	†	Δ
Disruption of socioeconomic systems	†	†	†	†	Δ

†, primary effect; Δ, secondary effect.

Looking at the consequences of disasters from the perspectives of humans, infrastructures, and societies can create a better understanding of the effects of disasters on nations. Table 15.4 summarizes the consequences of some of the most common natural disasters. A brief analysis of this table shows that besides casualties and the destruction of houses and infrastructures, disasters can also lead to socioeconomic disruptions such as unemployment, emigration and immigration, and the halting of day-to-day business and industrial activities. Hence, designing preventive measures and recovery plans for these consequences and other more obvious consequences are necessary.

15.3 Disaster Management System Cycle

Disaster management can be defined as the discipline of avoiding and dealing with risks [20]. In other words, disaster management is a set of processes designed to be implemented before, during, and after disasters to prevent or mitigate their effects. This discipline involves preparing for disasters, responding to them, and finally supporting and rebuilding the society after initial disaster relief operations have ended. Because disasters pose a permanent threat, disaster management systems and practices should be continually monitored and improved. Also, the success of these systems relies heavily on effective and efficient cooperation and coordination of organizations participating in relief operations.

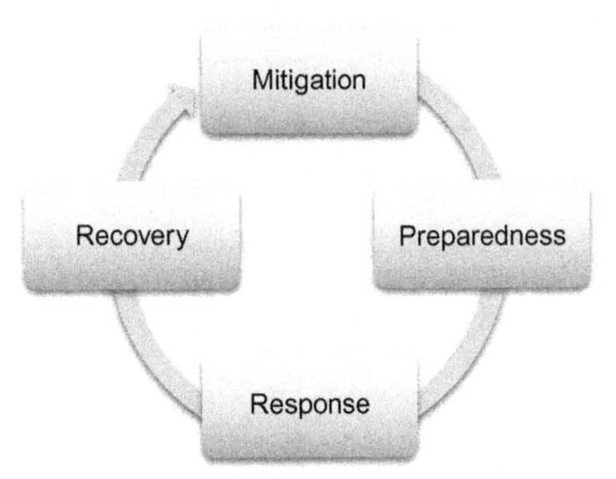

Figure 15.3 The four main phases of a disaster management system.

Any disaster management system consists of four main phases: mitigation, preparedness, response, and recovery (Figure 15.3); each will be discussed in the remainder of this section. Also, the main activities during each phase are shown in Table 15.5. A recent survey of articles on application of operations research (OR) and management science techniques in each of the four aforementioned phases are given in Altay and Green [21].

Finally, we must bear in mind that the success of every disaster management system is highly dependent on the characteristics of the region affected by a disaster in addition to the characteristics and intensity of each particular disaster. For example, in addition to countries' usual logistical preparedness, factors such as transportation and communication infrastructures and systems, environmental conditions, geographical conditions, and time of occurrence during the day and year have crucial impacts on the level of casualties and destruction caused by a disaster. Therefore, a comprehensive model for disaster management cannot be achieved without localizing the available models and best practices for each country and region.

15.3.1 Mitigation

Mitigation measures try to prevent hazards from turning into disasters or to reduce their destructive effects. This phase is different from the other three phases in requiring long-term planning and investment [22]. Because of the nature of preventive measures, mitigation is the most effective and important phase against disaster effects. The measures implemented in this phase can be categorized as *structural* and *nonstructural* [20]. Structural measures use technological advancement in order to mitigate the disasters effects (e.g., flood levees, strengthening existing buildings, and strengthening crucial links such as bridges in transportation networks). Examples of nonstructural measures are legislation, land-use planning, and insurance.

15.3.2 Preparedness

In this phase, various plans and solutions are devised in case a disaster occurs. Examples of these plans and solutions include various aspects of everything

Table 15.5 Major Activities of Disaster Management System Life Cycle [21]

Phase	Activity
Mitigation	– Establishing land-use planning and control to prevent occupation of high-hazard areas – Using technological advancement to mitigate disasters effects – Establish preventive measures to control developing situations – Improving disaster resistance of structures by enforcing building codes – Establishing tax incentives or disincentives – Ensuring application of proper methods in rebuilding buildings and infrastructures after disasters – Measuring potential for extreme hazards using risk-analysis techniques – Enforcing the use of insurance plans to reduce disasters' financial impacts
Preparedness	– Recruiting personnel for emergency services – Establishing community volunteer groups – Emergency planning – Logistical planning – Acquiring and stockpiling necessary items – Developing mutual aid agreements and memorandums of understanding with other organizations, NGOs, international organizations, and other countries – Providing training for both response personnel and concerned citizens – Performing threat-based public education – Budgeting – Acquiring necessary vehicles and equipments – Acquiring, stockpiling, and maintaining emergency supplies – Constructing central and regional emergency operations centers – Developing communications systems – Planning regular disaster exercises to train personnel and test capabilities
Response	– Activating emergency operations plan – Activating emergency operations centers – Evacuating disaster areas – Opening shelters and providing mass care – Providing emergency rescue and medical care – Firefighting – Performing search and rescue – Providing emergency infrastructure protection and recovering lifeline services – Establishing fatality management – Ensuring the security of affected areas by deploying police or military forces
Recovery	– Providing disaster debris cleanup – Providing financial assistance to individuals and governments – Rebuilding roads, bridges, and key facilities – Providing sustained mass care for displaced people and animals – Reburying displaced human remains – Fully restoring lifeline services – Providing mental health and pastoral care

about the disaster management system such as preplanning the logistics of relief operations (e.g., locating necessary facilities, stockpiling necessary items, and transporting people, equipments, and other items), establishing communication plans, defining the responsibilities of each participating relief organization, coordinating operations, and training relief personnel.

15.3.3 Response

This phase requires the immediate dispatching of the necessary personnel, equipment, and items to the disaster area. Generally, a combination of medical units, police or military forces, firefighters, and search units with the necessary vehicles and equipment are deployed right after a disaster occurs, depending on its intensity and extent. The next waves usually include backup human resources and equipments for the aforementioned groups as well as necessary items (e.g., primary supplies such as food, drinking water, clothing, tents and temporary building structures, and medicine), voluntary forces, and nongovernmental organizations (NGOs). The preparation of an effective response plan for coordinating relief forces and operations is critical to the success of a disaster management system.

15.3.4 Recovery

The main purpose behind the recovery phase is restoring the areas affected by disasters to their previous state. This phase is mainly concerned with secondary needs of people such as restoring and rebuilding houses and city facilities. One of the main opportunities of this phase is to enhance the infrastructures and conditions of the affected area by using fundamental mitigation techniques [23].

15.4 Humanitarian Logistics

Humanitarian logistics is a branch of logistics dealing with the preparedness and response phases of a disaster management system. Humanitarian logistics can be defined as:

> ... the process of planning, implementing and controlling the efficient, cost-effective flow and storage of goods and materials, as well as related information, from the point of origin to the point of consumption for the purpose of alleviating the suffering of vulnerable people. The function encompasses a range of activities, including preparedness, planning, procurement, transport, warehousing, tracking and tracing, and customs clearance [4].

Humanitarian logistics is crucial to the effectiveness and speed of relief operations and programs [24]. These logistics systems are usually required to procure, store, and transport food, water, medicine, and other supplies as well as human resources, necessary machinery and equipments, and the injured during the pre- and postdisaster periods.

The variety of logistical operations in disaster relief are so extensive that they make humanitarian logistics the most expensive part of disaster relief operations, accounting for about 80% of them [25]. Also, relief operations require deploying a huge number of logistical vehicles, equipments, and personnel. For example, in the Wenchuan earthquake on May 12, 2008, in China [26], 6 cargo-transport planes and 19 helicopters were sent to the region in 24 hours. About 5800 military medical and rescue personnel and 150 tons of supplies were conveyed to the affected area. The effective and efficient implementation of such a huge operation, considering the chaotic nature of the situation (e.g., public panic and the destruction of transportation and communication infrastructures), is actually a complex and difficult one.

In the remaining parts of this section, we first see brief comparison of humanitarian logistics systems and commercial supply chains in Section 15.4.1. Then the main stages of a generic humanitarian logistics system and required items and equipments in relief operations are discussed in Sections 15.4.2 and 15.4.3, respectively. For more information on concepts, methods, and models in the fields of humanitarian logistics and disaster operations management, see [3,21,24,25, 27,28].

15.4.1 Humanitarian Logistics Systems Versus Commercial Supply Chains

In practice, managing humanitarian logistics systems can be considered to be very different from managing their commercial counterparts. This is mainly because of different inherent characteristics of demand in each system. In commercial supply chains, the demand for the product is usually either estimated using proper forecasting techniques (i.e., push production system) or initiated by the customer (i.e., pull production system). Therefore, commercial supply chain managers try their best to eliminate the elements of uncertainty as much as possible. However, the nature of demand in humanitarian logistics is very uncertain because disaster time, location, and intensity—and hence exact relief requirements—are not known until after a disaster occurs. Based on the above explanations, the specific attributes of humanitarian logistics systems are as follows [24,25,27–32]:

1. The missions of not-for-profit organizations are different from profit-making entities (i.e., ensuring speedy and lifesaving responses instead of maximizing profits and reducing costs).
2. There are more complicated trade-offs of objectives because of different types of stakeholders, including governments, relief organizations, donors, and people affected by the disaster.
3. Complex characteristics of demand include:
   a. uncertainty of demand in features such as location, time, type, and quantity;
   b. suddenly occurring demand and therefore urgently shorter lead times;
   c. high stakes associated with adequate and timely delivery.
4. Complex operational conditions exist because of:
   a. the chaotic nature of events during the postdisaster period;

   b. a lack of resources (e.g., vehicles, equipments, food and water supplies, and medical supplies);
   c. a lack of proper access to vital infrastructures (e.g., transportation and communication);
   d. a lack of experienced and professional human resources;
   e. a lack of security in the regions affected by the disaster.
5. Coordination between organizations participating in relief operations is often lacking.
6. Relief organizations must act in accordance with humanity, neutrality, and principles of impartiality.
7. There is often a politicized environment in which it is difficult to maintain a humanitarian perspective to operations.
8. There is no way to punish ineffective organizations because of absence of the humanitarian logistics system final beneficiaries' voice in the performance appraisal and evaluation process. Since the affected people are not directly involved in this process, provided they are not dead, they usually cannot claim for more than their damages, which is usually paid by insurances and governments, whereas in a commercial supply chain, an ineffective member has to pay for its own inefficiencies.

It is worth noting that almost all of the items listed are serious challenges to the performance of any supply chain system, not just humanitarian logistics systems. For example, a lack of proper transportation infrastructures forces humanitarian relief teams to use various modes of transportation ranging from advanced modes (and usually more expensive) such as helicopters and cargo planes to more primitive modes such as animals (e.g., elephants and donkeys).

Although, until about 10 years ago, logistics was considered to be not necessary by many humanitarian organizations [24,25,27–32], but today many of them are trying to implement many of the concepts and practices used by commercial supply chains as advocated by researchers [24,30]. Also as mentioned in [24,25,29], commercial supply chains can learn a lot from humanitarian logistics systems, especially about important topics such as supply chain risk and disruption management. Therefore, mapping the application of best practices of each side to the other is important ongoing research in the field of supply chain management.

15.4.2 Humanitarian Logistics Chain Structure

The humanitarian logistics chain structure consists of three main stages (Figure 15.4): supply acquisition and procurement, pre-positioning and warehousing, and transportation [25]. The first stage in any humanitarian logistics chain is acquisition and procurement of necessary items and equipments. Any relief organization is required to obtain its necessary items and equipments from local or global suppliers using various procurement techniques such as direct purchasing and tenders. The main challenges in this stage are reducing the purchasing costs (considering the possible inflation of prices in local markets after disasters), ensuring the availability of supplies during the necessary times, reducing lead times, and coordinating in-kind donations with respect to other acquired items [29].

After acquiring necessary items and equipments for the predisaster and postdisaster periods, the responsible relief organizations are obliged to pre-position and

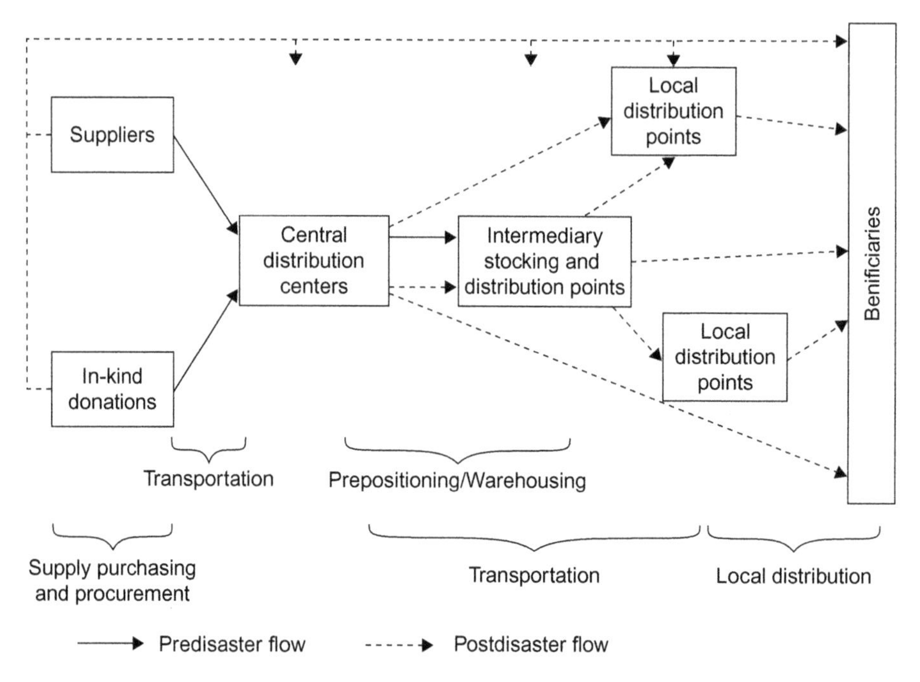

Figure 15.4 Humanitarian logistics chain structure [29].

store their items and equipments in suitable locations considering the location of disaster-prone areas. Challenges of this stage include the high costs of opening and operating permanent warehouses [29], inventory holding costs, and possible deterioration of items. Also, there is a high risk that warehouses will be destroyed during disasters, so those used for humanitarian logistics should have higher resistance against disasters and be located wisely.

Finally, transportation is the last important stage of any humanitarian logistics chain in which human personnel, equipment, and necessary items are sent to predefined central distribution centers (CDCs), distribution intermediary points, local distribution centers, and finally regions affected by the disaster. Transportation during the postdisaster period is somehow the most difficult stage of humanitarian logistics even if different kinds of preventive measures and plans have been taken into account [27]. This is mainly because transportation infrastructures and equipment are usually damaged and in poor condition after a disaster. Also, the geographical, weather conditions, and insecurities of the affected regions might restrict the types of transport vehicles and their usage methods.

15.4.3 Required Items and Equipments in Humanitarian Logistics

Usually after a disaster in a region, there is a high demand for various items and equipments for facilitating the relief operations. Based on information from the

Pan American Health Organization and the World Health Organization [33], an extended list of required items and equipment includes but is not limited to the following:

1. Food
2. Water and sanitary items
3. Environmental health equipments and items (e.g., water-treatment equipment and items)
4. Medicine (including both general pharmaceutical products and specific pharmaceutical products in possible cases of epidemics)
5. Health kits and supplies for supporting health-care processes
6. Field hospitals
7. Clothing and blankets
8. Items associated with infants and children (e.g., instant milk, diapers, formula, and toys)
9. Shelters and temporary housing facilities (e.g., tents)
10. Electrical power generating equipment
11. Fuel (e.g., coal, gas, or oil)
12. Field kitchen equipment and utensils
13. Cleaning supplies
14. Agricultural commodities and livestock
15. Specialized equipment for handling hazardous materials
16. Communication equipment
17. Firefighting equipment
18. Debris-removal equipment and vehicles
19. Construction equipment and vehicles

Keep in mind that the list is general, so different items and various amounts of them would be required based on the characteristics of each disaster and its specific situation. The urgency level of each item differs from others and is based on each specific situation; some of them are top priority and should be delivered during the early stages of postdisaster period. For example, in case of an earthquake during cold winter, the timely delivery of enough clothing, blankets, and fuel is critical. Also, in some countries specific items are required according to their cultural rules. For example, delivering enough chadors after a disaster is an important feature of disaster relief operations in Muslim countries. Finally, another important aspect of required items is the perishability of some items such as food and medicine. This characteristic calls for specifically designed ordering and inventory systems for such items because they cannot be feasibly stocked for longer terms.

15.5 Humanitarian Logistics Problems

One crucial aspect of humanitarian logistics systems is the importance of effective planning of both preparedness processes and response operations. This is because of the uncertain nature of disasters, where many factors such as type, time, location, and intensity are unknown before the disaster occurs. In addition, many decisions regarding types of operations and their requirements have to be taken either in advance (preparedness phase) or shortly after the disaster (response phase). Finally, the chaotic nature of events during the postdisaster period makes common human judgment and decision making subject to more costly mistakes. Therefore, advanced analytical decision-making techniques are important assets to the decision-making process in humanitarian logistics systems.

Among various available analytical decision-making techniques, OR is a great set of tools for better decision making about humanitarian logistics systems problems, and consequently improving the performance of disaster management systems. OR can be simply defined as an interdisciplinary field for "applying advanced analytical methods to help make better decisions."[4] The common analytical tools used in OR include but are not limited to mathematical modeling, optimization, simulation, probability and statistics, stochastic processes, queuing systems, game theory, forecasting, data mining, multicriteria decision making, and system dynamics.

Remaining parts of this section introduce some of the fundamental models and their solution techniques, which can be used in modeling and optimizing humanitarian logistics systems.

15.5.1 Location Models

One important aspect of a humanitarian logistics system is determining the location of various facilities and infrastructures, including but not limited to central warehouses of relief items, local warehouses, permanent relief facilities such as major hospitals and positioned relief equipment and vehicles, and temporary relief facilities such as mobile hospitals. To deal with locating these kinds of facilities, location science can be of great assistance. A simple but useful model for locating facilities is the p-median problem, which locates p facilities and allocates the demand nodes to them while minimizing the total weighted transportation costs. Without loss of generality, we assume that the facilities are CDCs, whereas the demand nodes are previously located regional distribution centers (RDCs). To model this problem, the following sets and parameters are required:

CDC: set of central distribution-center nodes
RDC: set of regional distribution-center nodes
F_i': available supplies of relief items in the ith CDC
h_j: required quantities of relief items at the jth RDC
CR_{ij}: transportation costs for a unit of the relief item between the ith CDC and the jth RDC

The required variables for the p-median problem are as follows:

x_{ji}: binary variable for allocating the demand of the jth RDC to the ith CDC
y_i: binary variable for opening the ith CDC

The mathematical model of the capacitated p-median problem is as follows [34]:

$$\text{Minimize} \sum_{i \in \text{CDC}} \sum_{j \in \text{RDC}} h_i d_{ij} x_{ji} \tag{15.1}$$

[4] http://www.scienceofbetter.org/

subject to

$$\sum_{i \in \text{CDC}} x_{ji} = 1 \quad \forall j \in \text{RDC} \tag{15.2}$$

$$\sum_{i \in \text{CDC}} y_i = p \tag{15.3}$$

$$\sum_{j \in \text{RDC}} h_j x_{ji} \leq F_i' y_i \quad \forall i \in \text{CDC} \tag{15.4}$$

$$x_{ji} \leq Y_i \quad \forall i \in \text{CDC},\ j \in \text{RDC} \tag{15.5}$$

$$x_{ji} \in \{0, 1\} \quad \forall i \in \text{CDC},\ j \in \text{RDC} \tag{15.6}$$

$$y_i \in \{0, 1\} \quad \forall i \in \text{CDC} \tag{15.7}$$

In the above model, objective function (15.1) minimizes the total weighted transportation costs. Constraint (15.2) allocates the demand of each RDC to exactly one CDC, whereas constraint (15.3) selects exactly p new CDCs to be opened. Constraint (15.4) limits the total demand allocated to each CDC, and constraint (15.5) ensures that demand of an RDC can be allocated only to a CDC, providing that CDC is opened. This formulation could be easily extended to consider the fixed cost of opening new facilities, FC_i', by replacing objective function (15.1) with the following objective function:

$$\text{Minimize} \sum_{i \in \text{CDC}} \text{FC}_i' y_i + \sum_{i \in \text{CDC}} \sum_{j \in \text{RDC}} h_i d_{ij} x_{ij} \tag{15.8}$$

The problem (15.8) and (15.2)–(15.7) is known as the capacitated fixed-charge facility location problem. Although both discussed problems are nondeterministic polynomial-time hard (NP-hard) [35], numerous effective solution techniques have been proposed to solve them, including benders decomposition, Lagrangian relaxation, and various metaheuristics such as genetic algorithm, tabu search, simulated annealing, and neural network. For a recent review of metaheuristics applied to solve the p-median problem, see [36].

The nature of the objective function in the p-median and fixed-charge location problems is a pull objective function that tries to locate the facilities as near as possible to the demand nodes. This characteristic is not always desirable because on-time delivery is a critical part of many relief operations. In such situations, it is most desirable that every demand node is reachable (or covered) in a specific amount of time. The famous model for dealing with such situations is the set covering location model. In this model, a demand node is considered to be coverable by a facility if and only if it is within a specified distance of the facility. If we denote the covering of demand node j by facility i with a binary parameter such as a_{ji}, then the set covering location model can be stated as follows [37]:

$$\text{Minimize} \sum_{i \in \text{CDC}} \text{FC}'_i y_i \tag{15.9}$$

subject to

$$\sum_{i \in \text{CDC}} a_{ji} y_i \geq 1 \quad \forall j \in \text{RDC} \tag{15.10}$$

$$y_i \in \{0, 1\} \quad \forall i \in \text{CDC} \tag{15.11}$$

In the formulation stated above, objective function (15.9) minimizes the total cost of opening CDCs, whereas constraint (15.10) ensures that each RDC is covered by at least one opened CDC. The set covering location problem is considered to be an integer-friendly problem [38] because its linear programming relaxation solution is usually integer; if not, the branch-and-bound method can solve it with limited search effort.

The model (15.9)–(15.11) is not perfect and has two major drawbacks:

1. Usually, locating all the needed facilities cannot be justified economically.
2. The model does not distinguish between demand nodes with different weights (e.g., population in the context of humanitarian logistics).

The maximal covering location problem can be considered an improvement on the set covering location problem because of eliminating the above drawbacks. This model locates at most p facilities in order to maximize the total covered demand. If z_j is defined as the binary variable for denoting the covering of RDC node j by at least one facility, the mathematical model of the maximal covering problem can be stated as follows [35]:

$$\text{Minimize} \sum_{j \in \text{RDC}} h_j z_j \tag{15.12}$$

subject to

$$\sum_{i \in \text{CDC}} y_i \leq p \tag{15.13}$$

$$z_j - \sum_{i \in \text{CDC}} a_{ji} y_i \leq 0 \quad \forall j \in \text{RDC} \tag{15.14}$$

$$y_i \in \{0, 1\} \quad \forall i \in \text{CDC} \tag{15.15}$$

$$z_j \in \{0, 1\} \quad \forall j \in \text{RDC} \tag{15.16}$$

In the above model, objective function (15.12) maximizes the total demand covered by the opened facilities. Constraint (15.13) ensures locating at most p facilities, whereas constraint (15.14) links the covering of the demand node variable to the location decision variable. Although this problem is NP-hard, it can be still solved effectively via various heuristics. For example, one can relax constraint (15.14) using Lagrangian relaxation and then solve the resulting problem using the subgradient optimization technique [39].

Note that because location models with median objectives are deciding about the allocation of demand nodes to facilities in addition to decisions on facilities' locations, they implicitly cover the transportation and inventory decisions as well. Therefore, explicit consideration of inventory and transportation structures and costs can be a great enhancement of location models in the context of humanitarian logistics. One model with such feature is discussed in great detail in Section 15.5.4.

Finally, other location models such as p-center can be discussed in the context of humanitarian logistics, but complete treatment of such models is out the scope of this chapter. The reader is referred to the following sources for a complete treatment of other types of location models: Daskin [34,35], Revelle and Eiselt [37], Drezner and Hamacher [40], and ReVelle et al. [41]. Also, the reader is referred to these additional sources for more familiarity with more advanced location models in humanitarian logistics literature: Balcik and Beamon [27]; Adivar et al. [42]; and Najafi et al. [43].

15.5.2 *Transportation and Distribution Models*

Another aspect of operating a humanitarian logistics system is the transportation of relief items from suppliers and major warehouses to local operation bases, and then distributing them among the people in need. The first transportation part, which is usually a long-haul type of transportation, can be dealt with using various transportation models available in the literature such as the famous basic transportation model discussed in numerous OR books such as Taha [44] and Hillier and Lieberman [45]. This problem can be defined on a bipartite graph $G = (N, E)$, where N consists of supply (suppliers and major warehouses) and demand (local operation bases) nodes. For the sake of simplicity, we assume that supply nodes are, in fact, CDCs and the demand nodes are RDCs as previously introduced in Section 15.5.1. Finally, E includes the edges connecting the nodes in each part of the graph to the other part. The only required decision variable is x_{ij}, which is the amount of relief items to be transported between the ith CDC and the jth RDC.

Considering the previously introduced notation, the mathematical model of the transportation problem would be as follows:

$$\text{Minimize} \sum_{i \in \text{CDC}} \sum_{j \in \text{RDC}} \text{CR}_{ij} x_{ij} \tag{15.17}$$

subject to

$$\sum_{i \in \text{CDC}} x_{ij} = h_j \quad \forall j \in \text{RDC} \tag{15.18}$$

$$\sum_{j \in \text{RDC}} x_{ij} \leq F_i^{'} \quad \forall i \in \text{CDC} \tag{15.19}$$

$$x_{ij} \geq 0 \quad \forall i \in \text{CDC}, \ j \in \text{RDC} \tag{15.20}$$

In the above formulation, objective function (15.17) minimizes total transportation costs, whereas constraints (15.18) and (15.19) ensure demand satisfaction and supply capacity limit, respectively. Because the technology matrix of the above problem is a totally unimodular matrix (TUM), the linear programming relaxation of constraints (15.17–15.20) will always yield an all-integer solution providing that all of the right-hand side coefficients are integers [46]. Techniques such as network simplex [47] can effectively solve this problem. Also, various assumptions can be added to this model, including considering multiple commodities of relief items, multiple periods of decision making, multiple modes of transportation, and transportation of human resources and wounded people to represent more complex situations and problems; see [48] for more information on implementing such assumptions.

The transportation problem has one inherent assumption regarding the type of deliveries—that each delivery requires a vehicle at full capacity, hence a round trip must be made each time for each delivery (Figure 15.5A). This assumption is not consistent with various situations where the quantities of delivered items to each demand node are less than vehicles' capacities, and multiple demand nodes can be served in each trip (Figure 15.5B). This is especially true when short-haul transportations are required to deliver necessary relief items. Hence, another aspect of a humanitarian logistics system is the distribution of relief items among people in need using short-haul transportation systems. The main mathematical model for dealing with such situation is the vehicle routing problem (VRP) model, which was originally introduced by Dantzig and Ramser [49]. The main assumption of this model is serving every relief demand node (RDN) in a tour (i.e., exactly one vehicle should enter and exit each demand node).

Considering the previously introduced notation, the remaining required sets and parameters are as follows:

RDN: set of relief demand nodes
0: index of the RDC acting as the depot of the vehicles
N_1: set of all nodes belonging to the set RDN $\cup$ $\{0\}$

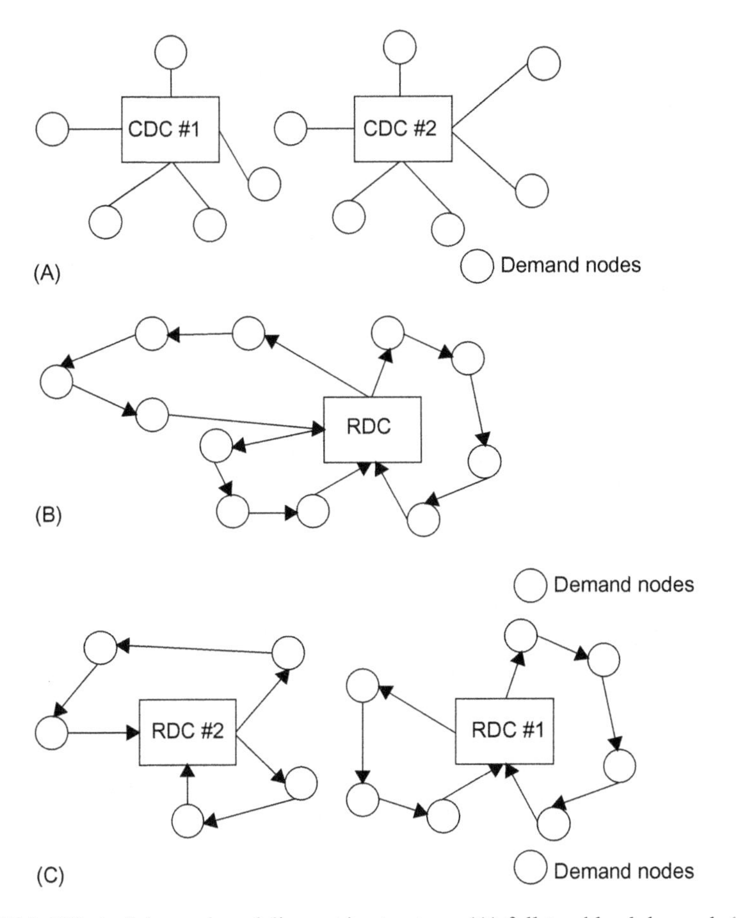

Figure 15.5 Effect of demand on delivery trip structure: (A) full-truckload demand; (B) less-than-truckload demand in a single-depot system; (C) less-than-truckload demand in a multiple-depot system.

L: set of all available vehicles
α_{km}: travel cost between the nodes k and m, k and $m \in N_1$
$r(S)$: minimum number of vehicles required to serve set S, $\forall S \subseteq \text{RDN}, \quad S \neq \varnothing$

Also, a binary variable, v_{km}, is required for denoting whether edge (k, m), where k and $m \in N_1$ $(k \neq m)$ is traversed by a vehicle. Considering the introduced notation, the mathematical model of the VRP is as follows [50]:

$$\text{Minimize} \sum_{m \in N_1} \sum_{k \in N_1} \alpha_{km} v_{km} \tag{15.21}$$

subject to

$$\sum_{k\in N_1} v_{km} = 1 \quad \forall m\in \text{RDN} \tag{15.22}$$

$$\sum_{m\in N_1} v_{km} = 1 \quad \forall k\in \text{RDN} \tag{15.23}$$

$$\sum_{k\in \text{RDN}} v_{k0} = |L| \tag{15.24}$$

$$\sum_{k\in \text{RDN}} v_{0k} = |L| \tag{15.25}$$

$$\sum_{k\notin S}\sum_{m\in S} v_{km} \geq r(S) \quad \forall S\subseteq \text{RDN}, \ S\neq \varnothing \tag{15.26}$$

$$v_{km}\in \{0, 1\} \quad \forall k, m\in N_1 \tag{15.27}$$

In the above formulation, objective function (15.21) minimizes the total distribution costs. Constraints (15.22) and (15.23) ensure that exactly one vehicle enters an RDN and one vehicle leaves it. Similarly, constraints (15.24) and (15.25) ensure that exactly $|L|$ vehicles leave the RDC and then return to it. Finally, constraint (15.26) acts as a subtour elimination constraint as well as a capacity constraint. Similar to the location problems discussed in Section 15.5.2, the VRP is NP-hard [51], therefore the variety of its solution techniques ranges from exact algorithms such as Lagrangian relaxation and branch and price to various metaheuristics. The reader is referred to [52,53] for a complete overview of the VRP, its extensions, and solution techniques. Also, the reader is referred to [54–56] for examples of applying variations of the vehicle-routing model to optimization of disaster relief operations.

At the end of this section, it is of great interest to pay more attention to an important aspect of relief items' distribution. During a disaster, various regions are usually affected with varied disaster strength levels. Hence, the condition and urgency level of various demand nodes might be completely different from one another. According to Sheu [57], different demand zones can be clustered into different groups based on time passed from the last delivery to each zone, percentage of casualties, percentage of children and elders, and level of damage to buildings.

15.5.3 Inventory Models

Humanitarian logistics systems are usually forced to keep some of their required relief items and equipment in stock, in order to increase their levels of preparedness against sudden disasters. However, similar to commercial supply chains, high levels of inventory holding costs could be a burden on humanitarian organizations because of their limited funds and operating resources. Therefore, designing effective inventory systems for humanitarian organization can be of great importance.

To demonstrate the modeling of an inventory system for a humanitarian logistics system, we discuss a simple deterministic multiperiod model in detail. Consider a set of CDCs that respond to the demand for a set of relief items, K, ordered by a set of RDCs during various relief operations in multiple consecutive periods of planning, T. Each relief item has a limited supply in each period, s_{kt}, $k \in K$, $t \in T$; and the distribution centers have limited capacities, σ_i, $i \in$ CDC. The demand of each RDC for each type of relief item is denoted by d_{jkt}, $j \in$ RDC, $k \in K$, $t \in T$. Also, consider p_{kt}, h_{kt}, and π_{kt} to be the purchasing, holding, and shortage costs of each relief item in each planning period. The pure inventory level, surplus inventory level, shortage inventory level, and purchasing quantities for each item in each period and CDC are denoted by $I_{kit}, I_{kit}^{+}, I_{kit}^{-}$, and q_{kit}, respectively. Also, the percentage of the demand of an RDC for a relief item during a period allocated to a specific CDC is denoted by x_{jkit}. Considering the above notations and assuming that the initial inventory level for each item in each CDC is known and equal to $I_{ki,0}$, the multiobjective mathematical model of such inventory system for a humanitarian logistics system is as follows:

$$\text{Minimize} \sum_{t \in T} \sum_{k \in K} \left\{ p_k \sum_{i \in \text{CDC}} q_{kit} + h_k \sum_{i \in \text{CDC}} I_{kit}^{+} \right\} \tag{15.28}$$

Minimize

$$\max_{t \in T} \left\{ \sum_{k \in K} \pi_k \sum_{i \in \text{CDC}} I_{kit}^{-} \right\} \tag{15.29}$$

subject to

$$I_{kit} - I_{ki(t-1)} = q_{kit} - \sum_{j \in \text{RDC}} d_{jkt} x_{jkit} \quad \forall i \in \text{CDC},\ k \in K,\ t \in T \tag{15.30}$$

$$\sum_{i \in \text{CDC}} x_{jkit} \leq 1 \quad \forall j \in \text{RDC},\ k \in K,\ t \in T \tag{15.31}$$

$$\sum_{i \in \text{CDC}} q_{kit} \leq s_{kt} \quad \forall k \in K,\ t \in T \tag{15.32}$$

$$\sum_{k \in K} I_{kit}^{+} \leq \sigma_i \quad \forall i \in \text{CDC},\ t \in T \tag{15.33}$$

$$I_{kit} = I_{kit}^{+} - I_{kit}^{-} \quad \forall i \in \text{CDC},\ k \in K,\ t \in T \tag{15.34}$$

$$x_{jkit} \in \{0,1\} \quad \forall i \in \text{CDC},\ j \in \text{RDC},\ k \in K,\ t \in T \tag{15.35}$$

$$q_{ki} \geq 0 \quad \forall k \in K,\ i \in \text{CDC} \tag{15.36}$$

In the above formulation, objective function (15.28) minimizes the total purchasing and holding costs for total planning horizon, whereas objective function (15.29) minimizes the maximum of the shortage costs in every planning period. Though objective function (15.29) is nonlinear, its linearization is fairly straightforward. Constraint (15.30) ensures an inventory balance of each relief item for each CDC in each period. Constraint (15.31) ensures demand for each item from each RDC is at most fulfilled completely. Constraints (15.32) and (15.33) ensure the supply limit of each item and capacity limit of each CDC in each period, respectively. Finally, constraint (15.34) relates the pure inventory level variable to its respective surplus and shortage inventory level variables. The proposed model is a multiobjective, mixed-integer linear programming model that requires efficient multiobjective optimization techniques such as multiobjective evolutionary algorithms for creating nondominated solutions.

It is noteworthy since humanitarian logistics systems mostly belong to nonprofit organizations, considering shortage cost and possibility of not satisfying an RDC are somehow questionable and against humanitarian principles. However, the proposed multiobjective model can be useful in reducing the negative effects of such assumptions. Also, it must be noted that in uncertain disaster environments, shortages can and will happen. Therefore, using various modeling techniques such as stochastic programming and robust optimization can be of great help in extending the above model to embrace such complex situations. Demonstration of such extension is beyond the scope of this chapter, and the reader is referred to [58,59] for an overview of the aforementioned modeling techniques.

At the end, it is noteworthy that research regarding effectively designing inventory systems for humanitarian logistics systems is still limited in comparison with other aspects of humanitarian logistics systems, and much research is needed. The reader is referred to [60] for a discussion of an advanced inventory model in the context of integrated logistics models (see Section 15.5.4).

15.5.4 Integrated Logistics Models

One important approach to logistics problems is the integration of decisions concerning different levels of decision making and then simultaneously solving the respective problem. This concept considers the effects of facilities location, transportation and routing, inventory control, and production planning and scheduling decisions on each other [61]. This approach to logistics problems prevents from local optimization of dependent problems such as location and routing problems [62]. Integrated logistics problems consist of different problems such as location-routing [63], inventory-location [64], queuing-location [65], and inventory-routing [66] problems.

In the context of humanitarian logistics, two of the most important decisions are the location of RDCs and the delivery of relief items such as medicine and medical equipments to various relief demand points in tours (because of less-than-vehicle capacity demand quantities; see Figure 15.5C). These two decisions can be modeled and solved simultaneously using a location-routing model.

Now, consider a two-echelon relief-chain structure consisting of three layers, including CDCs of relief items (which are preestablished and might not be near the regions affected by the disaster), RDCs of relief items (which might be set up temporarily), and finally the relief demand points. This problem can be defined on an undirected graph $G = (N, E)$ in which its nodes, N, consist of three previously introduced entities including CDCs, RDCs, and RDNs. Also, the edges of this graph, E, are composed of edges linking CDCs to RDCs, RDCs to RDNs, and RDNs to RDNs. Triangle inequality is assumed to be valid for edges linking RDCs to RDNs and any RDN to another RDN. Also, capacities of the CDCs, RDCs, and homogeneous vehicle fleet are deterministic and known. Demand of each RDN is deterministic and cannot be split among various RDCs. Finally, it is assumed that delivering relief items to each RDN must happen exactly in a specific period of time known as the hard time window. The hard time window constraints indicate the level of urgency for delivering the required items to each demand node.

This problem is known as the two-echelon location-routing problem with hard time windows constraints (2ELRPHTW) in the literature. Considering the previously introduced notations, the remaining required set and parameters of the 2ELRPHTW are as follows:

RDC: set of regional distribution centers
FC_j: fixed cost of opening jth RDC
VC_j: variable cost of operating the jth RDC for a unit of the commodity
F_j: capacity of the jth RDC
nv_j: maximum number of vehicles assignable to the jth RDC ($nv_j = \lceil F_j/\sigma \rceil$)
CV: fixed cost of operating a vehicle
σ: vehicle capacity
τ: maximum allowable route duration
D_k: demand of the kth RDN
t_{km}: travel time between nodes k and m, k and $m \in N_1$
$[a_k, b_k]$: acceptable time interval for serving the kth RDN

Considering the previously introduced variables, the required variables for the 2ELRPHTW are as follows:

y_j: binary variable for opening the jth RDC
v_{kml}: binary variable for traveling edge (k,m) by the lth vehicle for all k and $m \in N_1$ $(k \neq m)$
u_{jl}: binary variable for assigning the lth vehicle to the jth RDC
z_{kj}: binary variable for assigning the kth RDN to the jth RDC
w_k^l: arrival time of the lth vehicle to the kth RDN

Based on [67], the mathematical model of the two-echelon location-routing problem with hard time window constraints is as follows:

$$\text{Minimize} \sum_{j\in\text{RDC}} \text{FC}_j y_j + \sum_{i\in\text{CDC}} \sum_{j\in\text{RDC}} \text{CR}_{ij} x_{ij} + \sum_{j\in\text{RDC}} \text{VC}_j \sum_{k\in\text{RDN}} D_k z_{kj} + \text{CV} \sum_{j\in\text{RDC}} \sum_{l\in L} u_{jl} + \sum_{l\in L} \sum_{m\in N_1} \sum_{k\in N_1} \alpha_{km} v_{kml} \tag{15.37}$$

subject to

$$\sum_{j\in\text{RDC}} x_{ij} \le F_i' \quad \forall i\in\text{CDC} \tag{15.38}$$

$$\sum_{i\in\text{CDC}} x_{ij} \le F_j y_j \quad \forall j\in\text{RDC} \tag{15.39}$$

$$\sum_{k\in\text{RDN}} D_k z_{kj} - \sum_{i\in\text{CDC}} x_{ij} \le 0 \quad \forall j\in\text{RDC} \tag{15.40}$$

$$\sum_{l\in L} \sum_{m\in N_1} v_{kml} = 1 \quad \forall k\in\text{RDN} \tag{15.41}$$

$$\sum_{m\in N_1} \sum_{k\in N_1} d_{km} v_{kml} \le \tau \quad \forall l\in L \tag{15.42}$$

$$\sum_{m\in N_1} v_{mkl} - \sum_{m\in N_1} v_{kml} = 0 \quad \forall k\in N_1, \ \forall l\in L \tag{15.43}$$

$$\sum_{m\in N_1} v_{kml} + \sum_{h\in N_1} v_{jhl} - z_{kj} \le 1 \quad \forall k\in\text{RDN}, \ \forall j\in\text{RDC}, \ \forall l\in L \tag{15.44}$$

$$\sum_{k\in\text{RDN}} v_{kjl} = u_{jl} \quad \forall j\in\text{RDC}, \ \forall l\in L \tag{15.45}$$

$$\sum_{k\in\text{RDN}} v_{jkl} = u_{jl} \quad \forall j\in\text{RDC}, \ \forall l\in L \tag{15.46}$$

$$\sum_{l\in L} u_{jl} \le nv_j y_j \quad \forall j\in\text{RDC} \tag{15.47}$$

$$w_m^l \ge w_k^l + t_{km} - M(1 - v_{kml}) \quad \forall k, m\in\text{RDN}, \ l\in L \tag{15.48}$$

$$a_m \le w_m^l \le b_m \quad \forall m\in\text{RDN}, \ l\in L \tag{15.49}$$

$$x_{ij} \ge 0 \quad \forall i\in\text{CDC}, \ \forall j\in\text{RDC} \tag{15.50}$$

$$y_j \in \{0, 1\} \quad \forall j \in \text{RDC} \tag{15.51}$$

$$u_{jl} \in \{0, 1\} \quad \forall j \in \text{RDC}, \ \forall l \in L \tag{15.52}$$

$$v_{kml} \in \{0, 1\} \quad \forall k \in N_1, \ \forall m \in N_1, \ \forall l \in L \tag{15.53}$$

$$z_{kj} \in \{0, 1\} \quad \forall k \in \text{RDN}, \ \forall j \in \text{RDC} \tag{15.54}$$

In the above model, objective function (15.37) includes RDCs opening fixed costs, CDCs to RDCs commodity transportation costs, RDCs variable costs, fixed costs of the vehicle, and, finally, routing costs. Constraint (15.38) restricts the amount of outgoing commodity from each CDC to its capacity, whereas the incoming commodity into each RDC is limited to its capacity by constraint (15.39). Constraint (15.40) balances the incoming and outgoing commodity volume at each RDC. Constraint (15.41) requires each RDN to be assigned to the route of exactly one vehicle. Constraint (15.42) limits the capacity of each vehicle.

Constraint (15.43) ensures the conservation of flow in each RDC and RDN node. Constraint (15.44) assigns an RDN to an RDC if a vehicle enters that RDN and leaves the RDC itself at first. Constraints (15.45) and (15.46) ensure that if a vehicle is assigned to an RDC, it both enters and leaves that RDC. Constraint (15.47) limits number of vehicles assigned to an RDC to its vehicle capacity. Constraint (15.48) calculates the arrival time of vehicles to RDNs and also eliminates subtours [68]. Finally, constraint (15.49) defines the hard time window domain for each RDN.

Solving medium- and large-sized instances of this problem via exact methods is a challenging and difficult task because of the NP-hard nature of its subproblems [51,69]. Among the effective solution techniques proposed for solving the location-routing problems, various techniques such as decomposition [70], Lagrangian relaxation [71], tabu search [72], and simulated annealing [73] have been proposed. For recent reviews of models and solution methods proposed for various types of location-routing problems, the reader is referred to [61,63].

Besides location-routing problem, various integrated models can be used in the context of humanitarian logistics. For example, Barbarosoglu et al. [74] have addressed the problem of hierarchically making decisions about assigning helicopters and pilots to air bases (tactical level) as well as routing helicopters to serve the affected regions (operational level). Because the complete introduction of such models is beyond the scope of this chapter, the reader is referred to [63,66] for reviews of inventory-routing and location-inventory problems, respectively.

15.6 Coordination of Humanitarian Logistics Systems

After a disaster occurs in a region, different international organizations and NGOs, besides the responsible governmental organizations, are involved in carrying out the humanitarian relief operations. This is mainly because of a lack of sufficient recourses, equipments, and (professional) human recourses. For example, more than 44 countries participated in Bam's earthquake rescue and relief operations [75]. Efficient and effective use of these many groups, coming from different organizational structures and cultures, requires a close cooperation and coordination between them. However, the chaotic nature of postdisaster relief operations, lack of necessary resources, and presence of different and numerous relief organizations have contributed to a lack of teamwork and consequently the failure of many relief operations [76,77]. Hence, coordination of humanitarian logistical operations is one of the most important aspects of humanitarian logistics, especially during preparedness and response phases. In the remainder of this section, supply chain coordination, important factors in coordinating humanitarian logistics operations, and coordination mechanism in humanitarian logistics are discussed in brief overviews.

15.6.1 Supply Chain Coordination

To manage a supply chain effectively and efficiently, a coordinated execution of supply chain operations is required [78]. However, because of conflict of interest between the supply chain participants (i.e., different objectives and each participant's self-serving focus), the necessary level of coordination cannot be achieved. This is mainly because of the inefficiency of locally optimal decisions of each supply chain participant regarding the whole supply chain efficiency [79]. Supply-chain coordination tries to increase the whole supply chain profit, efficiency, and effectiveness through global decision-making models instead of using individually local optimal decisions of each participant. Hence, coordination mechanisms seek to encourage supply chain members to follow globally optimal decisions for the whole supply chain [80]. Broadly speaking, supply chain coordination can be classified into two main classes, namely, vertical coordination and horizontal coordination. Vertical coordination refers to the coordination of an organization with its downstream and upstream supply chain participants, whereas horizontal coordination refers to the coordination of same-level supply chain participants.

Supply chain coordination mechanisms can be classified based on five characteristics: level of formality, resource-sharing structure, decision style, level of control, and risk and reward sharing scheme [81,82]. Therefore, to achieve coordination in a supply chain, various techniques may be employed such as information and resource sharing, central decision making, conducting joint projects, various types of contracts such as revenue sharing [78], regional division of tasks, or a cluster-based system.

15.6.2 Important Factors in Coordinating Humanitarian Logistics Operations

Because of the severe effects of disasters, usually multiple organizations are engaged in relief operations. However, as mentioned earlier, the characteristics of

Table 15.6 Characteristics of Potential Relief-Chain Coordination Mechanisms [29]

Coordination Mechanism	Coordination Cost	Opportunistic Risk Cost	Operational Risk Cost	Technological Requirements	Beneficial to Relief Environment	Potential for Implementation
QR[a], CR[b], VMI[c], CVMI[d]	Low	High	High	High	No	Higher for large NGOs but generally low
Collaborative procurement	Low, especially if supported by an umbrella organization	Low	Low when no contracts; high in competitive environment	Low	Yes	High (currently observed)
Warehouse standardization	High	Varies	High	Medium	Yes	Low
Third-party warehousing (umbrella organization)	Low	Low	Low	Low	Yes	High (currently observed)
Third-party warehousing (private-sector partner)	Medium	Medium	Varies	Low	Yes	High (currently observed)
Transportation						
Shipper collaboration	High	Varies	High	Medium	No	Low
4PL	High	High	High	Medium	No	Low

[a]Quick response.
[b]Continuous replenishment.
[c]Vendor-managed inventory.
[d]Consignment vendor-managed inventory.

postdisaster relief operations can have a direct negative effect on the successful coordination of cooperative operations between different organizations. Some of these characteristics originally discussed in [29] are as follows:

1. *Number and diversity of organization involved*: Usually, because of a disaster's severe effects, various organizations with different policies, training, and operational procedures and methods are required to be involved in relief operations. Although the participating organizations have the very same goal (i.e., helping people), the diversity in methods, organizational culture, and policies [24]; the challenges of communicating between organizations with different speaking languages [83]; and the lack of coordination experience in affected countries [84] may lead to create more barriers in coordination of relief operations.
2. *Donor expectations and funding structure*: Most of the time, dealing with disasters requires huge amount of money, food, clothing, and other relief items that might not be readily available during initial days after the disaster. As a result, different international and private aid agencies along various countries usually provide the affected national governments and their corresponding organizations with necessary items and money. Because aid agencies are heavily dependent on donations [84], they are usually forced to act in accordance with the donors' requirement such as using the donations in a short amount of time or types of allowed relief activities [83,85]. These kinds of restrictions are considered as a barrier against the coordination of relief operations [29].
3. *Unpredictability*: Disaster environments are actually very uncertain because of various factors such as the location, timing, and intensity of disasters; regional infrastructures; preparedness of relief organizations; and, finally, the political environment of the affected country [29].
4. *Resource scarcity and oversupply*: During the pre- and postdisaster humanitarian logistics operations, deploying enough human resources and equipment as well as delivering the right amount of necessary items is crucial. However, because of the inherent uncertainties of disasters, insufficiency or oversupply of human resources, equipment, and other items might still happen. Hence, matching demand and supply is an important aspect of relief operations [29].
5. *Costs of coordination*: Coordinating the relief operations, whether during the pre- or postdisaster periods, requires various trips to different countries and regions as well as holding meetings between the corresponding organizations [86]. These trips and meetings impose higher staff salaries and travel expenses, and funding these costs might be impossible for smaller relief organizations and aid agencies because of their limited resources [83].

15.6.3 Humanitarian Coordination Mechanisms

As mentioned in Section 15.4, the main stages of any humanitarian logistics chain can be classified into several stages: acquisition and procurement, pre-positioning and warehousing, and transportation. Based on the results of Balcik et al. [29], characteristics of the most prominent commercial supply chain coordination mechanisms that are currently being practiced in humanitarian logistics chains or can be applied to them are given in Table 15.6. These supply chain coordination mechanisms are compared with one another based on six factors, including coordination cost (i.e., direct costs associated with coordination), opportunistic risk cost (i.e., costs associated with reduced or lost bargaining power or resource control), operational risk cost (i.e., costs associated with partner's poor performance), technological requirements, being beneficial to relief environment, and finally the potential for implementation [29,81]. The results of Balcik et al. [29] show that collaborative procurement and third-party warehousing are currently the most widely used coordination mechanisms because of their inherent low costs and technological requirements. Also, the authors pointed out that the use of more costly mechanisms such as shipper collaboration and 4PL[5] service providers requires new types of relationships between relief organizations and aid agencies.

15.7 Performance Measurement of Humanitarian Logistics Systems

Every organization is accountable for its decisions and actions to its customers, employees, suppliers, and shareholders according to its mission, vision, and goals. Hence, performance measurement and management, either working for profit or not, is an important part of analyzing and aligning an organization's performance path with its original vision and goals as well as shareholders' views and needs. Bear in mind, however, that balancing the performance of an organization in various performance measures might be a challenging task because some performance measures are usually in conflict with one another. For example, increasing the flexibility of a process might require higher levels of investment and operating costs. Performance measurement in supply chain management is also an important aspect of an organization's performance measurement mainly because of its important role in management of physical and information flows in the logistics systems of the organization.

Considering the previous discussions, performance measurement of humanitarian logistics systems and improving them and gaining experience from previous disaster relief operations are of great importance. This importance can be attributed to two main factors, including dependency of human lives on the effectiveness and responsiveness of humanitarian logistics systems as well as limited available resources for relief operations after a disaster occurs (e.g., unpreparedness of relief organizations and destruction of vital infrastructures). Also as mentioned by Schulz [87], the continuous use of a performance measurement system in a relief organization can lead to continuous improvement and more efficiency.

Besides the above reasoning, many organizations that participate in relief operations are not-for-profit organizations, and performance measurement is an important aspect of their systems as required by their donors and supporters. As pointed out by Poister [88], "effective performance measurement systems can help nonprofit managers make better decisions, improve performance, and provide more accountability." Moreover, when they are designed and implemented effectively, performance measures provide feedback on agency performance, and motivate managers and

[5] Fourth-party logistics.

employees to work harder and smarter to improve performance. They can also help allocate resources more effectively, evaluate the efficacy of alternative approaches, and gain greater control over operations, even while allowing increased flexibility at the operating level.

Usually, in evaluating a supply chain performance, different performance metrics such as cost, quality, time, responsiveness of the whole supply chain as well as its processes, reliability of processes, process flexibility, level of resource sharing, visibility, innovativeness, and trust between different actors of a supply chain are considered [89–93]. However, the types of metrics needed in humanitarian logistics systems are different and unique because of the inherent characteristics of humanitarian logistics systems mentioned in Section 15.4.1. This is highlighted in the following facts [32,94]:

1. Many of the services offered in humanitarian relief operations are intangible and therefore hard to quantify.
2. Outcomes are unknown.
3. The performance in each mission is hard to quantify and hence hardly measurable.
4. Interests, goals, and standards of stakeholders are different.
5. Accuracy and reliability of available data is not satisfactory.

Considering the above facts, we can conclude that the process of designing and implementing a performance measurement system in humanitarian systems would be a challenging and complicated task. However, as mentioned by Schulz and Heigh [95] based on experiences of designing such system for the International Federation of Red Cross and Red Crescent Societies (IFRC), this process can and should be kept as simple as possible. The success factors of such a process are the integration of key stakeholders throughout the process as well as simplicity and user friendliness of the methods and tools [95].

In this section, two examples of performance metric systems proposed by Beamon and Balcik [32] and Davidson [96] are explained. Davidson has proposed a metric system based on the following four core metrics:

1. *Appeal coverage*: This measures the performance of an organization in meeting its appeal for an operation in terms of both finding donors and delivering items. This metric is measured via two submetrics, including percentage of appeal coverage and percent of items delivered.
2. *Donation-to-delivery time*: This is a measure of how long it takes for an item to be delivered to the destination after a donor has promised its donation.
3. *Financial efficiency*: As a measure of the financial aspects of organizational performance, this metric is measured via three submetrics: (a) absolute ratio of the forecasted prices to the actual prices paid for items delivered in the operation, (b) relative ratio of the forecasted prices to the actual prices paid for items delivered in the operation, and (c) ratio of the total transportation costs incurred over the total costs for delivered items at a specific point in time.
4. *Assessment accuracy*: This measures how much the operation's final budget changed over time from the original budget. It is measured as the ratio of revised budget to the original budget.

Beamon and Balcik [32] also have proposed a framework for evaluating a humanitarian logistics system performance based on the three main metric classes, namely, resource, output, and flexibility, originally proposed by Beamon [97] for commercial supply chains. The resulting performance measures are available in Table 15.7.

At the end, it is noteworthy that, in comparison with the number of investigations done on the planning aspects of humanitarian logistics systems (see Sections 15.5 and 15.6), only limited research has been done on the performance measurement of these systems [32,94–96,98]. Therefore, one of the main research areas in the field of humanitarian logistics systems is to develop more comprehensive performance measurement frameworks, especially considering the intangible aspects of humanitarian logistics systems.

Table 15.7 Proposed Humanitarian Logistics System Performance Measures [32]

Resource	Output	Flexibility
Total cost	Total amount of delivered disaster supplies	Amount of individual units of tier 1 supplies that can be provided in time period T_c
Overhead costs	Total amount of delivered disaster supplies of each type	Minimum response time
Total transportation and handling cost	Total amount of delivered disaster supplies to each region	Variety of different types of supplies that the logistics system can provide in a specified time period
Value of held inventory	Amount of delivered disaster supplies to each recipient	
Inventory obsolescence and spoilage	Target fill rate achievement	
Ordering and setup costs	Average item fill rate	
Inventory holding costs	Probability of stock out	
Cost of supplies	Percentage of back orders	
Number of relief workers employed per aid recipient	Number of stock outs	
Percentage of value-added hours	Average back-order level	
Money spent per aid recipient	Average response time	
Donations received per time period	Minimum response time	

15.8 Case Studies and Learned Lessons

This section presents case studies and learned lessons regarding the humanitarian logistics system as presented through available reports, documents, and articles. The three cases include the decentralized approach of the IFRC to humanitarian logistics following the 2006 Indonesian earthquake in Yogyakarta, the performance evaluation of disaster logistics systems during hurricane Katrina in 2005, and the key success factors of natural disaster management systems in the Asian tsunami of 2004.

15.8.1 The Yogyakarta Earthquake, 2006

On May 27, 2006, an earthquake of magnitude 6.3 occurred in the Indian Ocean 20 km south-southwest of the Indonesian city of Yogyakarta [99]. The earthquake caused 5782 deaths, injured 36,299 people, damaged 135,000 houses, and left an estimated 1.5 million people homeless [100]. The quake also destroyed or severely damaged public buildings and facilities (e.g., railways, airports, and hospitals) [101]. Because Yogyakarta International Airport was closed, relief flights had to be rerouted to either Solo airport (60 km northeast of Yogyakarta) or Semarang airport (120 km north of Yogyakarta). Finally, water, electricity, transportation, and communications infrastructures in the regions affected by the quake were severely damaged. Immediately following the disaster, the Indonesia Red Cross Society mobilized more than 400 employees and volunteers and 10 mobile medical action teams to undertake assessment and relief operations [102]. The Indonesian president moved the army to the central Java province to aid rescue efforts and evacuate victims [100]. Also, about 20 countries and humanitarian organizations offered relief aid to the Indonesian government [100].

One prominent feature of the Yogyakarta earthquake was the exemplary performance of the IFRC during the relief operations compared to its previous performance in similar operations such as the Asian tsunami in 2004 and the Pakistani earthquake in 2005 [101]. Various improvements were observed such as lower total costs (cheaper relief chains), lower lead times (faster relief chains), and the number of families served by the relief teams (better relief chains) based on an internal IFRC case study [101]. For example, the Yogyakarta relief operations were 3 times and 6 times faster than the Pakistani earthquake and Asian tsunami relief operations, respectively. The main reason behind such performance is attributed to the decentralization of the IFRC humanitarian logistics chain.

Traditionally, the IRFC used to transport necessary items using transcontinental flights from various donors and national Red Cross and Red Crescent societies to local airports near a disaster. These kinds of operations required more money and were slow and time consuming (because of the bottlenecks created at local airports and warehouses). In the process of decentralizing their relief chain [101], IFRC chose three cities to be their regional logistics units (RLUs). These cities include Dubai (covering Africa, Europe, and the Middle East), Kuala Lumpur (covering

Asia and Australia), and Panama (covering the Americas). Hence, on receiving a help request from any national Red Cross or Red Crescent society, one of the RLUs is the main RLU responsible for coordinating and operating relief operations. Based on the results of Gatignon et al. [101], various factors contributed to such performance including the following:

1. *Pre-positioned, regional operational capacity*: Having regional infrastructures and stocked items helps increasing the speed of operations.
2. *Procuring required items from local and regional sources*: Having local and regional suppliers instead of relying heavily on long-distance donations leads to lower purchasing and transportation costs and of course, faster delivery.
3. *Local coordination of relief logistics operations*: Having a local coordinator more familiar with the region and local authorities instead of the previously global coordinator, positioned in Geneva, has increased the level of relief operations coordination.
4. *Standardization of processes and items*: Using various standardized logistics processes including needs assessments, procurement, warehousing, and fleet management leads to better flow of material management, maintaining the quality of procured items, and consequently lower costs.
5. *Adapting information systems for tracing goods*: Customizing their Humanitarian Logistics Software (HLS) according to the needs of the field ensures better tracking of material flows.
6. *Employing skillful and experienced personnel and training them*: Applying the knowledge and experience of such human resources can lead to better decision making in the field as well as better performance.

15.8.2 Hurricane Katrina, 2005

On August 23, 2005, hurricane Katrina formed over the Bahamas and crossed over southern Florida as a category 1 hurricane. Then it headed toward the Gulf of Mexico as a category 3 hurricane, causing massive destruction in southeast Louisiana on August 29 [103]. Because of Katrina's massive power, the levee system protecting New Orleans from Lake Pontchartrain and the Mississippi River failed and most of the city became flooded [104]. Considered to be the costliest hurricane of US history, hurricane Katrina caused about 1836 deaths and an economic damage of approximately $81.2 billion [105]. Besides deaths and the huge economic damage, the hurricane's environmental effects were considerable because of massive oil spills [106].

The US disaster management system in United States (currently known as the National Response Framework) has a hierarchical structure, so disasters are usually dealt with first at the lowest possible governmental level. Hence, a state government is supposed to be the last entity responsible for disaster relief. The scale of hurricane Katrina, however, led the state government to declare a state of emergency and call for national support. The disaster response started with various operations such as positioning relief teams across the inundated areas to assist and evacuate citizens and distribute relief items initiated by the Federal Emergency Management Agency (FEMA). The deployed forces included FEMA, state and local agencies, federal and National Guard soldiers, and volunteers [106]. After

Katrina's second landfall on August 29, the relief operations continued with various operations such as evacuating remaining citizens, distributing relief items, providing temporary housing, and removing debris. Besides the national capacity to respond to the disaster, many foreign countries and aid agencies offered to help the United States in relief operations (see [107] for a list of responders).

However, the quality and quantity of relief operations raised various questions (see [105] for a brief review of such criticisms). The weak performance of relief operations can be attributed to multiple factors, including the following:

1. A lack of central command and leadership
2. Insufficient local government, federal, and state capacities for dealing with a disaster of such magnitude, including enough trained personnel, funds, and resources
3. Lack of clarity in the national response plan
4. Late deployment of forces
5. Insufficient experience of the FEMA director and members
6. Poor performance of communication systems
7. Lack of collaboration and coordination
8. Inefficient budget expenditure system
9. Inconsistencies of the information systems
10. Poor logistics planning, including planning of transportation and distribution operations [105,108]

Iqbal et al. [108] have suggested the following for better disaster relief operations considering the event before, during, and after the disaster of hurricane Katrina:

1. Nationwide preparedness
2. Better and preplanned humanitarian logistics operations
3. Improved forecasting techniques used for disaster relief
4. Use a more flexible and transparent logistics system
5. Enhancing evacuation operations
6. Better pre-positioning of relief supplies in the country
7. More involvement of NGOs, volunteers, and private sector in various relief operations
8. Use of more reliable communication technologies and information technology structures for better collaboration and coordination
9. Use of modern technologies such as GIS and real-time tracking systems for better and more equitable distribution of disaster relief items

15.8.3 Asian Tsunami, 2004

On December 26, 2004, an underground earthquake of magnitude 9.1 struck in the Indian Ocean, off the west coast of Sumatra, Indonesia [109]. The severe earthquake created a series of tsunamis with waves as high as 30 m. The effects of this disaster on peoples' lives were catastrophic: 230,000 deaths, 125,000 injured, and about 1.69 million people left homeless in more than 15 countries [110]. The economic damages by the disaster were devastating. For example, the reported economic damage to Indonesia and Sri Lanka (the two most severely affected countries) were about $4451.6 and $970 billion, respectively [111]. Regarding the humanitarian actions, this disaster attracted donations worth more than $7 billion from around the world [110].

The responses of affected countries by the tsunami were different and diverse [112]. Among the affected countries, Maldives was the only country with no national disaster management system, so it relied mainly on its military forces capacities and appointed the minister of defense as the chief coordinator of relief operations. On the other hand, other countries were more dependent on their own national disaster management systems. However, the preparedness level of these countries varied greatly. For example, India and Thailand succeeded in assisting other affected countries in addition to attending to the needs of their own affected citizens. In Indonesia and Sri Lanka, the severities of damages were beyond their national capacities. Besides the national relief operations, many countries (such as Australia, Canada, Germany, Japan, the United States, and Pakistan) and international agencies and NGOs (such as IFRC, UNDP, UNICEF, and Islamic Relief Worldwide) participated in the relief operations [113].

Based on the study of various reports and research article on the disaster and interviews with some of the logistics managers after the disaster [114], the following weaknesses were observed:

1. Lack of preparedness
2. Lack of knowledge about local needs, capacities, and vulnerabilities among foreign aid teams
3. Lack of coordinated requirement assessments
4. Presence of competition between aid agencies
5. Lack of coordination and information sharing between foreign aid agencies
6. Lack of logistics professionals
7. Lack of long-term vision of the situation among donors and governments

Based on the above weaknesses, the following key factors regarding the success of disaster management systems were identified [114]:

1. Preparing for disaster in vulnerable regions
2. Cooperating with local organizations and NGOs as well as using local information about capacities and requirements in planning the relief operations
3. Involving local forces in assessment of the needs
4. Sharing information between participating organization in relief operations
5. Involving logistics professionals during planning and execution of relief operations
6. Improving the logistics knowledge of local officials and forces
7. Effectiveness of logistics systems
8. Importance of creating a national disaster plan
9. Necessity of using holistic approach to design disaster management systems

15.9 Conclusion

Disasters have always profoundly affected human lives in any country, whether developed, developing, or underdeveloped. As stated by Oloruntoba [115], the destructive effects of disasters not only affect underdeveloped and developing countries but also affect developed countries. Therefore, proactive planning to reduce negative impacts of disasters is a necessary step toward better living conditions for all human beings. Disaster management systems are part of humans' endeavors to better prepare themselves against disasters. However, despite some improvements in infrastructures, disasters still kill many humans. In such situations, conducting effective relief operations is of high importance.

A crucial aspect of disaster management systems are humanitarian logistics systems, which are responsible for "planning, implementing and controlling the efficient, cost-effective flow and storage of goods and materials, as well as related information, from the point of origin to the point of consumption for the purpose of alleviating the suffering of vulnerable people" [4]. Humanitarian logistics systems are usually required to procure, store, and transport food, water, medicine, other supplies as well as human resources, necessary machinery and equipment, and the injured during pre- and postdisaster periods. Although similar in concepts to commercial supply chains, humanitarian logistics systems differ from their commercial counterparts in the sense that demand for humanitarian services is highly uncertain in many aspects such as time, location, and type of services needed.

This chapter discussed basic concepts in disaster management and fundamentals of humanitarian logistics systems, including disasters and their effects, cycles of disaster management systems, definition and stages of humanitarian logistics, their comparison with commercial supply chains, basic mathematical models for optimizing humanitarian logistics operations, coordination mechanisms in humanitarian logistics, and performance measurements of such systems. Also, case studies were presented for more familiarity with important learned lessons and success factors of humanitarian logistics systems in practice.

At the end, various aspects of humanitarian logistics seem to need more attention from academics. These aspects include developing more advanced OR models for dealing with the complexities and uncertainties of disaster environment, designing specific inventory-control systems for humanitarian logistics by considering medium-term stocking of relief items and the perishability of relief items, devising better coordination practices and systems, and finally extending performance measurement systems to cover more tangible and intangible aspects of humanitarian logistics.

References

[1] G. Kovács, K.M. Spens, Humanitarian logistics in disaster relief operations, Int. J. Phys. Distrib. Logist. Manage. 37 (2007) 99–114.

[2] R. Oloruntoba, R. Gray, Humanitarian aid: an agile supply chain? Supply Chain Manage. 11 (2006) 115–120.

[3] G. Kovács, K. Spens, Identifying challenges in humanitarian logistics, Int. J. Phys. Distrib. Logist. Manage. 39 (2009) 506–528.

[4] Fritz Institute, in: A.S. Thomas, L.R. Kopczak (Eds.), From Logistics to Supply Chain Management: The Path Forward in the Humanitarian Sector, Fritz Institute, San Francisco, CA, 2005.
[5] UNDRO, An Overview of Disaster Management—Disaster Management Training Program, second ed., UNDP/UNDRO, New York, 1992.
[6] F.C. Cuny, Disasters and Development, Oxford University Press, Oxford, 1983.
[7] T. Russell, The Humanitarian Relief Supply Chain: Analysis of the 2004 South East Asia Earthquake and Tsunami, Massachusetts Institute of Technology, Cambridge, MA, 2005.
[8] A.R. Akkihal, Inventory Pre-Positioning for Humanitarian Operations, Massachusetts Institute of Technology, Cambridge, MA, 2006.
[9] ISDR, Terminology: basic terms of disaster risk reduction. International Strategy for Disaster Reduction (ISDR), Geneva, Switzerland.http://www.unisdr.org/eng/library/lib-terminology-eng%20home.htm, 2004 (accessed October 3, 2010).
[10] ISDR, Disaster statistics 1991–2005: disaster occurrence—Trends century. International Strategy for Disaster Reduction (ISDR), Geneva, Switzerland. http://www.unisdr.org/disaster-statistics/occurrence-trends-century.htm, 2006 (accessed October 3, 2010).
[11] ISDR, Disaster statistics 1991–2005: disaster occurrence—Type of disasters. International Strategy for Disaster Reduction (ISDR), Geneva, Switzerland. http://www.unisdr.org/disaster-statistics/occurrence-type-disas.htm, 2006 (accessed October 3, 2010).
[12] Université catholique de Louvain, EM-DAT: The OFDA/CRED International Disaster Database. EM-DAT, 11.08, Brussels, Belgium, 2009.
[13] ISDR, Disaster statistics 1991–2005: disaster impact—Economic damage. International Strategy for Disaster Reduction (ISDR), Geneva, Switzerland. http://www.unisdr.org/disaster-statistics/impact-economic.htm, 2006 (accessed October 3, 2010).
[14] Université catholique de Louvain, EM-DAT: The OFDA/CRED International Disaster Database. EM-DAT, 12.07, Brussels, Belgium, 2010.
[15] M. White, Deaths by mass unpleasantness: estimated totals for the entire 20th century. http://users.erols.com/mwhite28/warstat8.htm, 2005 (accessed October 3, 2010).
[16] I. Kelman, Disaster deaths. http://www.ilankelman.org/disasterdeaths.html, 2009 (accessed October 3, 2010).
[17] M. Aslanzadeh, E. Ardestani Rostami, L. Kardar, Logistics management and SCM in disasters, in: R. Zanjirani Farahani, N. Asgari, H. Davazani (Eds.), Supply Chain and Logistics in National, International and Governmental Environment: Concepts and Models, Springer, Berlin, 2009.
[18] K. Seneviratne, D. Amaratunga, R. Haigh, C. Pathirage, Knowledge management for disaster resilience: identification of key success factors, CIB2010 World Congress, The Lowry, Salford Quays, United Kingdom, 2010.
[19] National Disaster Management, Social, economic and health consequences of natural calamities, Vol. 2007, National Disaster Management Center of Ministry of Home Affairs of Government of India, North Block, New Delhi, 2007.
[20] G.D. Haddow, J.A. Bullock, Introduction to Emergency Management, Butterworth-Heinemann, Amsterdam, 2004.
[21] N. Altay, W.G. Green, OR/MS research in disaster operations management, Eur. J. Oper. Res. 175 (2006) 475–493.
[22] J.P. Wilson, Policy Actions of Texas Gulf Coast Cities to Mitigate Hurricane Damage: Perspectives of City Officials, Texas State University, San Marcos, TX, 2009.

[23] D. Alexander, Principles of Emergency Planning and Management, Terra Publishing, Harpenden, 2002.
[24] L.N. Van Wassenhove, Humanitarian aid logistics: supply chain management in high gear, J. Oper. Res. Soc. 57 (2006) 475–489.
[25] R. Tomasini, L.N. Van Wassenhove, Humanitarian Logistics, Palgrave Macmillan, Basingstoke, Hampshire, 2009.
[26] L. Li, S. Tang, An artificial emergency-logistics-planning system for severe disasters, IEEE Intell. Syst. 23 (2008) 86–88.
[27] B. Balcik, B.M. Beamon, Facility location in humanitarian relief, Int. J. Logist. Res. Appl. 11 (2008) 101–121.
[28] B.M. Beamon, Humanitarian relief chains: issues and challenges, in: Proceedings of the 34th International Conference on Computers and Industrial Engineering, San Francisco, CA, 2004, pp. 77–82.
[29] B. Balcik, B.M. Beamon, C.C. Krejci, K.M. Muramatsu, M. Ramirez, Coordination in humanitarian relief chains: practices, challenges and opportunities, Int. J. Prod. Econ. 126 (2010) 22–34.
[30] A. Thomas, L.R. Kopczak, Life-saving supply chains: challenges and the path forward, in: H.L. Lee, C.-Y. Lee (Eds.), Building Supply Chain Excellence in Emerging Economies, Springer, New York, NY, 2006, pp. 93–111.
[31] L. Kopczak, M.E. Johnson, Rebuilding confidence: trust, control and information technology in humanitarian supply chains, in: 2007 Academy of Management Annual Meeting, Philadelphia, PA, 2007.
[32] B.M. Beamon, B. Balcik, Performance measurement in humanitarian relief chains, Int. J. Publ. Sect. Manage. 21 (2008) 4–25.
[33] Pan American Health Organization and World Health Organization, Humanitarian Supply Chain and Logistics in the Health Sector, PAHO, Washington, DC, 2001.
[34] M.S. Daskin, Networks and Discrete Location: Models, Algorithms, and Applications, Wiley, New York, NY, 1995.
[35] M.S. Daskin, What you should know about location modeling, Naval Res. Logist. 55 (2008) 283–294.
[36] N. Mladenović, J. Brimberg, P. Hansen, J.A. Moreno-Pérez, The p-median problem: a survey of metaheuristic approaches, Eur. J. Oper. Res. 179 (2007) 927–939.
[37] C.S. Revelle, H.A. Eiselt, Location analysis: a synthesis and survey, Eur. J. Oper. Res. 165 (2005) 1–19.
[38] C.S. Revelle, Facility sitting and integer friendly programming, Eur. J. Oper. Res. 65 (1993) 147–158.
[39] R.D. Galvo, C. ReVelle, A Lagrangean heuristic for the maximal covering location problem, Eur. J. Oper. Res. 88 (1996) 114–123.
[40] Z. Drezner, H.W. Hamacher (Eds.), Facility Location: Applications and Theory, Springer-Verlag, New York, NY, 2002.
[41] C.S. ReVelle, H.A. Eiselt, M.S. Daskin, A bibliography for some fundamental problem categories in discrete location science, Eur. J. Oper. Res. 184 (2008) 817–848.
[42] B. Adivar, T. Atan, B.S. Oflaç, T. Orten, Improving social welfare chain using optimal planning model, Supply Chain Manage. 15 (2010) 290–305.
[43] M. Najafi, R. Zanjirani Farahani, M.P. de Brito, W. Dullaert, Location and Distribution Management of Relief Centers: A Genetic Algorithm Approach, Amirkabir University of Technology, Tehran, 2010.
[44] H. Taha, Operations Research: An Introduction, eighth ed., Prentice Hall, Upper Saddle River, NJ, 2007.

[45] F.S. Hillier, G.J. Lieberman, Introduction to Operations Research, seventh ed., McGraw-Hill, New York, NY, 2001.

[46] L.A. Wolsey, Integer Programming, Wiley, New York, NY, 1998.

[47] S. Bazaraa, J.J. Jarvis, Linear Programming and Network Flows, second ed., Wiley, New York, NY, 1977.

[48] R. Zanjirani Farahani, S. Sharahi, Bi-objective Genetic Algorithm for Multi-objective Logistics Planning in Disasters, Amirkabir University of Technology, Tehran, 2010.

[49] G.B. Dantzig, R.H. Ramser, The truck dispatching problem, Manage. Sci. 6 (1959) 80–91.

[50] P. Toth, D. Vigo, An overview of vehicle routing problem, in: P. Toth, D. Vigo (Eds.), The Vehicle Routing Problem, Society for Industrial and Applied Mathematics, Philadelphia, PA, 2002, pp. 1–26.

[51] J.-F. Cordeau, G. Laporte, M.W.P. Savelsbergh, D. Vigo, Vehicle routing, in: C. Barnhart, G. Laporte (Eds.), Handbook of Operations Research and Management Science—Transportation, Elsevier, Amsterdam, 2006, pp. 367–428.

[52] P. Toth, D. Vigo (Eds.), The Vehicle Routing Problem, Society for Industrial and Applied Mathematics, Philadelphia, PA, 2002.

[53] B. Golden, S. Raghavan, E. Wasil (Eds.), The Vehicle Routing Problem: Latest Advances and New Challenges, Springer, New York, NY, 2008.

[54] B. Balcik, B.M. Beamon, K. Smilowitz, Last mile distribution in humanitarian relief, J. Intell. Transport. Syst. Technol. Plan. Oper. 12 (2008) 51–63.

[55] A.M. Campbell, D. Vandenbussche, W. Hermann, Routing for relief efforts, Transport. Sci. 42 (2008) 127–145.

[56] L. Ozdamar, E. Ekinci, B. Kucukyazici, Emergency logistics planning in natural disasters, Ann. Oper. Res. 129 (2004) 217–245.

[57] J.-B. Sheu, An emergency logistics distribution approach for quick response to urgent relief demand in disasters, Transport. Res. Part E 43 (2007) 687–709.

[58] J.R. Birge, F. Louveaux, Introduction to Stochastic Programming, Springer, New York, NY, 1997.

[59] A. Ben-Tal, L. El Ghaoui, A. Nemirovski, Robust Optimization, Princeton University Press, Princeton, NJ, 2009.

[60] H.O. Mete, Z.B. Zabinsky, Stochastic optimization of medical supply location and distribution in disaster management, Int. J. Prod. Econ. 126 (2010) 76–84.

[61] H. Min, V. Jayaraman, R. Srivastava, Combined location-routing problems: a research directions synthesis and future, Eur. J. Oper. Res. 108 (1998) 1–15.

[62] S. Salhi, G.K. Rand, The effect of ignoring routes when locating depots, Eur. J. Oper. Res. 39 (1989) 150–156.

[63] G. Nagy, S. Salhi, Location-routing: issues, models and methods, Eur. J. Oper. Res. 177 (2007) 649–672.

[64] M.S. Daskin, C.R. Coullard, Z.-J.M. Shen, An inventory-location problem: formulation, solution algorithm and computational results, Ann. Oper. Res. 110 (2002) 83–106.

[65] H. Shavandi, H. Mahlooji, Fuzzy hierarchical queueing models for the location set covering problem in congested systems, Sci. Iranica 15 (2008) 378–388.

[66] F. Baita, W. Ukovich, R. Pesenti, D. Favaretto, Dynamic routing-and-inventory problem: a review, Transport. Res. Part A 32 (1998) 585–598.

[67] E. Nikbakhsh, S.H. Zegordi, A heuristic algorithm and a lower bound for the two echelon location-routing problem with soft time windows constraints, Sci. Iranica 17 (2010) 36–47.

[68] M. Desrochers, G. Laporte, Improvements and extensions to the Miller–Tucker–Zemlin subtour elimination constraints, Oper. Res. Lett. 10 (1991) 27–36.

[69] N. Megiddo, K.J. Supowit, On the complexity of some common geometric location problems, SIAM J. Comput. 13 (1984) 182–196.
[70] T.H. Wu, C. Low, J.W. Bai, Heuristic solutions to multi-depot location-routing problems, Comput. Oper. Res. 29 (2002) 1393–1415.
[71] Z. Özyurt, D. Aksen, Solving the multi-depot location-routing problem with Lagrangian relaxation, in: K. Baker, A. Joseph, A. Mehrotra, M.A. Trick (Eds.), Extending the Horizons: Advances in Computing, Optimization, and Decision Technologies, Springer, New York, NY, 2007.
[72] M. Albareda-Sambola, J.A. Díaz, E. Fernández, A compact model and tight bounds for a combined location-routing problem, Comput. Oper. Res. 32 (2005) 407–428.
[73] C.K.Y. Lin, C.K. Chow, A. Chen, A location-routing-loading problem for bill delivery services, Comput. Ind. Eng. 43 (2002) 5–25.
[74] G. Barbarosoglu, L. Ozdamar, A. Cevik, An interactive approach for hierarchical analysis of helicopter logistics in disaster relief operations, Eur. J. Oper. Res. 140 (2002) 118–133.
[75] USAID, Assistance for Iranian earthquake victims. U.S. Agency for International Development, Washington, DC. http://www.usaid.gov/iran/, 2004 (accessed October 3, 2010).
[76] G. Fenton, Coordination in the Great Lakes, Forced Mig. Rev. (2003) 23–24.
[77] N. Altay, Issues in disaster relief logistics, in: M. Gad-el-Hak (Ed.), Large-Scale Disasters: Prediction, Control and Mitigation, Cambridge University Press, New York, NY, 2008, pp. 120–146.
[78] G.P. Cachon, Supply chain coordination with contracts, in: A.G. de Kok, S.C. Graves (Eds.), Handbooks in Operations Research and Management Science: Supply Chain Management: Design, Coordination and Operation, vol. 11, Elsevier Science Publishers, Amsterdam, 2003.
[79] S. Whang, Coordination in operations: a taxonomy, J. Oper. Manag. 12 (1995) 413–422.
[80] I. Giannoccaro, P. Pontrandolfo, Supply chain coordination by revenue sharing contracts, Int. J. Prod. Econ. 89 (2004) 131–139.
[81] L. Xu, B.M. Beamon, Supply chain coordination and cooperation mechanisms: an attribute-based approach, J. Supply Chain Manage. 42 (2006) 4–12.
[82] J.E. McCann, J.R. Galbraith, Interdepartmental Relations, in: P.C. Nystrom, W.H. Starbuck (Eds.), The Handbook of Organizational Design, vol. 2, Oxford University Press, New York, NY, 1981, pp. 60–84.
[83] S. Moore, E. Eng, M. Daniel, International NGOs and the role of network centrality in humanitarian aid operations: a case study of coordination during the 2000 Mozambique floods, Disasters 27 (2003) 305–318.
[84] J. Seaman, Malnutrition in emergencies: how can we do better and where do the responsibilities lie? Disasters 23 (1999) 306–315.
[85] M. Stephenson, Making humanitarian relief networks more effective: operational coordination, trust and sense making, Disasters 29 (2005) 337–350.
[86] L. Minear, The Humanitarian Enterprise: Dilemmas and Discoveries, Kumarian Press, Bloomfield, CT, 2002.
[87] S.F. Schulz, Disaster Relief Logistics: Benefits of and Impediments to Horizontal Cooperation Between Humanitarian Organizations. Doctor of Economics, Berlin Institute of Technology, 2008.
[88] T.H. Poister, Measuring Performance in Public and Nonprofit Organizations, Jossey-Bass, San Francisco, CA, 2003.

[89] R. Bhagwat, M.K. Sharma, Performance measurement of supply chain management: a balanced scorecard approach, Comput. Ind. Eng. 53 (2007) 43–62.

[90] F.T.S. Chan, Performance measurement in a supply chain, Int. J. Adv. Manuf. Tech. 21 (2003) 534–548.

[91] Supply Chain Council, Supply Chain Operations Reference Model 9.0, Supply Chain Council, Cypress, TX, 2009.

[92] A. Gunasekaran, C. Patel, R.E. McGaughey, A framework for supply chain performance measurement, Int. J. Prod. Econ. 87 (2004) 333–347.

[93] J.P.C. Kleijnen, M.T. Smits, Performance metrics in supply chain management, J. Oper. Res. Soc. 54 (2003) 507–514.

[94] E.A. Van Der Laan, M.P. De Brito, D.A. Vergunst, Performance measurement in humanitarian supply chains, Int. J. Risk Assess. Manag. 13 (2009) 22–45.

[95] S.F. Schulz, I. Heigh, Logistics performance management in action within a humanitarian organization, Manag. Res. News 32 (2009) 1038–1049.

[96] A.L. Davidson, Key Performance Indicators in Humanitarian Logistics, Massachusetts Institute of Technology, Cambridge, MA, 2006.

[97] B.M. Beamon, Measuring supply chain performance, Int. J. Oper. Prod. Manag. 19 (1999) 275–292.

[98] G. Kovács, P. Tatham, Humanitarian logistics performance in the light of gender, Int. J. Prod. Perform. Manag. 58 (2009) 174–187.

[99] USGS NEIC, Magnitude 6.3—Java, Indonesia. U.S. Geological Survey, Reston, VA. http://earthquake.usgs.gov/earthquakes/eqinthenews/2006/usneb6/, 2010 (accessed October 3, 2010).

[100] Wikipedia Contributors, May 2006 Java earthquake. Wikipedia, The Free Encyclopedia. http://en.wikipedia.org/wiki/May_2006_Java_earthquake, 2010 (accessed October 3, 2010).

[101] A. Gatignon, L.N. Van Wassenhove, A. Charles, The Yogyakarta earthquake: humanitarian relief through IFRC's decentralized supply chain, Int. J. Prod. Econ. 126 (2010) 102–110.

[102] IFRC, Indonesia: Yogyakarta Earthquake—IFRC Information Bulletin No. 1, May 27, 2006.

[103] NASA, Hurricane Season 2005: Katrina. National Aeronautics and Space Administration, Washington, DC. http://www.nasa.gov/vision/earth/lookingatearth/h2005_katrina.html, 2005 (accessed October 3, 2010).

[104] M. Gad-el-Hak, The art and science of large-scale disasters, in: M. Gad-el-Hak (Ed.), Large-Scale Disasters: Prediction, Control and Mitigation, Cambridge University Press, New York, NY, 2008, pp. 5–68.

[105] Wikipedia Contributors, Hurricane Katrina. Wikipedia, The Free Encyclopedia. http://en.wikipedia.org/wiki/Hurricane_Katrina, 2010 (accessed October 3, 2010).

[106] Wikipedia Contributors, Hurricane Katrina disaster relief. Wikipedia, The Free Encyclopedia. http://en.wikipedia.org/wiki/Hurricane_Katrina_disaster_relief, 2010 (accessed October 3, 2010).

[107] Wikipedia Contributors, International response to Hurricane Katrina. Wikipedia, The Free Encyclopedia. http://en.wikipedia.org/wiki/International_response_to_Hurricane_Katrina, 2010 (accessed October 3, 2010).

[108] Midwest Transportation Consortium, in: Q. Iqbal, K. Mehler, M.B. Yildirim (Eds.), Comparison of Disaster Logistics Planning and Execution for 2005 Hurricane Season, 2007 Ames, IA, May 2007

[109] USGS NEIC, Magnitude 9.1—Off the West Coast of Northern Sumatra. U.S. Geological Survey, Reston, VA. http://earthquake.usgs.gov/earthquakes/eqinthenews/2004/usslav/, 2009 (accessed October 3, 2010).
[110] Wikipedia Contributors, 2004 Indian Ocean earthquake and tsunami. Wikipedia, The Free Encyclopedia. http://en.wikipedia.org/wiki/2004_Indian_Ocean_earthquake_and_tsunami, 2010 (accessed October 3, 2010).
[111] P.-C. Athukorala, B.P. Resosudarmo, The Indian Ocean Tsunami: Economic Impact, Disaster Management and Lessons, Australian National University, Economics RSPAS, Canberra, 2005.
[112] R. Samii, L.N. Van Wassenhove, The Logistics of Emergency Response: Tsunami Versus Haiti. INSEAD—Social Innovation Center, Fontainebleau Cedex. http://www.insead.edu/facultyresearch/centres/isic/documents/TsunamiversusHaiti.pdf, 2010 (accessed October 3, 2010).
[113] Wikipedia Contributors, Humanitarian response to the 2004 Indian Ocean earthquake. Wikipedia, The Free Encyclopedia. http://en.wikipedia.org/wiki/Humanitarian_response_to_the_2004_Indian_Ocean_earthquake, 2010 (accessed October 3, 2010).
[114] M. Perry, Natural disaster management planning: a study of logistics managers responding to the tsunami, Int. J. Phys. Distrib. Logist. Manag. 37 (2007) 409–433.
[115] R. Oloruntoba, An analysis of the cyclone Larry emergency relief chain: some key success factors, Int. J. Prod. Econ. 126 (2010) 85–101.

16 Freight-Transportation Externalities

Fatemeh Ranaiefar[1] *and Amelia Regan*[2]

[1]Institute of Transportation Studies, University of California, Irvine, CA, USA

[2]Computer Science and Institute of Transportation Studies, University of California, Irvine, CA, USA

16.1 Introduction

Freight transportation is a primary component of all supply-chain and logistics systems. However, the cost of moving commodities between cities and countries is borne not only by direct stakeholders (shippers, carriers, or consignees) but also by other members of society who may not benefit directly from these movements. In economics literature, this is referred to as the external cost of an activity. Air, noise, and water pollution; vegetation and wildlife destruction; and road accidents are some of the negative impacts of freight transportation. Freight movements and their associated negative impacts have been steadily increasing over the last few decades in most parts of the world. Negative environmental effects of freight transportation are a serious concern, because of the associated long-term direct and indirect impacts such as increasing greenhouse gases (GHGs) and global warming.

External or social costs of freight transportation have received increased attention recently in development strategies because of sustainability issues. A widely accepted definition of sustainable development is "development that meets the needs of the present without compromising the ability of future generations to meet their own needs" [1]. Strategies for sustainable transportation consider economic development, environmental preservation, and social development [2]. But as the negative impacts of freight transportation increase, achieving sustainability goals becomes ever more challenging.

In general economic terms, the external costs of a product or service can be a source of market failure because these occur outside the market. It is desirable to internalize all costs of transportation, first because the demand-and-supply equilibrium will occur at a more sustainable level, and second because providers are more likely to be more responsible for their decisions and customers will use the services more efficiently. It is also important for governments to know to what extent stakeholders cover the costs when designing taxation and regulation policies.

Logistics Operations and Management. DOI: 10.1016/B978-0-12-385202-1.00016-5

In recent years "greenness" ideas such as green ports, green transportation, and green logistics have been discussed in the literature to emphasize the environmental and sustainability aspects of freight transportation. The goal of green logistics is "planning freight logistics systems that incorporate sustainability goals with a primary focus on the reduction of environmental externalities" [3]. However, there are typically trade-offs or even conflicts between environmental and logistics systems requirements, which makes it hard to achieve both simultaneously [4].

In this chapter, first we investigate different types of freight-transportation externalities with a focus on road transportation. The results of related studies in the United States and Europe are presented and compared though consistent comparison is challenging because of differences in the times and locations of the studies as well as currency exchange rates and basic assumptions such as the statistical value of life. Later we present practical policies that have been introduced to reduce these externalities in the United States and some other parts of the world.

16.2 Freight-Transportation Trends and Costs

Trucks are the dominant mode of inland freight transportation. Based on the latest available data (2004–2007), 76.2% of freight ton-km transported in Europe [5] and 73.0% of total inland tonnage movements in the United States (28.6% of total ton-km) [6] was carried by road transportation. In the United States, truck transportation is growing at 3.5% annually, compared to 2.5% for all vehicles. Trucks now routinely approach 40% of the traffic mix on certain segments of interstate highways at peak times and are likely to increase in the future. At the same time, truck accidents and fatalities are a significant and continual public concern [7]. On average, trucks have higher costs, fuel consumption, and emissions per ton-km of freight transported than marine and rail modes. The US Federal Highway Administration report on air quality (2005) reveals that heavy-duty vehicles have the highest contribution to regional and global emissions: 33% of NO_x and 23.3% of fine particulate matter (PM_{10}) from all mobile source pollutions are produced by heavy-duty trucks [8]. Therefore, this review has truck transportation as its focus. To understand, different types of costs in truck transportation we have to classify the costs and discuss each class separately. In general, there are four nonexclusive classes of costs in each industry. These are shown in Table 16.1 [9].

Litman [9,10] proposed a detailed cost breakdown for passenger transportation. Here we present this list with modification for over-the-road freight transportation.

Not all of the categories listed in Table 16.2 have been estimated in the literature or in practice. Some cannot be easily estimated; even when estimated, the results might not be homogenous for all studies because of a diversity of cost structures. Some categories such as the barrier effect are well known to exist but are hard to estimate because there is no measure to quantify the mobility of nontruck users in roadways with or without trucks.

Table 16.1 General Classification of Costs

Cost Class	Definition	Example
Internal and external costs	Costs borne by users are internal, and those borne by others or society are external costs	Fare: internal Air pollution: external
Variable and fixed costs	Short-term costs proportional to service being used are variable, but long-run costs not related to amount of service being used are fixed costs	Fuel: variable Insurance: fixed
Market and nonmarket costs	Market costs involve goods that are regularly traded in a competitive market. Nonmarket costs involve entities that are not regularly traded in the market	Fuel: market Noise pollution: nonmarket
Perceived and actual costs	Costs that the users are aware of and make an estimation of are perceived costs, whereas actual costs include all costs that may be ignored or underestimated by users. The greater the difference between perceived and actual costs, the less efficient the system	Expected travel time versus actual travel time due to delays in congested traffic

In the SOFTICE [11] project, about 40 major European shippers were surveyed to identify the main parameters affecting freight cost structure. Drivers' wages and fuel cost were identified as the most significant costs factors. For collection and distribution, truck drivers' wages are, on average, 50.4% of the total operating cost. The cost of administration (11.6%), depreciation (10.9%), and fuel (10.3%) are the next three largest costs. For long-distance haulage, trucks drivers' wages and fuel cost are, respectively, 33.0% and 20.4% of total operating cost. These statistics do not include the external costs, but the study presents simulation results of several scenarios in which users have to pay the full costs of freight transportation. External, fixed, long-term, nonmarket, and indirect costs tend to be undervalued by users. This skews user and society decisions and results in an inefficient and unsustainable transportation systems.

As shown simply in Figure 16.1, the demand for freight transportation might change if marginal (total) cost increases because of including external costs. D2 is the quantity demanded, considering full freight-transportation costs as part of the private total cost, whereas D1 is the quantity-demanded level with only users' costs. In the next section, we review different classes of external costs related to freight transportation on roads.

Table 16.2 Freight Transportation Costs Specification

Cost	Description	E/I	V/F	M/NM
Vehicle ownership	Vehicle expenses that are not proportional to the distance that the vehicle is driven	I	F	M
Vehicle operation	User expenses that are proportional to vehicle use	I	V	M
Operating subsidies	Vehicle expenses not paid by the user	E	V	M
Reliability risk	Costs associated with likelihood of delay	E	V	M
Accidents	Vehicle accident costs borne by users	I	V	M
	Vehicle accident costs not borne by users	E	V	NM
Handling facilities and terminals	Cost for providing cranes, loaders for loading and unloading facilities, land for intermodal terminals	I	V	M
	Terminal costs not borne by users	E	F	M
Congestion	Increased delay, vehicle costs, and stress an additional truck imposes on other road users	E	V	NM
Road facilities	Road construction, maintenance, and operating expenses not borne by road users	E	V	M
Roadway land value	Opportunity cost of land used for roads	E	F	NM
Municipal services	Public services devoted to vehicle traffic	E	V	M
Equity/option value	Reduced mode options for rural areas	E	V	NM
Air pollution	Costs of motor-vehicle emissions	E	V	NM
Noise	Costs of motor-vehicle noise	E	V	NM
Resource consumption	External costs resulting from the consumption of petroleum and other natural resources	E	V	NM
Barrier effect	The disutility truck traffic imposes on pedestrians or other vehicles' mobility. Also called *severance*	E	V	NM
Land-use impact	Economic, environmental, and social costs resulting from developing new facilities, terminals, and ports	E	V	NM
Water pollution	Water pollution and hydrologic impacts	E	V	NM
Waste disposal	External costs from motor-vehicle waste disposal	E	V	NM

E, external cost; F, fixed cost; I, internal cost; M, market cost; NM, nonmarket cost; V, variable cost.

16.3 Over-the-Road Freight-Transportation Externalities

Logistics systems have both physical and information infrastructure and regulatory components. Truck fleets, road network infrastructures, freight terminals, and loading and unloading facilities are the components of physical infrastructure, whereas communication systems make up the information infrastructure and road traffic regulations and other related laws comprise the regulatory component. All of these components affect system performance. The costs mentioned in Table 16.2 can be categorized by three primary sources: (1) infrastructure development and maintenance, (2) vehicle operations, and (3) system inefficiency or malfunction.

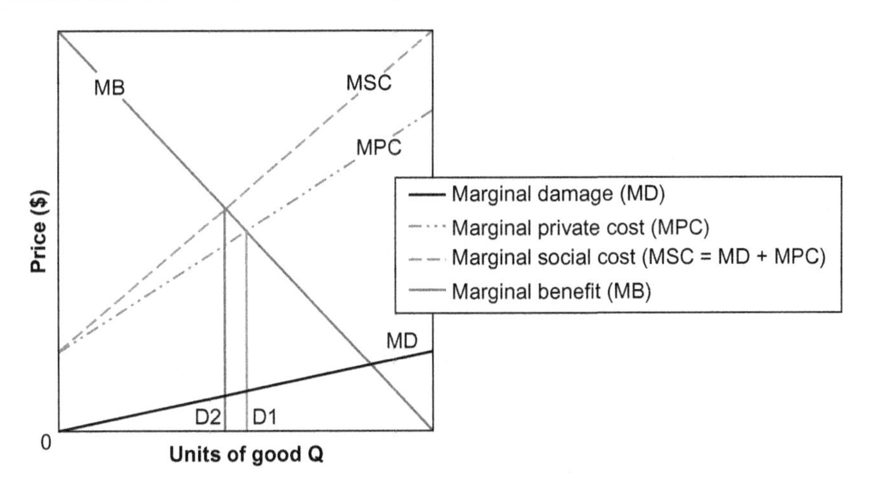

Figure 16.1 Economic diagram for freight-transportation demand and cost.

The first group results in some destruction of the natural environment. Clearly, however, the infrastructures are shared by both passenger and freight transportation, and the external costs imposed to society are generated by both types of users. In addition to the extensive resources required for road building, highway development also affect regional land-use patterns because cities tend to develop along access roads. The pollutants generated from infrastructure maintenance also are a burden to the environment.

The second source of cost includes extensive environmental and social externalities. Some are directly related to trucking operations, such as air pollution from engine exhaust, but some are because of the way drivers operate trucks, such as idling when they could be waiting with their engines off. The US Department of Energy estimated that more than 25 million barrels a year (>$2 billion) is consumed by trucks idling overnight (about 2% of heavy-vehicle fuel consumption in the United States). Each individual truck typically idles for about 1830 hours each year, and idling trucks consume 3157 million gallons of fuel (diesel and gas) each year. This is 8.5% of total commercial truck fuel consumption [12]. About 34% of engine run time for long-haul trucks is spent idling, and around 41% of truck drivers do not take any steps to reduce their idle time [13].

The third group of costs relates to inefficiencies or reduced productivity in operations. This is a topic of interest to both industry analysts and researchers. Increasing fuel efficiency is a crucial goal. The other issue is empty running. When a vehicle is traveling empty or with a low load factor, the operating external costs are imposed on the system with little or no benefit. About 20% and more than 30% of truck trips are empty flows (these numbers vary by type of truck) in the United States and Europe, respectively [14,15]. Empty movements are caused by geographical demand imbalances, scheduling constraints, and vehicle incompatibilities. Clearly, not all empty flows can be eliminated, but communication and scheduling technologies, along with better management, should lead to a decrease in these nonproductive movements.

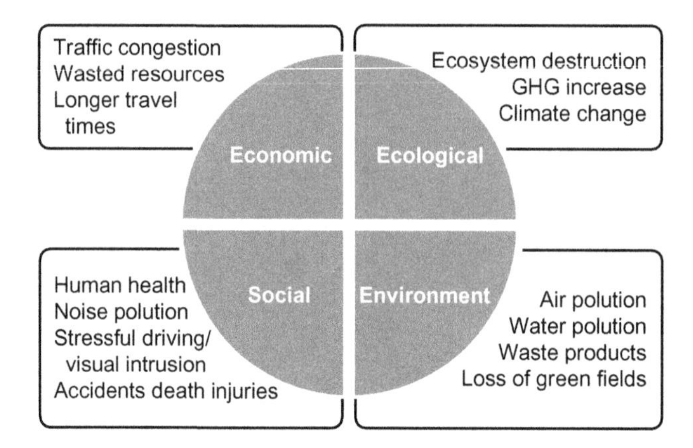

Figure 16.2 Truck-transportation negative impacts.

The other approach to classify the external costs is by the type of impacts they have on society as shown in Figure 16.2.

Different methods are used to quantify or estimate the costs of the negative impacts identified above, although not all of them can be easily measured or estimated. However, where such measurement or estimation is possible, two different economic approaches are generally applied: average cost estimation and marginal cost estimation. Marginal cost is the cost to society of one additional unit of freight transportation. This is a more accurate measure of the cost, but it can require disaggregate data to estimate. Such data may be difficult and expensive to acquire. On the other hand, average cost can be calculated using aggregate data such as number of fatalities and injuries associated with a unit distance of truck operation, average load factor, and average cost per ton-km, which are much more likely to be available to researchers and decision makers. Regardless of the technical approach, there are three general steps as shown in Figure 16.3.

Estimation of each type of externality requires special consideration. It should be noted that the monetary value of damages that are related to human health care or human life are highly dependent on the value of statistical life (VOSL). Different agencies and countries have different estimations, and they update their numbers annually based on inflation rates or other factors. This creates challenges for comparing findings across different studies. Some studies only discuss specific types of externalities [3,16–20], whereas others try to estimate or review the most important externalities related to transportation in general or freight transportation specifically [10,21–29].

16.3.1 Air Pollution

Air pollution causes extensive damage to the environment and society at both the local and global levels. Local-level effects can be seen on humans and animal health, in vegetation and crop damage, and in reduced enjoyment of outdoor

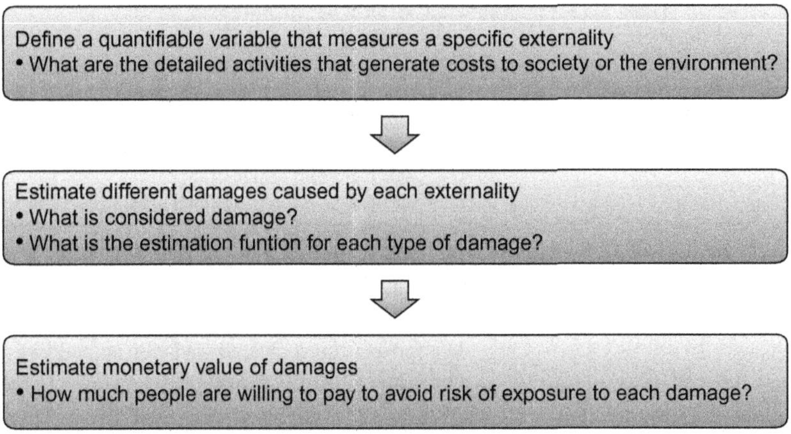

Define a quantifiable variable that measures a specific externality
• What are the detailed activities that generate costs to society or the environment?

Estimate different damages caused by each externality
• What is considered damage?
• What is the estimation funtion for each type of damage?

Estimate monetary value of damages
• How much people are willing to pay to avoid risk of exposure to each damage?

Figure 16.3 General steps in estimation of freight-transportation externalities.

activities. Global-level effects are primarily in climate change. Researchers estimate the social cost of releasing GHGs in the atmosphere separately from air-pollution costs. As mentioned before, if we want to quantify the environmental externalities of freight transportation, then we should estimate relevant emissions from the activities that generate those emissions. Then different damages and related monetary values should be estimated. This process includes estimation of values that rely on the existence of required data, an acceptable and efficient data-collection method, confidence in the quality of the data, and knowledge about the different effects of emissions. Therefore, all the methods have some degree of uncertainty. Further, data obtained in a single geographic region are typically not directly applicable to other regions because of differences in geographic layout and willingness to pay for mitigation measures (e.g., the greater Los Angeles basin vs. the greater Houston area). The environmental effects of different types of pollutants are shown in Table 16.3.

The magnitude of air pollution caused by trucks depends on vehicle characteristics (engine type and condition, fuel type, vehicle age), road characteristics (vehicle speed, traffic congestion level), and driver behavior (idling, average speed, acceleration profile). Different methodologies have been presented in the literature to estimate the environmental costs of freight transportation [10,22,24,27,31,32].

The Transportation Research Board report *Paying Our Way* presents a comprehensive methodology to calculate marginal external cost of freight transportation in the United States [27]. A five-step methodology is applied to estimate the air pollution external marginal cost for four different case studies, including different types of trucks, origin and destination, length of trip, and type of commodity. The steps are (1) estimate change in emissions, (2) resulting change in ambient air quality, (3) change in exposure of humans and property, (4) physical effects of the change in exposure, and (5) economic value of the effects. Their results show the average air-pollution costs for rural and urban areas to be

Table 16.3 Environmental Effects of Air Pollution [8,30]

Pollutant	Effects
VOC: volatile organic compounds (mainly hydrocarbons)	Produces ground-level ozone (O_3), which leads to regional smog production and impairs visibility and alters the taste and smell of air
SO_2: sulfur dioxide	Formation of acid rain, which can adversely affect vegetation, buildings, and humans
NO_x: nitrogen oxides	Produces ground-level ozone (O_3), which leads to regional smog production
	Formation of nitric acid (HNO_3), which causes paint deterioration, corrosion, degradation of buildings, and damage to agricultural crops
	Short-term health effects include acute irritation, neurophysiological dysfunction, and respiratory problems
	Long-term health effects are damage to lung tissue and possibly lung cancer
PM_{10}: particulate matter (10 μm)	Can cause severe health problems
	Increases GHG emissions
CO: carbon monoxide	CO can form ozone and has direct effect on global warming when reacting with hydroxyl (OH) radicals
CO_2: carbon dioxide	Concentration of CO_2 increases GHG effects

$0.017 and $0.025, respectively, per truckload per kilometer for intercity freight transportation.[1]

The European Commission has sponsored a series of studies related to the estimation of the transport air-pollution external cost [17,29,31,32]. Maibach et al. [32] presented a comprehensive report on the estimation of transport marginal external costs with case studies in European countries. The bottom-up "impact pathway approach" [24] is used to estimate the external cost of passenger transport. A passenger car unit (PCU) factor that varies between 2 and 3.5, depending on the type of freight vehicle, is then applied to estimate freight-transportation's air-pollution costs. However, top-down approaches are also discussed. Although the results of the approaches are similar in terms of their order of magnitude, the bottom-up approach aims at estimating marginal costs, whereas the top-down approach produces an average value. The bottom-up approach starts at the microlevel: marginal costs are estimated from the source of emission via quality changes in the air, water, and soil. Finally, monetary values of the impacts are estimated. In contrast, top-down approaches are more aggregate and consider average costs but can analyze the total market equilibrium when new policies are in effect, such as a modal shift because of a change in gas prices. Their results show that the average external

[1] Original calculations were based on 1995 dollars.

cost of air pollution for different types of trucks is between €0.05 and €0.012 per vehicle-km for urban areas and as much as €0.09 per vehicle-km for interurban freight transportation.[2]

Small and Kazimi [20] estimated the health (mortality and morbidity) cost from particulate matter and ozone for heavy-duty vehicles using what they termed the *direct estimation of damages model.* They developed several scenarios based on different statistical values of human life and different levels of emissions' impacts on human health. Their baseline scenario results for Los Angeles estimate the cost at $0.527 per vehicle-mile ($0.327 per vehicle-km),[3] which is significantly higher than other estimates in the literature. The authors attribute this difference to the geographic characteristics of the Los Angeles basin region, which traps pollutants between mountain barriers that surround the area.

Forkenbrock [24] took an average from 2233 US counties, using estimates provided in a range of different studies, and came up with $0.082 per vehicle-mile ($0.051 per vehicle-km).[4] His estimate is very low compared with other studies.

16.3.2 Global Climate Change

Although there is no consensus about the precise role of GHG emissions in climate change, it is clear that the transportation sector is the second greatest source of GHG (after industrial sources) in the United States [33]. Because CO_2 is the dominant GHG, the impacts of other emissions such as CH_4, CO, and NO_2 are also calculated based on CO_2 equivalent factors. The uncertainty of different cost-estimation methods associated with damages from truck GHG emissions is very high because of a wide potential range of GHG impacts that can be included in the models, the intensive data requirements for quantifying them, and exogenous variables used to estimate global effects. Thus, many researchers propose lower and upper bounds rather than a single marginal value or average cost per unit. Apart from the different impacts and range of GHGs included in a model, considering the emissions from the complete life cycle of freight-transportation activities or only from tailpipe emissions results in significant variation in final estimations.

ExternE [31] proposed two complementary methods. The first one is a bottom-up approach that estimates damage costs from the impacts of climate change. The second one is based on the avoidance costs. These avoidance costs are estimated as an equivalent for the preferences followed when focusing on a target policy (i.e., based on the motivation to follow the path to sustainable development). The latter approach is mainly developed for measuring the Kyoto Protocol target, which is an 8% reduction in GHG emissions by 2012, compared to 1990 for the European Union (EU) as a whole.

[2] Based on 2000 prices.
[3] Based on 1992 prices using a $4.87 million value of life.
[4] Based on 1994 prices using a $2.9 million value of life.

The first approach is fairly complicated and uses the Climate Framework for Uncertainty, Negotiation, and Distribution (FUND), and it is done separately for CO_2, CH4, and NO_2. Impacts of climate change with energy consumption, agriculture, coastal zones, forestry, unmanaged ecosystems, water resources, malaria and dengue fever, and cardiovascular health are estimated separately or reviewed from other sources. Their final estimation is €5–20 per ton of CO_2 equivalent.

Maibach et al. [32] also recommended using an avoidance cost method to reduce the uncertainties associated with assessing climate damage costs. He proposed that the marginal cost of GHG emissions can be present based on a fuel-consumption rate for each class of vehicles, including trucks. He concluded that the long-term external cost of GHG emissions is an increasing function starting from €20 per ton of CO_2 in 2010 to €85 per ton of CO_2 in 2050.

Delucchi and McCubbin [23] estimated the cost of GHG emission for the entire life cycle of freight transportation, including a wide range of GHGs, as $0.0003–0.0274 per ton-mile ($0.0002–0.0188 per ton-km) using 2006 prices.

Forkenbrock [24] considered only CO_2 as the main GHG related to trucking. He estimated the social cost of GHG emission from truck transportation based on an initial set of assumptions as follows: one gallon of diesel fuel releases 22.8 pounds of CO_2, the minimum cost of releasing CO_2 to the atmosphere for society is $10 per ton of CO_2, the average fuel efficiency for long-haul trucks is 5.2 miles per gallon, and the average payload is 14.8 tons[5] per vehicle-mile. He concluded that the cost of GHG emissions from truck transportation is $0.15 per ton-mile ($0.10 per ton-km).

16.3.3 Noise Pollution

Noise pollution is more of a concern in urban than rural transportation systems. Some studies even assume external costs of zero for noise pollution in rural areas [27]. Medium to heavy trucks are 10–18 decibels (dB) louder than passenger cars [27]. Truck noise is annoying to residents and pedestrians. Therefore, truck operations during evening and night hours are restricted or prohibited in some areas. Transport noise above a threshold can increase or cause health problems such as changes in heartbeat frequency, increases in blood pressure, hormonal changes, and sleeping problems. The external costs of noise pollution have been studied extensively in Europe and the United States [10,27,29,31,32].

The external cost of noise is mainly reflected in property values when people are less willing to pay for areas near highways. However, this is independent from the health-related costs of noise pollution.

The level of truck noise pollution is influenced by several factors such as vehicle speeds, traffic flow (free flow vs. stop and go), road surfaces, weather, and vehicle type and conditions. The index used for noise is the energy mean sound

[5] An American ton is a short ton, which is 0.90718474 metric ton. Thus, the American 1 ton-mile is 1.45997231821056 ton-km.

level, [dB(A)[6]]. It gives the average sound level over a given period. Noise has a logarithmic relationship with traffic volume. This means that marginal noise pollution decreases as traffic flow increases. In other words, if the current traffic is medium to high, then the marginal noise pollution is small and probably below average, but in low traffic the marginal level can be much higher than average. Time of day, population density exposed to the pollution, and existing noise level are the main drivers of noise cost [32]. Noise levels also decrease with an almost linear relationship with distance from the source. At about 1000 ft (0.3048 km) from the highway, the noise level reaches the background level [27].

In the United States, the first noise-estimation models were developed by the Federal Highway Administration (FHWA) and the National Cooperative Highway Research Program in the late 1960s. The most common model for vehicle traffic is the FHWA's software TNM 2.5 model. Haling and Cohen [16] provide a review of noise-estimation and -prediction models. They also estimate the noise cost produced by trucks of different sizes and carrying different loads using a hedonic price method. This method is based on the reduction of property values caused by vehicle noise emissions. In a hedonic price technique, the actual value of a residential property is dependent on both the physical characteristics of the property and environmental attributes such as pollution levels. Their results show a large variation of noise cost, depending on the type of vehicle, operating conditions, and location of the roadway in relation to residential areas. Haling and Cohen present the results of their estimation by truck type (number of axles and weight), traffic volume, and land development type (urban, rural). Each category is then classified by vehicle speed. The cost estimations vary from 0 to $0.1148 per vehicle-mile ($0.0713 per vehicle-km) in 1993 prices. Mayeres et al. [34] also used the same technique and applied it for Brussels with classification in the results. They estimated that the marginal cost of noise pollution in Brussels is €0.014 per vehicle-km during peak hours and €0.058 per vehicle-km during off-peak hours (using 2005 prices).

The European Commission studies used two approaches (marginal value and average value) to estimate the noise cost. The extensive report, ExternE [31], provides an estimate of the marginal cost of noise pollution. The noise costs for two simulated scenarios in which one has an additional vehicle are calculated. The second approach is based on willingness to pay to have a more quiet environment. This amount is then multiplied by the number of people exposed to the noise. This approach allows us to calculate the average cost of noise pollution. As mentioned before, in moderate traffic the marginal and average costs of noise are approximately the same, whereas in uncongested or heavily congested situations these costs can be very far apart. Nash [29] estimates that the noise cost of heavy good vehicles (HGVs) for daytime versus nighttime and urban versus rural separately. The daytime cost is between €0.08 and €0.26, and the nighttime cost is between

[6] "This index has a logarithmic scale, reflecting the logarithmic manner in which the human ear responds to sound pressure. Since the human ear is also more sensitive at some frequencies than at others, a frequency weighting is applied to measurements and calculations. The most common frequency weighting is the 'A weighting,' hence the use of dB(A)" [32].

€0.23 and €0.78 per vehicle-km for urban areas and the upper bound of €0.03–0.05 per vehicle-km for daytime and nighttime for interurban areas.

16.3.4 Congestion

External congestion costs are experienced by all system users because of the entrance of one additional user into the system when the system is approaching or exceeding capacity. There are two types of congestion costs. The first refers to the delay imposed by one vehicle on another (flow congestion), and the second refers to one vehicle preventing another from gaining access to the network. The other impact of congestion is a significant increase in pollution because of speed reduction. A detailed review of different types of congestion costs is presented in Maibach et al. [32].

Flow congestion is usually described and measured in the literature using speed-flow diagrams. This represents the relationship between congestion and travel time, which is not a linear function. Thus, on a congested road, a small decrease in traffic volume results in a relatively large decrease in delays. On the other hand, the larger the vehicle is, the greater its contribution to congestion. Trucks can have 1.5–2.5 times the impact on congestion that passenger cars have, depending on the roadway conditions because they require more road space and are slower to accelerate [35].

Litman [10] did a comprehensive literature review of transport congestion costs and compared the results of different studies in the United States and Canada. Maibach et al. [32] also reviewed different methods of estimation and internalization of congestion costs and present empirical results for European countries.

The common microsimulation procedure to calculate the marginal contribution of trucks to transport congestion cost is to calculate the difference in speed of other vehicles when a truck is added to the traffic flow [29,34]. Nash [29] estimated the congestion cost of HGVs at between €0.02 and €0.09 per vehicle-km for urban areas and €0.09–0.13 per vehicle-km for interurban areas.

16.3.5 Accidents

The US National Highway Traffic Safety Administration (NHTSA) system [36] reported that in 2008, large trucks were involved in 21.1% of fatal accidents, 4.7% of injury crashes, and 7.8% of crashes with only property damage. Light trucks, which includes many passenger vehicles (SUVs, vans, minivans, and so on in addition to package delivery and utility vehicles), were involved in 60.3%, 59.9%, and 63.1% of fatal accidents, injury crashes, and property-damage-only crashes, respectively.

Each accident has a monetary and nonmomentary cost. The monetary cost of accidents is paid by affected road users, so these are viewed as internal costs, including automobile insurance and medical and emergency service. However, the nonmonetary costs, including increased travel time and emissions, are borne by the larger society, and some individuals who are not beneficiaries of the transportation system bear the loss of family members or long-term pain from injuries. Estimation of the monetary value of the costs caused by freight trucks requires access to accident records that detail the role of trucks in accidents, information that is not generally readily available [24].

The cost of an accident can be viewed as the amount of money that people would pay to reduce the risk of that type of accident. Making this calculation is highly dependent on the availability of appropriate data. The first step is to measure accident rates. There are two typical approaches used to measure accident rates. The first approach uses accident records directly and calculates the accident rate per vehicle distance traveled [19,23,25,29]. The second approach is to estimate accident rates as function of traffic volume [27,29,34]. Other factors also affect the rate of accidents, such as road conditions, weather, and time of day, but it is almost impossible to incorporate them because of expensive data-collection issues.

The next step is to measure the cost associated with each type of accident. Lindberg [19] divided the cost of an accident into direct and indirect parts. The direct part is the tangible cost in the present or future, whereas the indirect one is the lost production capacity resulted from losing a member of society. The estimation of the indirect cost depends on the structure of the society, but Lindberg estimated the indirect cost as a percentage of VOSL for some European countries: 8% of VOSL for fatal accidents and 20% of VOSL for injury accidents. He also estimated an extra external cost of an accident to the relatives and friends of the person who suffered from the accident at 40% of VOSL.

Lindberg [19] reviewed two case studies to find out if the external accident cost increases with axle weight or not. In other words, is the accident rate for HGVs different from other road users? Do the characteristics of the accidents in which HGVs are involved differ significantly from other accidents? He defined three accident-risk measures to compare HGV and light-vehicle accidents: number of accidents per 1000 vehicles, number of accident per million vehicle-kilometers, and average proportion of internal cost to total cost of the accident. He concluded that there is no conclusive evidence on the relationship between truck configuration and accident risk. In a given traffic flow of other vehicles, the total number of accidents increases at a decreasing rate. The number of accidents per 1000 vehicles also strictly increases with increases in vehicle weight class in all case studies. However, the accident risk per vehicle-kilometer does not show a consistent pattern. The risk per kilometer for light vehicles is more than 3 times that of the heavy vehicles. This might be because of differences in the type of exposure as heavier vehicles operate less in urban areas. The ratio of internal cost of the accident to the total cost is between 0.03 (for the lightest vehicles) and 0.38 (for the heaviest vehicles) for the different weight classes, but it does not show a clear trend as vehicle weight increases.

Delucchi and McCubbin [23] estimated the external cost of an accident at between \$0.1 and \$2.0 per ton-mile (\$0.07–1.37 per ton-km) using 1991 US data. These lower and upper bounds are related to assumptions regarding the fraction of accident costs internalized by insurance liability premiums. Forkenbrock [24] estimated the cost at \$0.0059 per ton-mile (\$0.0040 per ton-km) using 1994 US data. The Transportation Research Board [27] estimated the marginal cost of accidents by examining six representative case studies. Within each case study, researchers presented their estimations for urban and rural areas separately. However, they did not arrive at any general conclusions about the difference of accident costs in rural and urban areas. Some case studies have higher cost in rural areas, and others have higher costs in urban areas. The average estimation across the different case studies is $0.0051 per ton-mile[7] (0.0035 per ton-km) using 1996 dollars. Some of the estimations from these studies differ by as much as an order of magnitude, making comparison and general conclusions difficult.

16.3.6 Construction and Maintenance

External cost of construction and maintenance of transportation facilities such as roads, bridges, tunnels, and terminals; the manufacturing of vehicles; and fuel provision, which requires petroleum exploration, refining, and distribution of fuel are often ignored as a source of cost borne by society. Dredging, landfilling, and clearing land for large freight terminals have significant environmental impacts, especially with respect to local wildlife [3].

It is very important how infrastructure costs are allocated between heavy and light vehicles. Litman [10] cited a study in Canada in 1994 in which trucks imposed an average marginal infrastructure cost of US$0.51 per ton-km. Heavy trucks make up only about 9% of Canadian vehicle traffic, but they account for about 25% of roadway costs.

The indirect emissions produced to facilitate road freight transportation are considered as part of the environmental life-cycle assessment (LCA) of freight transportation. The result of freight-transportation LCA in the United States demonstrates that significant emissions are produced outside the operational phase [18]. "In fact the majority of emissions of PM_{10}, SO_2, CO, and Pb are found to occur outside the operational phase for road freight transportation. In particular, PM_{10} and SO_2 are found to have significant emissions associated with infrastructure, comprising approximately 75% and 20% of the life-cycle emissions, respectively" [37]. Heavy trucks also cause most of the damage to road pavements because of the exponential relationship between one extra axle load and pavement damage. The American Association of State Highway and Transportation Officials (AASHTO) [38] estimated this effect roughly as the fourth power rule, which indicates that road damage is proportionate to axle weight raised to the power four.

16.4 Policies to Reduce Externalities

Clearly, no single policy can compensate for all externalities. Each group of strategies focuses on certain aspect of external costs, but the common goal is to either reduce truck travel or increase efficiency of the transportation system. If that can be done, then the total cost will be reduced because external costs are included in variable operations cost. User charges should be applied to the source of an externality to have maximum economic efficiency. Murphy and Poist [39] identified the

[7] Assuming, on average, 14.8 tons transported per vehicle.

following strategies that could be employed by firms to manage and respond to environmental issues related to freight transportation and logistics:

- Reduce consumption whenever possible
- Reuse materials whenever possible
- Recycle materials whenever possible
- Redesign logistical system components for greater environmental efficiency
- Reject suppliers who lack environmental concerns
- Increase the education and training of company personnel
- Encourage greater governmental involvement and regulation
- Publicize environmental efforts and accomplishments
- Promote industry cooperative efforts
- Conduct environmental audits
- Use outside or third parties to manage environmental issues
- Hire and promote environmentally conscious personnel

Policies for reducing freight-transportation externalities can be divided into two groups based on the level of their decision makers. At lower levels firms, manufacturers, and trucking companies can implement different policies to reduce the external costs of freight transportation. At the higher levels, local or national government policies can provide incentives or penalties for lower-level participants to behave in a more sustainable manner. Yano and Katsuhiko [40] listed some of these policies as follows:

- Shift from truck to rail or ships
- Shift from private trucks to for hire trucks
- Promote efficient delivery schedules
- Use back hauling
- Reconsider delivery routes based on real-time traffic conditions
- Promote cooperative deliveries
- Drive with ecology in mind (e.g., reduce idling or maintain constant speed)
- Reconsider locating distribution centers to decrease delivery distances
- Decrease the number of deliveries by reconsidering lot size or delivery frequency
- Change to larger trucks
- Introduce environmentally friendly trucks

In this section, we will discuss these policies in more detail.

16.4.1 *Urban Freight Strategies*

The economic success of a city depends on the efficient movement of goods and services as well as people. The main purpose of urban freight transport is the distribution of goods at the end of the supply chain to the final users, thus many smaller deliveries are made with high frequency.

The environmental impacts of freight movements are more problematic in urban areas than in rural ones. Urban areas are affected more by negative impacts of freight transportation, because quality of life in urban areas is more vulnerable. External costs of public health impacts because of emissions, injuries, and death caused by accidents, noise pollution, and visual intrusion are much higher in cities.

There are two groups of solutions to reduce external costs of urban freight transportation: (1) government policies and regulations and (2) company initiatives [21]. In practice, it is necessary to apply both to get desired advantages. Some potential solutions for reducing urban freight external costs are shown in Table 16.4.

16.4.2 Vehicle-Technology Improvements

New and tightened standards to regulate pollutants (carbon monoxide, hydrocarbons, oxides of nitrogen, and particulate matter) have forced heavy vehicle manufacturers to improve vehicle designs to be less destructive to the environment. Table 16.5 classifies different technology improvements in trucks with respect to different external costs.

A detailed investigation of truck technology development is beyond the scope of this study, but making new technologies economical for carriers is a critical way to reduce the external cost of emissions and noise.

16.4.3 Intelligent Transportation Systems

Information technologies and intelligent transportation systems (ITSs) offer some promise for reducing freight transportation's external costs. Intelligent freight transportation uses advanced technologies and intelligent decision making to make existing infrastructure for freight transportation more efficient [41]. ITS has two key elements: intelligence and integration. The first is characterized by knowledge discovery made possible by better access to data and advanced data-analysis techniques, and the second by an understanding of how to use that data to manage the elements of the system more efficiently. Freight ITS benefits from reduced delay and congestion costs because of the development of integrated systems [42]. Containerized cargo is the main focus in some intelligent freight transportation because it reduces handling cost and loading and unloading time, increases storage efficiency, and increases average payloads [43].

Crainic et al. [44] reviewed freight ITS, including advances in two-way communication, location and tracking devices, electronic data interchange (EDI), advanced planning, and operation decision-support systems and classified them into two groups: commercial vehicle operations (CVO) and advanced fleet management systems (AFMS). The first group includes systemwide, regional, national, or continental applications, and the second group is more concerned about the operations of a particular firm or group of firms. Both groups require e-business activities to be partially integrated across firms, and both are the result of adopting EDI in freight transportation. EDI is a two-way communication and vehicle and cargo location and tracking technology. Of course, the original motivation for development of these technologies was not reducing externalities but increasing the efficiency of the freight-transportation system, which can indirectly decrease externalities.

Table 16.4 Solutions for Reducing External Costs of Urban Transportation

Solution	Challenges	Benefits
Out-of-hours delivery	Staffing Quality control Noise Security Increased costs resulting from longer operating hours	Reduced congestion Increased access to curbsides Parking and maneuvering for other vehicles in congested area during daytime Reduced accidents
Consolidation and transshipment centers	Large land required with good location and access to multiple transport modes Sufficient handling facilities required and high initial investment and operating cost Not applicable for all types of goods (highly perishable, hazardous, high security) Increased air and noise pollution in area adjacent to terminal	Reduced VMT/VKT will reduce emissions, pollution, accidents, and congestion levels Increased load factors will increase efficiency Reduced emissions result from increased use of rail mode
Cooperative operation of private sector (city logistics) [41]	Hard to mange Requires cooperation among retailers and logistics companies rather than normal competition Requires higher level of logistics control in supply chain High initial investment	Avoids duplication and inefficiencies Reduces total operation cost Reduced VMT/VKT reduces emission, pollution, accidents, and congestion
Improve operational efficiency	Increase load factors (vehicle size, shipment size) Increase fuel efficiency Using routing and scheduling software Improve collection and delivery system (material handling technology, unitization of loads, coordination between shipper, carrier, customer) Using effective in-vehicle communication systems	Improve economic efficiency or increase market share by being more environmentally responsible

Table 16.5 Technological Improvements to Reduce External Costs

External Cost	Technology Improvement
Noise	Reduction of engine noise
	Air brake silencers
	Reduced exhaust noise levels
	Better body design
Safety and accident	Better brakes
Emission	Reduction of exhaust emissions
	Alternative fuels
Energy efficiency	Larger vehicles

16.4.4 Pricing Strategies

Efficient pricing is very important for sustainable freight transport. Underpricing has long-term effects on land-use patterns and commodity movements. Pricing strategies have been suggested or implemented in some countries in the form of extra fuel taxes or road tolls for trucks [11]. Fair pricing strategies require a reasonable estimation of the external costs generated by different users. The average cost and marginal cost approaches discussed in previous sections affect the magnitude of pricing strategies, especially in cases with high variation between marginal and average costs such as noise pollution. The significant concern used to justify increasing prices is *equity*. Equity becomes a very important issue when the extra revenue generated by such strategies is to be allocated. If the revenues are used to compensate for the harm caused by freight transportation, then equity is increased. But if the revenue is spent on roadway development or new freight terminals' buildings and equipment, then the equity is decreased.

Nonpricing strategies such as increasing fuel efficiency might lead to increase other environmental externalities because of so-called "rebound effect" of fuel efficiency. For example, increasing fuel efficiency makes operation costs cheaper so idle engine or empty truck flows might increase [45].

Fuel Taxes

Fuel taxes are the most common pricing strategy to internalize freight-transportation externalities. However, this is not the most effective strategy because higher fuel prices do not affect the time of travel, routing, or commodity flow distribution or provide incentives for trucking companies to buy low-pollutant trucks or encourage industries to make their supply chains more efficient. Because of the low elasticity of freight-transportation demand to fuel prices, increasing fuel taxes in the long run probably will not reduce congestion, accidents, or noise and air pollution. The other issue is the international transport or transit flows that allow external cost spillovers to other regions. In transportation systems such as the United States or Europe, where different tax rates are implemented in neighboring regions, it is well known that drivers fill their tanks in the lower tax-rate region before crossing into the higher tax regions. De Borger et al. [46] investigate the potential of tax competition between regions in this situation and present a model for the second-best solution considering the externalities spillover. However, they also analyzed the implication of price discrimination between domestic and international transport on a regional territory and concluded that the optimum tax level depends on the presence or absence of possible discrimination between domestic and international flows. De Borger et al. did not solve their model empirically, but Parry [47] estimated that the optimal (second-best) diesel fuel tax is $1.12 per gallon in the United States.

In Europe, fuel tax is the main policy to internalize external costs of road freight transport, which is roughly estimated as €0.26 per vehicle-km. However, the degree of internalization varies from 30% in Poland, Greece, and Luxembourg to 88% in England [48].

Congestion Pricing

Congestion pricing or dynamic road pricing based on traffic congestion can also internalize congestion costs but not other externalities such as pollution and accidents. However, less congestion on roadways will indirectly reduce pollution and the number of accidents. There are two approaches for congestion pricing in the literature. These are referred to as the first-best and second-best solutions. The first-best policy for a congested road network applies tolls equal to marginal external costs on each individual link. This approach has limitations in practice because of the implementation and monitoring cost of each toll point. However, it is often used as a theoretical benchmark. In the second-best solution, tolls are set for a subset of links in the network only. Tolls can be set on lanes or cordons. The objective is to maximize social welfare given some set of constraints—for example, on a number of toll points. The resulting policy, which is not the best efficient solution, is called the *second-best policy* [49].

The main goals for tolling are demand reduction, traffic allocation, and revenue generation, so it is important to study the impacts of different tolling regimes. Time-varying versus static road pricing is one of the aspects of these regimes. The trade-offs are between the costs, revenue, and network efficiency. Dynamic toll pricing tends to generate higher revenue and improved network efficiency but has higher operating and infrastructure costs than static pricing. De Palma et al. [50] studied different models of road pricing, including flat rate (time-independent tolling), fine tolling (tolls vary over time), and step tolling (a base toll for off-peak time and an extra toll during peak hours) and concluded that time-varying tolls will increase the efficiency of a network.

Kleist and Doll [51] studied the impacts of tolling for HGVs by analyzing the change in the behavior of freight-transportation actors (shippers, carriers and forwarders, and consignees). They found that carriers may (1) shift to secondary roads, (2) shift to more cost-effective vehicles, or (3) improve their productivity. The first impact is a negative consequence of tolling, because "secondary roads are usually more sensitive to pavement degradation, congestion, environmental and noise pollution than the highways." The second impact is ambiguous. Carriers may switch to lighter vehicles to pay fewer or no tolls or to heavier vehicles because they tend to be more cost-effective. The third impact is the best result of tolling because some carriers may decrease their vehicle-miles (or -kilometers) traveled (VMT or VKT) by improving routing plans to reduce empty travel or by increasing load factors. It also encourages firms to merge in order to access larger markets as well as load and vehicle pooling. Shippers and consignees may also react in two different ways to increases in transport costs: (1) reducing the quantity of shipments and (2) reducing travel distances. These are possible through relocation of the firms, suppliers, or consignees and in-house production instead of purchased goods. As the transportation cost is very small compared to production cost, tolling HGV usually does not affect shipper or consignees behavior significantly. Kleist and Doll also comprehensively studied the relative influence of different HGV tolling schemes on the economy and on freight-transport flows for interurban goods movement with a microsimulation model and a four-step travel demand model across all of Europe. The toll rates were set to between €0.13 and €0.20 per km for all or some European highways. They estimated that 5–25% of HGV traffic would shift from highways to secondary roads under different scenarios. The modal shift was estimated to be insignificant after road tolling (less than 1% of total market share), but the impacts were highly dependent on commodity type and trip distance. Unitized goods show the maximum modal shift toward rail transport for long hauls because of how easily these can be transferred. Further, their study found that the average load factor increased between 0.2% and 1.8% for various commodities in different scenarios.

16.4.5 Intermodal Transportation

Intermodal freight terminals provide opportunities to consolidate loads from at least two modes before these loads are delivered to customers. These terminals are usually dominated by trucks as the primary mode, connecting to either rail transportation or waterways. Because the loads are consolidated and shifted to modes other than truck, intermodal transports are more sustainable and have less external costs than single-truck load transportation.

Janic [26] proposed a framework to compare the full cost of intermodal and road freight transportation. He applied the model in European intermodal rail–truck transport and to an equivalent road freight transport network using EU data. The results show that although intermodal network has an economies-of-scale property (the average full costs decrease at a decreasing rate as the quantity of loads rises), the road network has a constant return to scale. Intermodal networks also have greater economies of distance than road networks. In other words, the full costs of both networks decrease more than proportionally as door-to-door distances increase, but the rate is greater for intermodal networks.

Janic's [26] study suggests,

> If the full costs are to be used as the main basis for pricing, the breakeven distance will increase for intermodal transport and thus push it to compete in longer distance markets, with increasingly diminishing demand. However, intermodal transport can neutralize the effects of the higher prices associated with internalizing by increasing the service frequencies in medium-distance markets (around 600–900 km) to meet the large demand there. (p. 43)

However, despite an increasing awareness among policy makers about the benefits of intermodal transportation, the highest estimated share of intermodal transportation in Europe is only 10% [28]. The numbers in the United States are no higher. This poor market share does not create incentives for transportation planners or policy makers to allocate large investments of public funds for required intermodal infrastructures. Ricci and Black [28] reviewed the factors that lead to inefficiencies in intermodal freight transportation in Europe and propose innovative policies and incentives to increase the market share and hence the productivity of intermodal facilities. They also break down the activities in an intermodal chain into 11 groups and discuss external and internal costs associated with each set of activities. The external costs of intermodal transport are generated by the movement of vehicles (locomotives, wagons, trains, trucks, barges, ships) or by the use of machinery and equipment to physically transship from one vehicle to another similar vehicle of the same mode or various modes. Calculating the marginal cost of externalities requires an integrated accounting system and may highly vary across the regions because of various parameters including efficiency of operations, load factors, average length of a journey leg, characteristics of the journey leg (e.g., average speed) and the scale of movements (economies of scale) [28]. They used the ExternE [31] methodology to calculate the external costs of intermodal transport for three major European corridors and compared it with all-road transport. On average, for transporting a 40-ft container, the marginal external costs of all-road transport is 2–3 times higher than intermodal transport.

16.4.6 Strategies to Reduce Empty Travel

As mentioned earlier, between 20% and 30% of truck travel involves repositioning empty vehicles. The associated external costs of empty travel are borne by society, whereas carriers (directly) and shippers and consignees (indirectly) pay for operational costs without gaining any direct benefits. Thus, there are incentives for all participants in freight transportation to reduce empty running vehicles. Empty movements or very low load factors result mainly from the imbalance of flows between different regions. However, as the average length of journey increases, the load factors and back loading tend to be higher because operators have stronger incentives to find return loads and consolidate loads. McKinnon and Ge [15] discussed the factors that constrain back loading practically:

- Lack of coordination between purchasing and logistics departments
- Unreliability of collection and delivery operations
- Priority given to the outbound delivery service
- Inadequate knowledge of available loads

- Incompatibility of vehicles and products
- Resource constraints

Despite the above constraints, decreasing empty running is still possible, and it is an effective way to reduce the external cost of freight transportation. As an example, the empty kilometers traveled by HGVs in England decreased from 33.7% of total truck VMT in 1973 to 26.5% in 2003, while average load factors remained almost stable during most of the period [15]. This change occurred gradually over three decades, but considering the growth in road freight movement, this reduction is equal to saving £1.3 billion in total road freight-transportation cost and not emitting 1.1 million tons of CO_2 into the atmosphere by trucks in 2003. McKinnon and Ge [15] investigated the reasons that led to this reduction and concluded that the following changes were effective in declining empty running:

- Outsourcing of road haulage operations from in-house to third-party transport
- Increasing geographical balance in traffic flow through wider sourcing of commodities, centralization of production and inventory, and greater regional specialization
- Increasing the average length of haulage
- Increasing the direct costs of traveling (fuel, labor) per kilometer
- Increasing the average number of stops (for collection or delivery) per trip
- Developing reverse logistics (packaging waste, handling equipment and product into existing logistics networks)
- Developing load-matching services through online freight-exchange data sources
- Adopting new management initiatives

From the preceding list, we can conclude that reducing operational costs is the main incentive for reducing empty distances traveled by trucks. Thus, if the actors in freight transportation (shippers, carriers, and consignees) have to pay the true cost of their actions, they will increase their performance and efficiency.

16.5 Conclusion

The intent of this study is to review the social costs of freight transportation and the policies being used to reduce these costs. Identifying sources of externalities in logistics systems is very important. First, these costs are nonmarket costs and can lead to market failure. Second, although the users do not pay for these costs, they are sources of inefficiency in the system; they cause damage to society and environment without creating any benefit. Third, in general, reducing externalities will increase the sustainability of the system. Finally, users can increase their long-term benefits by reducing external costs.

We focused on freight transported by trucks because trucks carry the highest share of goods (in ton-kilometer) relative to other transportation modes while having the highest external costs as well. Predictions for the future in United States and Europe are that this mode share will remain essentially unchanged in the future.

We categorized external costs of freight transportation based on their impacts in four groups, namely social, economic, environmental, and ecological. We reviewed

Table 16.6 Summary of Freight Transportation External Costs from Different Studies

Cost	Study					
	Forkenbrock [24]	**Beuthe [22]**	**Paying Our Way [27]**	**ExternE [31]**	**UNITE [29]**	**Delucchi [23]**
Base year prices	1994	1995	1996	2000	2003	2006
Method	Average	Marginal	Marginal	Marginal	Marginal	Average
Unit	$ct/ton-km	€ct/ton-km	$ct/v.km	€ct/v.km	€ct/v.km	$ct/ton-km
Congestion	—	2.108	1.528–4.207	1.05–3.15	2.0–13.0	0.370
Accident	0.404	0.937	5.477–13.414	0.09–17.05	0.084	0.075–1.370
Air pollution	0.055	1.82	1.650–2.069	2.05–28.9	2.09–17.52	0.068–12.88
Climate change	0.103	—	—	1.425–2.05	2.0–3.28	0.014–4.041
Noise pollution	0.027	0.665	0.0–4.039	0.06–30.98	0.0–78.25	0.0–3.630
Water pollution	—	—	—	1.05	—	0.021–0.034
Energy security	—	—	0.444–0.888	—	—	0.151–0.576
Infrastructure	0.171	0.204	11.00–14.919	—	3.62–5.17	—
Total	0.76	5.734	20.099–39.536	5.725–83.18	9.794–117.304	0.329–22.459

the literature for different estimation methods. Not all of the negative impacts of freight transportation have been quantified and estimated in previous literature, because of data availability constrains and the cost of gathering required data. The major costs include congestion, accidents, air pollution, climate change, noise pollution, and construction and maintenance of infrastructures. Water-pollution and energy-security costs are also reviewed in some studies, but we did not discuss them here. Table 16.6 presents a summary of estimations from some of the reviewed studies. Note that the initial assumptions of these studies are not the same, which explains part of the differences in the results. The results are presented directly from the original studies without any adjustment for currency or discount rate for different years.

Acknowledgments

We would like to thank Professor Kenneth Small for his helpful comments and suggestions on our work. Any errors or omissions are, of course, solely the responsibility of the authors.

References

[1] National Research Council (U.S.), Policy Division, Board on Sustainable Development, Our Common Journey: A Transition Toward Sustainability, National Academy Press, Washington, DC, 1999. p. 23.

[2] C.M. Jeon, A. Amekudzi, Addressing sustainability in transportation systems: definitions, indicators and metrics, J. Infrastruct. Sys. 11 (2005) 31–50.

[3] N. Sathaye, Y. Li, A. Horvath, S. Madanat, The Environmental Impacts of Logistics Systems and Options for Mitigation, University of California, Berkeley, CA, 2006UCB-ITS-VWP-2006-4.

[4] J.P. Rodrigue, B. Slack, C. Comtois, Green Logistics, in: A. Brewer, K.J. Button, D. Hensher (Eds.), Handbook of Logistics and Supply Chain Management, Pergamon, Amsterdam, 2001.

[5] Eurostat, Online. http://epp.eurostat.ec.europa.eu/portal/page/portal/product_details/dataset?p_product_code=TSDTR220, cited August 12, 2010.

[6] U.S. Department of Transportation. National Transportation Statistics, Online. http://www.bts.gov/publications/national_transportation_statistics/, cited August 15, 2010.

[7] Transportation Research Board, Freight Transportation Research Needs Statements, TRB, Washington, DC, 2002.

[8] J. Ang-Olson, S. Ostria, Assessing the Effects of Freight Movement on Air Quality at the National and Regional Level, Office of Natural and Human Environment, U.S. Federal Highway Administration, Washington, DC, 2005.

[9] T. Litman, Transportation Cost Analysis for Sustainability, Victoria Transport Policy Institute, Victoria, BC, 1999.

[10] T.A. Litman, Transportation Cost and Benefit Analysis Techniques, Estimates and Implications, Second ed., Victoria Transport Policy Institute, Victoria, BC, 2010. Cited online: September 15, 2010. http://www.vtpi.org/tca/

[11] A. Musso, Survey on Freight Transport Including a Cost Comparison for Europe (SOFTICE), European Commission, 1999.
[12] L. Gaines, T. Levinson, Clean Cities Program: Idle Reduction, U.S. Department of Energy, 2009. Cited online: June 27, 2010. http://www1.eere.energy.gov/cleancities/pdfs/idle_reduction.pdf.
[13] N. Lutsey, C.J. Brodrick, D. Sperling, C. Oglesby, Heavy-duty truck idling characteristics: results from a nationwide truck survey, Transport. Res. Record 1880 (2004) 29–38.
[14] U.S. Census Bureau, Service Annual Survey: transportation and warehousing 2008, 2010, p. 14.
[15] A.C. McKinnon, Y. Ge, The potential for reducing empty running by trucks: a retrospective analysis, Int. J. Phys. Dist. Logist. Manag. 36(5) (2006) 391–410.
[16] D Haling, H. Cohen, Residential noise damage costs caused by motor vehicles, J. Transport. Res. Board 1559 (1996) 84–93.
[17] J. Bates, Economic Evaluation of Emissions Reductions in the Transport Sector of the Bottom-Up Analysis, European Commission, 2001.
[18] C. Facanha, A. Horvath, Environmental assessment of freight transportation in the U.S., Int. J. Life Cycle Assess. 11 (2006) 229–239.
[19] G. Lindberg, Accidents, Res. Transport. Econ. 14 (2005) 155–183.
[20] K.A. Small, C. Kazimi, On the costs of air pollution from motor vehicles, J. Transport Econ. Pol. 29 (1995) 7–32.
[21] M. Browne, M. Piotrowska, A. Woodburn, J. Allen, Literature review WM9: part I—urban freight transport, Green Logist. (2007) 277–290.
[22] M. Beuthe, J.-F. Geerts, F. Degrandsart, B. Jourquin, External costs of the Belgian interurban freight traffic: a network analysis of their internalization, Transport. Res. Part D 7 (2002) 285–301.
[23] M. Delucchi, D. McCubbin, External Cost of Transport in the U.S, in: A. de Palma, R. Lindsey, E. Quinet, R. Vickerman (Eds.), Handbook of Transport Economics, Edward Elgar Publishing Ltd, Cheltenham, UK, 2010.
[24] D. Forkenbrock, External costs of intercity truck freight transportation, Transport. Res. Part A 33 (1999) 505–526.
[25] INFRASS/IWW, External cost of transport, Final report, 2004.
[26] M. Janic, Modeling the full cost of an intermodal and road freight transport network, Transport. Res. Part D 12 (2007) 33–44.
[27] Transportation Research Board, Paying Our Way: Estimating Marginal Social Costs of Freight Transportation, TRB, Washington, DC, 1996. Special Report 246.
[28] A. Ricci, I. Black, The social costs of intermodal freight transport, Res. Transport. Econ. 14 (2005) 245–285.
[29] C. Nash, UNITE (UNIfication of accounts and marginal costs for Transport Efficiency) Final Report, 5th Framework RTD Programme, 2003.
[30] U.S. Environmental Protection Agency, Health Assessment Document for Diesel Engine Exhaust, Office of Research and Development. National Center for Environmental Assessment, U.S. EPA, Washington, DC, 2002.
[31] P. Bickel, R. Friedrich, Externalities of Energy, European Commission, 2005.
[32] M. Maibach, C. Schreyer, D. Sutter, H.P. van Essen, B.H. Boon, R. Smokers, et al., Handbook on Estimation of External Costs in the Transport Sector, European Commission, 2008.
[33] U.S. Environmental Protection Agency, Inventory of U.S. Greenhouse Gas Emissions and Sinks: 1998–2008, U.S. EPA, Washington, DC, 2010.
[34] I. Mayeres, S. Ochelen, S. Proost, The marginal external costs of urban transport, Transport. Res. Part D 1 (1996) 111–130.

[35] Highway Capacity Manual, Transportation Research Board, National Research Council, Washington, DC, 2000.
[36] NHTSA Traffic Safety Facts, A Compilation of Motor Vehicle Crash Data from the Fatality Analysis Reporting System and the General Estimates System, U.S. Department of Transportation, 2008.
[37] C. Facanha, A. Horvath, Evaluation of life-cycle air emission factors of freight transportation, Environ. Sci. Tech. 41 (2007) 7138–7144.
[38] Highway Research Board, The AASHTO road test, Report 7, Summary report, Special Report 61G, Publication no. 1061, National Academy of Sciences, National Research Council, Washington, DC, 1961.
[39] P.R. Murphy, F.P. Richard, Green perspectives and practices: a comparative logistics study, Supply Chain Manage. An. Int. J. 8 (2003) 122–131.
[40] Y. Yano, K. Hayashi, Typology of efforts by Japanese companies to address logistics related environmental issues, in: E. Taniguchi, R.G. Thompson (Eds.), Innovations in City Logistics, Nova Science Publishers, Inc., Hauppauge, NY, 2008.
[41] H. Quak, R.V. Duin, J. Visser, City logistics over the years ... lessons learned, research directions and interests, in: E. Taniguchi, R.G. Thompson (Eds.), Innovations in City Logistics, Nova Science Publishers, Inc, Hauppauge, NY, 2008, pp. 37–53.
[42] E. Taniguchi, R. Thompson, T. Yamada, R.V. Duin, City Logistics, Pergamon, Oxford, 2001.
[43] P.A. Ioannou, Intelligent Freight Transportation, CRC Press, Boca Raton, FL, 2008.
[44] T.G. Crainic, M. Gendreau, J.Y. Potvin, Intelligent freight-transportation systems: assessment and the contribution of operations research, Transport. Res. Part C 17 (2009) 541–557.
[45] F. Ruzzenenti, R. Basosi, The rebound effect: an evolutionary perspective, Ecol. Econ. 67 (2008) 526–537.
[46] B. De Borger, C. Courcelle, D. Swysen, Optimal pricing of transport externalities in a federal system: identifying the role of tax competition, tax exporting and externality spillovers, J. Transport. Econ. Pol. 37 (2003) 69–94.
[47] I. Parry, How should heavy-duty trucks be taxed? Resource for the future, 2006.
[48] M. Piecyk, A. McKinnon, Internalizing the External Costs of Road Freight Transport in the UK, Heriot-Watt University, Edinburgh, 2007.
[49] K.A. Small, E.T. Verhoef, The Economics of Urban Transportation, Routledge, 2007.
[50] A. De Palma, R. Lindsey, E. Quinet, Time-varying road pricing and choice of toll locations, in: G. Santos (Ed.), Road Pricing: Theory and Evidence, Elsevier, Amsterdam, 2004.
[51] L. Kleist, C. Doll, Economic and environmental impacts of road tolls for HGVs in Europe, Res. Transport. Econ. 11 (2005) 153–192.

17 Robust Optimization of Uncertain Logistics Networks

Sara Hosseini[1] *and Wout Dullaert*[2,3]

[1]Petrochemical Industries Development Management Co., Tehran, Iran
[2]Institute of Transport and Maritime Management Antwerp, University of Antwerp, Belgium
[3]Antwerp Maritime Academy, Antwerp, Belgium

Logistics, as comprehensively defined by Riopel et al. [1] "is that part of the supply-chain process that plans, implements, and controls the efficient, effective forward and reverse flow and storage of goods, services, and related information between the point of origin and the point of consumption in order to meet customers' requirements." So it is obvious that planning and managing this vast range of processes would be extremely complicated. In other words, decision makers should concentrate on managing any probable risk of the logistics system, starting from the design phase. As a result, unforeseen conditions during implementation will be less likely to invalidate the basic design plan or disturb the performance targets.

One of the main difficulties of the logistics management problems is how to wisely consider the uncertainty about the future in the modeling phase. In the real world, the existence of noisy, incomplete, or erroneous information and data is an unavoidable fact that widely affects the efficiency of the logistics network processes (e.g., location of logistics centers, distribution plans, and customer demands) at the implementation phase, so not correctly modeling for these inherent uncertainties might result in impractical plans. Hence, one critical role of logistic managers relates to the way of facing noisy and uncertain environments in order to obtain more effective networks with less replanning.

The importance of this subject has caused a considerable growth in the number of studies dedicated to the uncertainty in supply-chain and logistic networks and their associated modeling approaches to optimize the design and performance of the networks under uncertainty.

Two general approaches have been used in the face of uncertainty in the logistics studies and planning: reactive and proactive.

1. *Reactive approaches*: They are postoptimal, so they cannot provide any direct mechanism to control the sensitivity of decisions to the uncertainties. Sensitivity analysis is categorized in this group.
2. *Proactive approaches*: In contrast to the previous group, these practices are applied to yield solutions that are less sensitive to the uncertainty. Stochastic programming is a conventional optimization method with probabilistic data that belong to this category.

Logistics Operations and Management. DOI: 10.1016/B978-0-12-385202-1.00017-7

One shortage of stochastic programming is its limitation to handle decision makers' preferences or risk aversion. More recently, an improved stochastic programming called *robust programming* has been developed with the capability of tackling this shortage. Owing to the flexible modeling qualifications allowed by robust optimization (RO), it is believed that this approach can provide a credible methodology for real-world uncertain logistics problems. In other words, the simplicity of implementation of this method enables decision makers to manage and control the logistics system without having to learn complicated programming procedures.

17.1 A Literature Review on RO

Klibi et al. [2] present a critical review of the supply-chain network (SCN) design problem under uncertainty and of the available proposed models for properly formulizing the uncertainty at the design phase. Several definitions of robustness, literature of it in the supply-chain context, and the necessity of SCN robustness to ensure sustainable value creation are reviewed and discussed in their paper. The theory, methodology, and main approaches to cope with optimization problems under uncertainty are reviewed by Sahinidis [3]. One of these approaches is a special type of stochastic nonlinear programming called RO introduced by Mulvey et al. [4]. Bai et al. [5] have made some applications for RO models in which the traditional stochastic linear programming fails to identify a robust solution. Gutierrez et al. [6] have addressed a robustness approach in the formation of regret model to a standard MIP formulation of uncapacitated network design problems. Therein, some algorithms that are adaptations of Bender's decomposition methodology have been developed to generate robust network designs. Vladimirou and Zenios [7] have presented a RO model to obtain a trade-off between the stability of recourse decisions and the expected cost of a solution. Yu and Li [8] have developed a new RO model for stochastic logistic problems. Their proposed formulation needs only adding half of the variables toward the formulation of Mulvey et al. [4] for transferring a nonlinear robust model to a linear program. Landeghem and Vanmaele [9] have proposed the concept of robust planning, which applies risk assessment within uncertain demand and supply chains at the tactical level. Also, they indicated the value of robust planning in a European chemical enterprise. Leung et al. [10] have presented a RO model for an uncertain cross-border logistics problem with fleet composition to determine an optimal long-term transportation strategy under different economic growth scenarios. More recently, Leung et al. [11] have applied the improved RO formulation proposed by Yu and Li [8] to solve a multisite production planning problem in an uncertain environment. An RO problem in the form of regret model is studied by Baohua and Shiwei [12] for logistics center location and allocation in an uncertain environment. By numerical experiments, it has been shown that the results of this method are better than those of the stochastic optimization model. Yin et al. [13] have developed three robust improvement schemes (sensitivity based, scenario based, and min-max) for road networks under future uncertain demand, applying different techniques to model uncertainty with different prospective on robustness.

17.2 Optimization Under Uncertainty

17.2.1 Uncertainties in the Logistics Networks

One of the main challenges for mathematical modeling of an effective logistics network that adapts with the real-world data is how to logically incorporate uncertain and noisy data in the network design phase. In fact, making any mistake in this phase may result in serious malfunction of the logistics network in the implementation phase, so it probably fails to achieve some of targeted goals. As a consequence, additional costs would be imposed on the system to penalize the lack of precise network design.

In recent decades, notable studies, but certainly not enough, have been published that focus on conveniently recognizing and formulizing the uncertainties in network design problems. For example, Davis [14] introduced uncertainty as a major factor affecting the supply chain that is not adequately handled by managers. He identified supplier performance, manufacturing processes, and customer demands as major sources of uncertainty. In a similar way, Landeghem and Vanmaele [9] listed the main sources of uncertainty regarding the degree of their effects at different levels of the supply chains in Table 17.1.

Table 17.1 Sources of Uncertainty in the Supply Chain: Low, Medium, and High Leverage of Decisions [9]

Sources of Uncertainties	Operational	Tactical	Strategical
Exchange rates	●●●	●●	
Supplier lead-time	●	●●●	●
Supplier quality	●●	●	
Manufacturing yield	●●	●●	
Transportation times	●●	●●	●
Stochastic costs	●	●●●	●●
Political environment			●●
Customs regulations	●	●●	●●●
Available capacity	●●	●●	●
Subcontractor availability	●●●	●●	
Information delays	●●●	●●	
Stochastic demand	●	●●●	●●
Price fluctuations	●	●●●	●

Low (●); medium (●●); high (●●●).

17.2.2 Optimization Approaches Under Uncertainties

As a good classification, Sahinidis [3] categorizes and reviews the main optimization approaches under uncertainty into three groups: (1) stochastic programming (recourse models, robust stochastic programming, and probabilistic models); (2) fuzzy programming (flexible and possibilistic models); and (3) stochastic dynamic programming.

In stochastic programming, it is assumed that the probability distribution functions of the uncertain parameters are known and that decision makers try to find an optimal solution that minimizes the expected value of objective. One development of stochastic programming is RO, which describes the uncertain parameters by the discrete scenarios or a continuous range and is capable of handling decision makers' favored risk aversion. The goal in this approach is obtaining a series of solutions that are less sensitive to any realizations of the uncertain parameters. In fact, the term *robustness* has emerged in statistics and become popular in the field of control theory since the 1970s. This word is used when the control system is influenced by unavoidable fluctuations of parameters.

17.2.3 Robust Optimization

In general, RO consists of two types of constraints: structural and control. The input data in the first group are free of noise; the second ones are influenced by noisy data. Also, two sets of variables are defined: design and control. The control variables are subjected to uncertainty, unlike the design variables.

The scenario-based RO involves a set of scenarios $\Omega = \{1,2,\ldots,S\}$. Under each scenario $s \in \Omega$, the coefficients of the control constraints will become $\{d_S, B_S, C_S, e_S\}$, with a fixed given probability of occurrence of a scenario s, $P_s\left(\sum_S P_S = 1\right)$. Let y_s be the control variable $\forall s \in \Omega$ and δ_s be the error vector that shows the value of the allowed infeasibility in the control constraints under scenario s. Then a mathematical formulation for the RO model is as follows:

$$\text{Min} \;\; \sigma(x, y_1, y_2, \ldots, y_s) + \omega\rho(\delta_1, \delta_2, \ldots, \delta_s) \tag{17.1}$$

subject to

$$Ax = b \tag{17.2}$$

$$B_s x + C_s y + \delta_s = e_s \quad \text{for all} \;\; s \in \Omega \tag{17.3}$$

$$x \geq 0 \;\; , \;\; y_s \geq 0 \quad \text{for all} \;\; s \in \Omega \tag{17.4}$$

The first term of the objective function would be a moment of the $\xi_s = c^T x + {d_s}^T y_s$with probability P_s under scenario s. This term represents the solution's robustness. The optimal solution of this model will be called robust if it remains *close* to *optimality* for any realization of the scenario $s \in \Omega$.

The second term of the objective function prepares a feasibility penalty value to penalize the probable violations of the control constraints. This term is referred as *model robustness*. Each solution will be robust if it remains *almost feasible* for any realization of the scenario $s \in \Omega$.

The weight ω is used to obtain a good trade-off between the *solution robustness* (optimality) and the *model robustness* (feasibility) under the conception of the multi-criteria decision making.

The RO includes two major model formulations [12]: a regret model and a variability model.

Regret Model

In this model, the regret value of a scenario is referred to as the *absolute* or *relative difference* between the objective value of the feasible solution and the best objective function.

Let S denote the set of scenarios and x be the feasible solution of the deterministic model P_s, $\forall s \in S$. $Z_s(x)$ is the objective of P_s and Z_S^* is the optimal objective of it. Also, let given constant $\omega \geq 0$ be the regret coefficient. $Z_s(x) - Z_S^*$ is the absolute regret value, and $[Z_s(x) - Z_S^*]/Z_S^*$ is the relative regret value. The overall framework of this model is discussed below.

$$\text{Min} \sum_S q_S Z_S(x)\sqrt{a^2 + b^2} \tag{17.5}$$

subject to

$$[Z_S(x) \leq (1 + \omega)Z_S^*] \quad x \in \Omega \tag{17.6}$$

If $[Z_s(x) - Z_S^*]/Z_S^* \leq \omega \quad \forall s \in S$, then x is the robust solution of P_s. Maybe there exist several robust solutions. The optimal solution of the above model is the best robust solution.

Variability Model

In this model, the higher moments of the distribution of the objective value are used to reduce the sensitivity of model to uncertain data—i.e., the standard deviation and variance.

The following nonlinear formulation was proposed by Mulvey et al. [4] for the first term of RO:

$$\sigma(x, y_1, y_2, \ldots, y_s) = \sum_{s \in \Omega} P_s \xi_s + \lambda \sum_{s \in \Omega} P_s \left(\xi_s - \sum_{s' \in \Omega} P_{s'} \xi_{s'} \right)^2 \tag{17.7}$$

To reduce the complexity of this formulation, Yu and Li [8] proposed the following form:

$$\sigma(x, y_1, y_2, \ldots, y_s) = \sum_{s\in\Omega} P_s\xi_s + \lambda\sum_{s\in\Omega} P_s\left|\xi_s - \sum_{s'\in\Omega} P_{s'}\xi_{s'}\right| \tag{17.8}$$

Then this form is converted to a linear form using two nonnegative deviational variables. The overall framework of their model is:

$$\sigma(x, y_1, y_2, \ldots, y_s) = \sum_{s\in\Omega} P_s\xi_s + \lambda\sum_{s\in\Omega} P_s\left[\left(\xi_s - \sum_{s'\in\Omega} P_{s'}\xi_{s'}\right) + 2\theta_s\right] \tag{17.9}$$

subject to

$$\xi_s - \sum_{s\in\Omega} P_s\xi_s + \theta_s \geq 0 \tag{17.10}$$

$$\theta_s \geq 0 \tag{17.11}$$

Besides these two main models—regret and variability—there exist other formulations of RO such as the worst-case analysis, which contains two principles named *minimax* and *maximin*. Here, it does not matter for decision makers how much the system performance changes above the level, as long as it achieves a certain acceptable level. This asymmetric effect could result in a performance improvement in the worst case, whereas the average performance is poorer. When applying this method, it is not necessary to consider all possible realizations of uncertain data in the uncertainty set.

17.3 RO of Logistics Networks

As mentioned before, in the real world, the logistics networks frequently encounter uncertain data. Ignoring each of them might result in resource waste and low network efficiency. The studies show the helpful role of RO to logistics managers and decision makers to wisely solve the uncertain logistics problems. Moreover, the obtained results from the sets of real-world data indicate that the RO model is more realistic in dealing with future economic conditions. This section presents applications of RO in logistics networks.

17.3.1 A Variability Formation of RO for the General Logistics Problem [8]

An efficient RO model has been presented by Yu and Li [8] for solving an uncertain logistics problem. First, they reviewed a general deterministic logistics problem and then adjusted this model in three RO approaches introduced by Mulvey et al. [4] and Mulvey and Ruszczynski [15]. They studied the limitations of each approach such as having to add many extra variables and constraints. After that, they presented a novel RO formulation for the general logistics problem. Finally, two logistics examples (a wine company and an airline company) prove that the proposed method is more computationally efficient than the conventional methods because it contains a lower number of variables or constraints.

We formulate a general framework of a deterministic logistics problem that aims to minimize the costs associated with the production and distribution of products under a variety of constraints as follows:

$$\text{Min} \sum_{i}\sum_{k} c_{ik}x_{ik} + \sum_{k} c_k x_k + \sum_{k}\sum_{j} c_{kj}x_{kj} \tag{17.12}$$

subject to

$$Ax \geq b \tag{17.13}$$

$$\sum_{k} x_{kj} - D_j = g_j \quad \forall j \tag{17.14}$$

$$\text{All } x_{ik}, x_k, x_{kj}, D_j, g_j \geq 0 \tag{17.15}$$

Let x_{ik} be the amount of raw material shipped from location i to plant k, and c_{ik} be the unit cost for their shipment. Let x_{kj} be the amount of product x shipped from plant k to market j, and c_{kj} be the unit cost for their shipment. Also, x_k denotes the amount of product x produced at plant k, and c_k represents the unit cost for their production. D_j denotes the demand of market j for this product, and g_j represents the safety stock at marketplace j, which is specified by the decision maker.

Here, the objective of Eqn (17.12) is to minimize the total cost of transportation, production, and inventory. Equation (17.13) represents the general constraints associated with flow balance, workers, materials, funds, and other resources requirements in related locations, plants, and markets. Equation (17.14) expresses the constraints of supply and demand in markets.

Now the novel RO formulation of the general logistics after transforming it to a standard linear problem, introduced by Yin et al. [13], is presented as follows:

$$\begin{aligned}
\text{Min} & \sum_{s\in S} P_s\left(\sum_{i}\sum_{k} c_{ik}x_{ik} + \sum_{k} c_{sk}x_k + \sum_{k}\sum_{j} c_{kj}x_{kj}\right) \\
& + \lambda \sum_{s\in S} P_s \left[\begin{array}{l} \left(\sum_{i}\sum_{k} c_{ik}x_{ik} + \sum_{k} c_{sk}x_k + \sum_{k}\sum_{j} c_{kj}x_{kj}\right) \\ - \sum_{s'\in S} P_{s'}\left(\sum_{i}\sum_{k} c_{ik}x_{ik} + \sum_{k} c_{s'k}x_k + \sum_{k}\sum_{j} c_{kj}x_{kj}\right) + 2\theta_s \end{array} \right] \\
& + \omega \sum_{s\in S} P_s \left(\omega_{sj}^{+} \left(\sum_{k} x_{kj} - D_j - g_{sj} + \delta_{sj} \right) + \omega_{sj}^{-}\, \delta_{sj} \right)
\end{aligned} \tag{17.16}$$

subject to

$$Ax \geq b \tag{17.17}$$

$$\sum_{s \in S} P_s \left(\sum_i \sum_k c_{ik} x_{ik} + \sum_k c_{sk} x_k + \sum_k \sum_j c_{kj} x_{kj} \right) - \left(\sum_i \sum_k c_{ik} x_{ik} + \sum_k c_{sk} x_k + \sum_k \sum_j c_{kj} x_{kj} \right) - \theta_s \leq 0 \tag{17.18}$$

$$\sum_k x_{kj} + D_j + g_{sj} - \delta_{sj} \leq 0 \quad \forall \ j \text{ and } s \tag{17.19}$$

$$\text{All } x_{ik},\ x_k,\ x_{kj},\ D_{sj},\ gs_j,\ \theta_s,\ \delta_{sj} \geq 0 \tag{17.20}$$

where θ_s denotes the deviation for violation of the mean, and δ_{sj} represents the deviation for violations of the control constraints.

This new linear formulation just needs to add $n+m$ deviation variables (where n and m are the number of scenarios and total constraints, respectively), whereas the proposed model by Mulvey et al. [4] requires adding $2n+2m$.

17.3.2 A Regret Formation of RO for the Logistic Center Location and Allocation [12]

Logistics centers have a very important role in the logistics networks. If the existent uncertainty in the parameters of these systems is ignored through the design phase, then some impractical location plans might be obtained. Moreover, it is recommended to consider the distribution plan when working on the location of the logistics centers.

This field is concentrated under uncertain demand in Baohua and Shiwei [12]. They present a RO model using the regret formation and solve it by both the enumeration method and the genetic algorithm. Finally, it has been shown by numerical experiments that the RO model can obtain better solutions than the stochastic optimization model via reducing the risk of decisions.

The formulation of the regret model of logistics center location and allocation model with uncertain demand is presented as follows.

Let S denote the set of scenarios. $\forall s \in S$, ρ_s denotes the probability of scenario s. V denotes the set of nodes. Q denotes the set of supply nodes. L represents the set of possible locations of logistics center. D is the set of demand nodes. A represents the set of arcs. $\forall i \in A$, o_i is the origin of arc i, and d_i is the destination of arc i. P is the set of products. q_j^p denotes the supply capacity of product p in the supply node j, $j \in Q$ and $p \in P$. η_j is the operation capacity of the possible location j, $j \in L$. r_j^p denotes the capacity needed for unit product p in possible location j, $j \in Q$ and $p \in P$. n denotes the largest number of logistics center. ω denotes the regret coefficient. In this chapter, the regret coefficients are assumed to be identical for all scenarios.

Also, let y_{ip}^{s} denotes the flow of product p on the arc i under scenario s, which is the integer decision variable. x_j denotes the location of logistics centers and binary decision variable. $x_j = 1$ if the logistics center is going to be built at site j; otherwise, $x_j = 0$.

$$P: \quad \text{Min } Z = \sum_{s \in S} \rho_s Z_s \tag{17.21}$$

subject to

$$Z_s = \sum_{j \in L} W_j x_j + \sum_{s \in S} \sum_{p \in P} \sum_{i \in A} c_i^p y_{ip}^s \tag{17.22}$$

$$\sum_{i \in A | d_i = j} y_{ip}^s = \sum_{i \in A | o_i = j} y_{ip}^s \quad \forall s \in S, \ j \in L, \ p \in P \tag{17.23}$$

$$\sum_{i \in A | d_i = j} y_{ip}^s \geq d_{jp}^s \quad \forall s \in S, \ j \in D, \ p \in P \tag{17.24}$$

$$\sum_{i \in A | o_i = j} y_{ip}^s \leq q_j^p \quad \forall s \in S, \ j \in Q, \ p \in P \tag{17.25}$$

$$\sum_{p \in P} r_j^p \left(\sum_{i \in A | d_i = j} y_{ip}^s \right) \leq \eta_j x_j \quad \forall s \in S, \ j \in L \tag{17.26}$$

$$\sum_{j \in L} x_j \leq n \tag{17.27}$$

$$\left(Z_s(x) - Z_s^* \right) / Z_s^* \leq \omega \quad \forall s \in S \tag{17.28}$$

$$Y_{ip}^s \in Z^+ \quad \forall s \in S, \ i \in A, \ p \in P \tag{17.29}$$

$$x_j \in (0, 1) \quad \forall j \in L \tag{17.30}$$

The objective function of Eqn (17.21) aims to minimize the total average cost under all scenarios. Equation (17.22) provides the total cost under each scenario s. Equation (17.23) is the flow conservation constraint of each logistics center node. Equation (17.24) ensures that the demand should be satisfied at each demand node. Equation (17.25) ensures that the total supply of each supply node should not exceed its capacity. Equation (17.26) is the operation capacity constraint of each logistics center node. Equation (17.27) ensures that the number of logistics centers is restricted to a given number. Equation (17.28) ensures that the feasible solution of model P should meet the requirement of the robust solution. Equations (17.29) and (17.30) are logical constraints of the decision variables.

17.3.3 A Min-Max Formation of RO for Road Networks [13]

Three robust improvement schemes are presented by Yin et al. [13], including sensitivity based, scenario based, and min-max for road networks under future uncertain demand. In this chapter, the improvement schemes are called robust if the resulted solutions are less sensitive to any realizations of uncertain demands (in the sensitivity-based and scenario-based schemes) or if the system performs better against the worst-case or high-consequence demand scenarios (in the min-max scheme). They formulate these different schemes as robust programs and present convenient solution algorithms. Finally, they validate their proposed models by numerical examples and simulation tests.

The min-max scheme for road networks is studied as follows.

Consider a network $G=(N,A)$, where N is the set of nodes, and A is the set of links. Let W be the set of all origin–destination (O–D) pairs in the network. Denote travel demand between all O–D pairs as a vector q, which is assumed to be unknown but bounded by an uncertainty set Q. R_w is the set of routes between O–D pair $w \in W$, and q_w is the demand between O–D pair w. $\delta_{ar}^{w}=1$ if route r between O–D pair w uses link a and 0 otherwise. Denote V as the set of feasible link flow vectors (v). Denote the travel time for each link $a \in A$ as $t_a(v_a, c_a)$. v_a is the traffic flow, and c_a is the capacity of link $a \in A$. c_a^{+} is the continuous capacity increase of link a. $h_a(c_a^{+})$ is the construction cost function that is generally assumed to be nonnegative, increasing, and differentiable; B is the available budget; $c_a^{\max}$ is the upper limit of the capacity increase; and c^0 is the vector of the original link capacities.

$$\underset{c^{+},v}{\text{Min}} \; \underset{q \in Q}{\text{Max}} \sum_{a} v_a \cdot t_a\left(v_a,\; c_a^0 + c_a^{+}\right) \tag{17.31}$$

subject to

$$\sum_{a \in A} h_a(c_a^{+}) \leq B \tag{17.32}$$

$$0 \leq c_a^{+} \leq c_a^{\max} \quad a \in A \tag{17.33}$$

$$v_a = \sum_{w \in W} \sum_{r \in R_w} q_w \cdot \delta_{ar}^{w} \cdot p_r^{w}(t^{w}(v,\; c^0 + c^{+})) \quad a \in A,\;\; v \in V,\;\; q \in Q \tag{17.34}$$

To solve this RO model, they propose a heuristic algorithm that includes an iterative procedure to obtain move directions and to generate a sequence of solutions until a convergence criterion is met.

At the end of their study, they conclude some facts about the situations in which the use of each improvement scheme is preferred. Upon their conclusion, the sensitivity-based model is preferred when the fluctuations of the uncertain parameters are believed to be nonsignificant, or when the robustness approach is considered as a side improvement effort. In the other side, when it is intended to make the system performance more stable, the preferred model will be the scenario-based one. Finally, when the system performance is measured under the worst-case or high-consequence scenarios, the min-max model will be appropriate to be applied by decision makers. This model does not need any prior information about distributions of uncertain parameters.

17.4 Challenges of RO

In spite of the simplicity of implementation of this method and its applicability in modeling real-world cases, it cannot be ignored that there are several limitations in this approach. Two major shortcomings of the scenario-based RO are (1) how to determine the number of scenarios that should be included in the model to find the robust solution and (2) how to generate those scenarios and specify their related probabilities [13]. In this way, some studies have been done to overcome these limitations. For example, variance-reduction methods can be used to generate the representative scenarios.

However, we believe that the merits of developing RO would encourage the decision makers to incorporate uncertainty into the logistics networks' design phase. Besides, this field is attractive enough for further research.

References

[1] D. Riopel, A. Langevin, J.F. Campbell, The network of logistics decisions, in: A. Langevin D. Riopel (Eds.), Logistics Systems: Definition and Optimization, Springer, New York, 2005.

[2] W. Klibi, A. Martel, A. Guitouni, The design of robust value-creating supply chain networks: a critical review, Eur. J. Oper. Res. 162 (2009) 4–29.

[3] N.V. Sahinidis, Optimization under uncertainty: state-of-the-art and opportunities, Comput. Chem. Eng. 28 (2004) 971–983.

[4] J.M. Mulvey, R.J. Vanderbei, S.A. Zenios, Robust optimization of large-scale systems, Oper. Res. 43 (1995) 264–281.

[5] D. Bai, T. Carpenter, J.M. Mulvey, Making a case for robust optimization models, Manage. Sci. 43 (1997) 895–907.

[6] G.J. Gutierrez, P. Kouvelis, A.A. Kurawarwala, A robustness approach to uncapacitated network design problems, Eur. J. Oper. Res. 94 (1996) 362–376.

[7] H. Vladimirou, S.A. Zenios, Stochastic linear programs with restricted recourse, Eur. J. Oper. Res. 101(1) (1997) 177–192.

[8] C.S. Yu, H. Li, A robust optimization model for stochastic logistic problems, Int. J. Prod. Econ. 64 (2000) 385–397.
[9] V.H. Landeghem, H. Vanmaele, Robust planning: a new paradigm for demand chain planning, J. Oper. Manage. 20 (2002) 769–783.
[10] S.C.H. Leung, Y. Wu, K.K. Lai, A robust optimization model for a cross-border logistics problem with fleet composition in an uncertain environment, Math. Comput. Model. 36 (2002) 1221–1234.
[11] S.C.H. Leung, S.O. Tsang, W.L. Ng, Y. Wu, A robust optimization model for multi-site production planning in an uncertain environment, Eur. J. Oper. Res. 181 (2007) 224–238.
[12] W. Baohua, H.E. Shiwei, Robust optimization model and algorithm for logistics center location and allocation under uncertain environment, J. Transp. Syst. Inf. Technol. 9(2) (2009) 69–74.
[13] Y. Yin, S.M. Madanat, X. Lu, Robust improvements schemes for road networks under demand uncertainty, Eur. J. Oper. Res. 198 (2009) 470–479.
[14] T. Davis, Effective supply chain management, Sloan Manage. Rev. 34 (1993) 35–46.
[15] J.M. Mulvey, A. Ruszczynski, A new scenario decomposition method for large-scale stochastic optimization, Oper. Res. 43 (1995) 477–490.

18 Integration in Logistics Planning and Optimization

Behnam Fahimnia[1], Reza Molaei[2] and Mohammad Hassan Ebrahimi[3]

[1]School of Management, Division of Business, University of South Australia, Adelaide, Australia
[2]Department of Technology Development, Iran Broadcasting Services (IRIB), Tehran, Iran
[3]Terminal Management System Department, InfoTech International Company, Tehran, Iran

18.1 Logistics Planning and Optimization Problem

A logistics system (LS) is a network of organizations, people, activities, information, and resources involved in the physical flow of products from supplier to customer. An LS may consist of three main networks or subsystems:

1. *Procurement*: The acquisition of raw material and parts from suppliers and their transportation to the manufacturing plants.
2. *Production*: The transformation of the raw materials into finished products.
3. *Distribution*: The transportation of finished products from plants to a network of stocking locations (warehouses) and from there to end users.

Logistics planning (LP) is the process of integrating and utilizing suppliers, manufacturers, warehouses, and retailers so that products are produced and delivered at the right quantities and at the right time while minimizing costs and satisfying customer requirements [1]. Implementation of LS has crucial impacts on a company's financial performance and LP optimization is essential to achieve globally optimized operations. The six major cost components that form the overall logistics costs are: (1) raw material costs, (2) costs of raw material transportation from vendors to manufacturing plants, (3) production costs at manufacturing plants, (4) transportation costs from plants to warehouses, (5) inventory or storage costs at warehouses, and (6) transportation costs from warehouses to end users (Figure 18.1). In a logistics optimization model, the overall systemwide costs are to be minimized through effective procurement, production, distribution, and inventory management. It is widely acknowledged that many benefits can be achieved by treating a logistics network as a whole (integration in LS) for optimization purposes, which requires the simultaneous minimization of all systemwide costs [2].

Logistics Operations and Management. DOI: 10.1016/B978-0-12-385202-1.00018-9

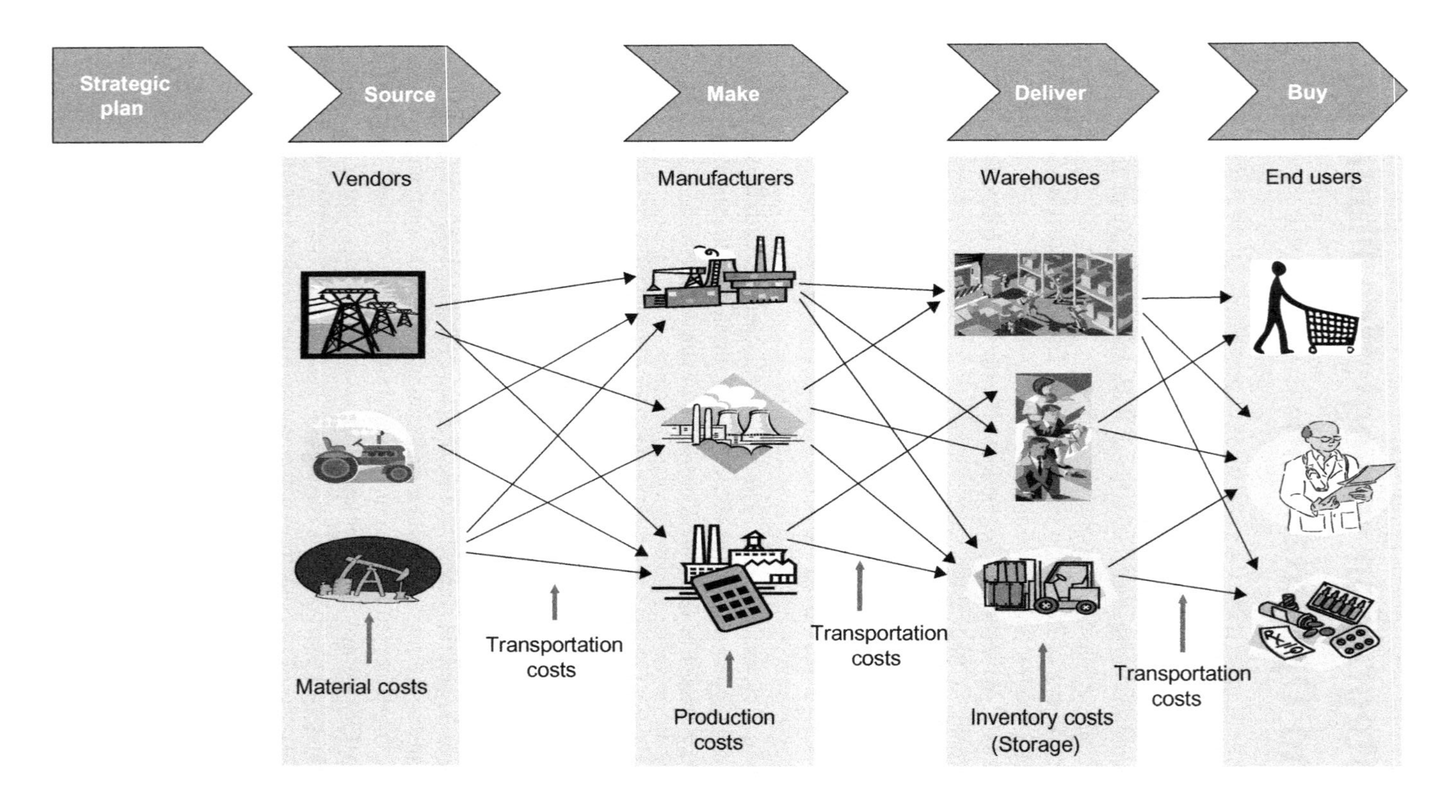

Figure 18.1 Participants of a logistics system.

18.2 Significance of Integrated LP

A new approach to the analysis of LSs has been recently proposed based on the integration of decisions of different functions in production and distribution networks into a single optimization model [3]. In fact, optimization of activities in individual subsystems of LS does not guarantee the global optimization, and this has been the driving reason why researchers have changed direction toward more integrated approaches. Efficient and effective planning and control of activities within a logistics network offer opportunities in terms of cost and lead-time reductions as well as improved quality [4]. Two issues—profitability and a quicker response to market changes—can justify the call to implement effective and efficient integrated logistics optimization models.

18.2.1 Profitability

Manufacturing, distribution, and service industries have realized the magnitude of savings that can be achieved through better planning and management of complex LSs [5]. Research shows that a great portion of such financial improvements are achieved when the associated decision makings are integrated and coordinated among like-minded entities participating in the logistics network [6]. Concurrent reduction in production costs, distribution costs, and inventory holding costs can only be achieved through the effective integration of procurement, production, and distribution activities across a logistics network.

18.2.2 Quicker Response to Market Changes

Lead-time reduction is an unavoidable element of today's time-based competitions. Sajadieh et al. [7] refer to this point and cites that one important benefit of coordination in a logistics network is a more efficient management of inventories across the entire supply chain (SC) that would consequently contribute to a shorter lead time. Further, the integrated management of a logistics network can improve information flow, which would naturally lead to improved product flow and thereby shorter lead times. Many potential benefits can be obtained from lead-time reductions, including better responsiveness to market changes, more accurate forecasts, significant reductions of bullwhip effects throughout a logistics network, smaller order sizes, reduction in work-in-progress (WIP) inventory and inventory of finished goods, and improved customer satisfaction [8–10].

18.3 Issues in Integrated LP

A logistics plan integrates the procurement plan, production plan, and distribution plan. A typical integrated logistics plan aims to deal with the following problems simultaneously: (1) quantity of raw material transported from vendors to manufacturing plants; (2) quantity of each product produced at each plant during each period; (3) quantity of each product outsourced during each period; (4) WIP inventory amount stored at each plant at the end of each period; (5) inventory amount of finished products stored at stack buffers at each plant at the end of each period; (6) quantity of each product shipped from stack buffers to warehouses during each period; (7) quantity of each product shipped from warehouses to end users during each period; (8) quantity of each product shipped directly from stack buffers to end users during each period; (9) inventory amount of finished products stored at warehouses at the end of each period; and (10) quantity of each product backordered (i.e., shortage or backlogged amount for failing to satisfy the customer demand at one period) at end users at the end of each period.

An integrated logistics plan covers the planning of activities in a vast scope from raw material suppliers to manufacturers and warehouses through to end users. This large planning scope with multiple players makes the LP problem complex containing several decision variables and constraints. The problem presented by the analysis of LSs is so complex that optimal solutions are very hard to obtain [3]. The difficulties associated with this type of decision making can be further amplified by the complex maze of the network, geographical span of the SC, limited visibility, and involvement of varied entities with conflicting objectives [6]. For this reason, simplification of a real-life scenario becomes unavoidable in developing an LP model [11,12].

Most of the LP problems are classified under the category of nondeterministic polynomial-time hard (NP-hard) problems, which are very difficult to solve using ordinary planning and optimization techniques. Literature on LP and optimization indicates that past research is subject to *oversimplification* of real-life scenarios. Oversimplification may preclude a logistics model from functioning effectively in real-world scenarios. Hence, there is a need to extend the scope of the proposed models to perform the optimization of the detailed aggregated logistics plan. The attempt to replicate the real scenarios as closely as possible makes the LP a challenging problem [12].

Various techniques have been used to solve small- and medium-sized LP problems ranging from mathematical models, heuristics, simulation, and knowledge-based systems to the latest fuzzy programming approaches [12]. However, finding the optimal solution in a complex LP problem using the presented approaches in the literature is impossible or subject to heavy computing overheads. Consequently, there is also a need to enhance the quality and precision of the solutions for the optimization of complex real-life LP problems.

18.4 An Integrated LP Model

This section aims to formulate an LP problem consisting of multiple production plants producing different product types during several time periods and distributing the finished products from plants to various end users located in different geographical locations through a number of warehouses. Mixed-integer programming (MIP) is used for this purpose. The following subsections will discuss the key performance indicators used, assumptions, parameters and decision variables, and finally the MIP formulation of the objective function.

18.4.1 Key Performance Indicators

The first issue in constructing a mathematical model is to determine the appropriate key performance indicator of the system. A number of performance indicators can be suggested for evaluating the performance of an LS, such as overall system costs, inventory management, delivery performance, and network responsiveness [13]. Literature on the LP models indicates that the cost-based value characteristics (e.g. total cost, profit, setup cost, delivery cost, and penalty cost) have been the most popular performance measures [12]. Cost-based optimization has a direct financial implication on system performance and clearly reflects the efficiency of the LS. The proposed model in this section is based on cost trade-off analysis, and therefore the objective functions aim to minimize the overall LS costs as the main performance measure.

18.4.2 Assumptions

There are a set of assumptions to be considered in the proposed LP model in this chapter. The procurement activities (including the raw material acquisition and its transportation to manufacturing plants) are disregarded in this model. Therefore, the proposed model is concerned with the production of multiple products in different manufacturing plants and the distribution of finished products from plants to end users (via a number of warehouses). In addition, demand is deterministic, and the aggregate demand for all types of final products in the concerned periods is assumed to be known for several periods in the near future. The aggregate demand at each end user is the total demand for each product that might have been ordered by several individuals and retailers at end users. Other assumptions include the following.

- Variety of products (i) to be produced is known.
- Number, location, and capacity of plants (m) and warehouses (w) are known.
- Number and location of end users (e) are known.
- All demands for each product have to be satisfied, sooner or later, during the planning horizon. A penalty cost will be incurred if the demand for a certain product at one period is decided to be backordered. The backordered demands are to be satisfied in the next periods before the end of the planning horizon.
- Production and distribution capacity limitations, capacity of raw material supply, and limitations in storage capacity at stack buffers and warehouses are known.
- To simplify the inventory management issues, a zero switch role is used in this model. This implies that the inventory levels of all products (at stack buffers and warehouses) are to be zero at the beginning and the end of the planning horizon. WIP inventory holding at manufacturing plants is disregarded in this model.
- Transportation costs are proportional to transportation distances.
- End users or customer zones are the locations where products are delivered to the final customers and have no holding capacity to store the products.

18.4.3 Parameters and Decision Variables

Before formulating the objective function of a model, all system inputs and decision variables must be clearly defined. Indices used for the purpose of mathematical modeling in this chapter include i for product types, m for manufacturing plants, b for stack buffers, w for warehouses, e for end users, and t for time-periods. Parameters represent the input data for a system. Therefore, a parameter is a variable with a fixed given value that is used as an input to the optimization system [12]. The following parameters are used in our model.

D_{iet} = forecasted demand for product i at end user e in period t
O_m = fixed costs of opening and operating plant m for the next planning horizon T
O'_w = fixed costs of opening and operating warehouse w for the planning horizon T
H_{ibt} = unit holding cost for finished product i at stack buffer b in period t
H'_{iwt} = unit holding cost for finished product i at warehouse w in period t
HC_{ibt} = holding capacity (maximum units) at stack buffer b for product i in period t
HC'_{iwt} = holding capacity (units) at warehouse w for product i in period t
T_{ibwt} = unit transportation cost for product i from stack buffer b to warehouse w in t
T'_{iwet} = unit transportation cost for product i from warehouse w to end user e in t
T''_{ibet} = unit transportation cost for product i directly from stack buffer b to e in t
P_{imt} = unit production cost of product i at plant m in period t
OS_{imt} = unit outsourcing cost of product i ordered by plant m in period t
SC_{iet} = unit backordering (shortage) cost for product i at end user e in period t
$S^{\max}_{iet}$ = maximum amount of shortage permitted (maximum backordering) for product i at end user e in period t
λ_{igmt} = capacity hours for the production of product i on g at plant m in t
γ_{imt} = capacity units of raw material supply for product i at plant m in period t
E_{ibt} = the distribution capacity at stack buffer b for product i in period t
E'_{iwt} = the distribution capacity at warehouse w for product i in period t

Decision variables are the outputs of the model or the variables in which the values need to be determined by the optimization model. The decision variables for the presented model in this chapter are listed below:

I_{imt} = quantity of product i produced at plant m in period t
I'_{imt} = quantity of product i outsourced by plant m in period t
J_{ibwt} = quantity of product i shipped from buffer b to warehouse w during period t
J'_{iwet} = quantity of product i shipped from warehouse w to end user e during period t
J''_{ibet} = quantity of product i shipped directly from stack buffer b to end user e during t
Y_{ibt} = inventory amount of finished product i left at buffer b at the end of period t
Z_{iwt} = amount of product i stored at warehouse w at the end of period t
S_{iet} = quantity of product i backordered at end user e at the end of period t

The integer variables include the following:

$$F_{ibwt} = \begin{cases} 1, & \text{If product } i \text{ is shipped from buffer } b \text{ to warehouse } w \text{ at period } t \\ 0, & \text{Otherwise} \end{cases}$$

$$F'_{iwet} = \begin{cases} 1, & \text{If product } i \text{ is shipped from warehouse } w \text{ to end user } e \text{ at period } t \\ 0, & \text{Otherwise} \end{cases}$$

$$F''_{ibet} = \begin{cases} 1, & \text{If product } i \text{ is shipped from buffer } b \text{ to end user } e \text{ at period } t \\ 0, & \text{Otherwise} \end{cases}$$

$$G_{mt} = \begin{cases} 1, & \text{If plant } m \text{ operates in period } t \\ 0, & \text{Otherwise} \end{cases}$$

$$G'_{wt} = \begin{cases} 1, & \text{If warehouse } w \text{ is open in period } t \\ 0, & \text{Otherwise} \end{cases}$$

$$d_{iet} = \begin{cases} 1, & \text{If demand for product } i \text{ at end user } e \text{ is not satisfied at period } t \\ 0, & \text{Otherwise} \end{cases}$$

18.4.4 Objective Function and Model Constraints

The objective function (i.e., cost function) in the LP problem under investigation minimizes the sum of production costs, outsourcing costs, inventory holding costs, transportation costs, and backlogging costs. The cost function for the proposed LP problem is the objective function of the model presented in Eqn (18.1). This equation consists of 10 cost components. Components 1 and 4 are the fixed costs of opening and operating plants and warehouses. These are independent of the rate and quantities of production and distribution at a plant or warehouse and may include the costs of building and facilities, amortizations of machines and tools, salaries of managers, annual insurance payments, and so on. Components 2 and 3 express production and outsourcing costs, respectively. Components 5 and 6 represent the inventory holding costs in stack buffers and warehouses, respectively. Components 7, 8, and 9 express the transportation costs for the distribution of items from plants to end users. This can be done directly from plants to end users (as in component 7) or indirectly from plants to warehouses and then from warehouses to end users (as in components 8 and 9). Component 10 stands for the shortage (penalty) costs incurred if backlogging occurs at the end users.

Min Z =

$$\begin{aligned} &\sum_m \sum_t G_{mt} O_{mt} + \sum_i \sum_m \sum_t I_{imt} \cdot P_{imt} + \sum_i \sum_m \sum_t I'_{imt} \cdot \mathrm{OS}_{imt} + \\ &\sum_w \sum_t G'_{wt} O'_{wt} + \sum_i \sum_b \sum_t H_{ibt} \cdot Y_{ibt} + \sum_i \sum_w \sum_t H'_{iwt} \cdot Z_{iwt} + \\ &\sum_i \sum_b \sum_w \sum_t J_{ibwt} \cdot T_{ibwt} \cdot F_{ibwt} + \sum_i \sum_w \sum_e \sum_t J'_{iwet} \cdot T'_{iwet} \cdot F'_{iwet} + \\ &\sum_i \sum_b \sum_e \sum_t J''_{ibet} \cdot T''_{ibet} \cdot F''_{ibet} + \sum_i \sum_e \sum_t S_{iet} \cdot \mathrm{SC}_{iet} \cdot d_{iet} \end{aligned} \tag{18.1}$$

The proposed model is subject to capacity constraints; demand and shortage constraints; balance constraints at stack buffers, warehouses, and end users; and variables constraints.

Capacity Constraints of Plants

Raw material supply capacity restrictions:

$$I_{imt} \leq \gamma_{imt} \quad \forall \ i, m, t \tag{18.2}$$

Demand satisfaction constraint: The total amount of production and outsourcing for every product at all plants must meet the forecast demand for that product at the end of planning horizon (i.e., complete satisfaction of all demands for every product at the end of planning phase):

$$\sum_{m}\sum_{t}\left(I_{imt} + I'_{imt}\right) = \sum_{e}\sum_{t} D_{iet} \quad \forall \ i \tag{18.3}$$

Capacity Constraints at Stack Buffers

Stack buffer capacity restriction:

$$Y_{ibt} \leq \mathrm{HC}_{ibt} \quad \forall \ i, b, t \tag{18.4}$$

Inventory balance at stack buffers:

$$Y_{ibt} = Y_{ib(t-1)} + \left[I_{imt} + I''_{imt}\right] - \left[\sum_{w} J_{ibwt} + \sum_{e} J''_{ibet}\right] \quad \forall \ i, b, m, t \tag{18.5}$$

Capacity Constraints of Warehouses

$$Z_{iwt} \leq \mathrm{HC}'_{iwt} \quad \forall \ i, w, t \tag{18.6}$$

Distribution Capacity Limits at Buffers

$$\sum_{w} J_{ibwt} + \sum_{e} J''_{ibet} \leq E_{ibt} \quad \forall \ i, b, t \tag{18.7}$$

Distribution Capacity Constraint at Warehouses

The distribution capacity limitation at warehouses:

$$\sum_e J'_{iwet} \le E'_{iwt} \quad \forall\ i, w, t \tag{18.8}$$

Inventory balance at warehouses:

$$Z_{iw(t-1)} + \sum_b J_{ibwt} \cdot F_{ibwt} = \sum_e J'_{iwet} \cdot F'_{iwet} + Z_{iwt} \quad \forall\ i, w, t \tag{18.9}$$

Backlogging Constraints at End Users

Maximum allowed shortage at end users:

$$S_{iet} \le S_{iet}^{\text{Max}} \quad \forall\ i, e, t \tag{18.10}$$

Balance equations at end users: The shipments of a product to an end user satisfy the demand for that product; otherwise some amount of shortage would appear.

$$\sum_w J'_{iwet} + \sum_b J''_{ibet} = D_{iet} - S_{iet} \cdot d_{iet} + S_{ie(t-1)} \cdot d_{ie(t-1)} \quad \forall\ i, e, t \tag{18.11}$$

Zero Switch Role

$$\sum_{t=0} Y_{ibt} = \sum_{t=T} Y_{ibt} = 0 \quad \forall\ i, b \tag{18.12}$$

$$\sum_{t=0} Z_{iwt} = \sum_{t=T} Z_{iwt} = 0 \quad \forall\ i, w \tag{18.13}$$

Nonnegativity Restriction for all Decision Variables

$$I_{imt} \ge 0 \quad \forall\ i, m, t \tag{18.14}$$

$$I'_{imt} \ge 0 \quad \forall\ i, m, t \tag{18.15}$$

$$J_{ibwt} \geq 0 \quad \forall \; i, b, w, t \tag{18.16}$$

$$J'_{iwet} \geq 0 \quad \forall \; i, w, e, t \tag{18.17}$$

$$J''_{ibet} \geq 0 \quad \forall \; i, b, e, t \tag{18.18}$$

$$Y_{ibt} \geq 0 \quad \forall \; i, b, t \tag{18.19}$$

$$Z_{iwt} \geq 0 \quad \forall \; i, w, t \tag{18.20}$$

$$S_{iet} \geq 0 \quad \forall \; i, e, t \tag{18.21}$$

18.5 Optimization Tools and Techniques

The concept of optimization refers to the process through which we minimize or maximize a function by means of systematically selecting the best values for the decision variables from a set of available alternatives. Many techniques have been proposed in the literature for the effective optimization of different integrated logistics plans, each with its own strengths and weaknesses. These tools and techniques can be classified into four categories: mathematical techniques, heuristics techniques, simulation, and genetic algorithms. This section briefly discusses the strengths and weaknesses of each technique.

18.5.1 Mathematical Techniques

Mathematical techniques are based on the representation of the essential aspects of an actual system using mathematical languages. Basically, a mathematical model needs to contain enough details to answer the questions for a certain problem [14]. Mathematical techniques may include linear programming, nonlinear programming, MIP, and Lagrangian Relaxation [15–17]. Different mathematical techniques have been adopted to solve logistics problems, including linear programming models [18–22], MIP models [23–39], and Lagrangian Relaxation models [40–43].

Mathematical programming models have been demonstrated to be useful analytical tools in optimizing decision-making problems such as those encountered in LP [44,45]. Linear programming was first proposed in 1947 and has been widely used in solving constrained optimization problems. "Programming" in this case is applicable when all of the underlying models of the real-world processes are linear

[17,46]. MIP is used when some of the variables in the model are real values and others are integer values (0, 1). Mixed-integer linear programming (MILP) occurs when objective function and all the constraints are linear in form; otherwise, it is mixed-integer nonlinear programming (MINLP), which is harder to solve [16]. The idea behind the Lagrangian Relaxation methodology is to relax the problem by removing the constraints that make the problem difficult to solve, putting them into the objective function, and assigning a weight to each constraint [47]. Each weight represents a penalty that is added to a solution that does not satisfy the particular constraint.

All of the mathematical techniques are fully matured and are thus guaranteed to produce the optimal solution (or near-optimal solutions) for a certain type of problem [12]. However, for two reasons this technique has limited application in solving complex logistics problems. First, mathematical equations are not always easy to formulate, and the associated complexities in the development of mathematical algorithms increase as the number of variables and constraints increase [12,48]. Because the majority of logistics networks are complex with the presence of large numbers of variables and constraints, mathematical methods may not be very effective in solving real-world LP problems [12,15]. Second, even if it is possible to translate a difficult LP problem into mathematical equations, the problem would become intractable or NP-hard because of the exponential growth of the model size and complexity [12,49]. The drawbacks of mathematical techniques may make it almost impossible to employ them for solving real-life, large-scale LP problems unless the problems are oversimplified.

18.5.2 Heuristic Techniques

The limitations of mathematical techniques have forced the use of heuristics in finding feasible solutions for large-scale LP problems. Heuristic methods are experience-based techniques that are generally used to rapidly find a solution that is hoped to be close to the optimal. Therefore, the upside of using heuristics is their relatively rapid response time in handling large problems. One popular heuristic that can be used for both discrete and continuous problems is simulated annealing [50]. Jayaraman and Ross [51] used simulated annealing methodology for distribution network design in SCs and demonstrated the effectiveness and usefulness of the solution approach to complex logistics problems. There have also been many other attempts in literature for solving various logistics problems using different heuristic techniques [52–58].

There are, however, few reasons why heuristics techniques are not always employed as an effective method for the optimization of complex integrated LP problems. The first problem with heuristic techniques is that they do not promise an optimal solution to the problem. In fact, in many cases they cannot even promise a near-optimal result [50]. The second problem is that heuristic techniques (e.g. simulated annealing) may not be very effective in locating the global optimal or near-optimal solutions in complex LP problems with a vast search space. Although the simulated annealing approach can often manage to make its way through the traps of local optima, its ability and efficiency in exploring the search space is highly limited by its characteristic of examining only one point of the space at a time [59].

18.5.3 Simulation Modeling

Simulation modeling in the area of LP is used to observe how a real system performs, diagnoses problems, predicts the effect of changes in the system, evaluates logistics activities, and suggests possible solutions for improvements [60]. Because of many influential sources of stochastic variation and interdependencies in today's LSs, simulation can be a highly effective tool in making operationally and economically sound business decisions [61]. Advancement of computing facilities and the development of user-friendly and easy-to-understand simulation software packages (e.g. AutoMod, Arena, and SIMFLEX) are the encouraging motivations toward the wider utilization of simulation modeling in SC analysis [12,14,61–65].

There are, however, two downsides for simulation modeling that can justify the limited application of this methodology for the optimization of complex logistics models [12,49]. First, it is difficult to search for an optimal value using simulation techniques. Like heuristic techniques, the first and main drawback of a simulation modeling is its inability to guarantee optimality of the developed solution. Second, it is costly and takes much time and effort to analyze the obtained results. Simulation software packages are generally very expensive to purchase and very time consuming to analyze the autogenerated reports and results.

18.5.4 Genetic Algorithms

Introduced by John Holland [66], a genetic algorithm (GA) is a stochastic algorithm categorized in the class of general-purpose search methods that simulate the processes in a natural evolution system [67,68]. GAs combine directed and stochastic search methods and are able to achieve a good balance between exploration and exploitation of the search space [68]. GAs have been proven to be highly effective in solving complex engineering and manufacturing problems, and some of their successful applications in the optimization of LP models have been proposed in the literature [22,49,69–80].

The advantages of using GA techniques for solving large optimization problems are their robustness, searching flexibility, and evolutionary nature [59]. GAs are able to search large, complicated, and unpredictable search spaces that facilitate this technique to locate the optimal solution demonstrated by the convergence of the fitness function as the number of evolutions increases [12,66,68,81,82]. GA produces a large population of solutions, for each of which the evaluation of the fitness function is sought; hence, a parallel computer is required when running GA for very large-scale optimization problems [83].

There are, however, a number of challenges when designing a customized GA procedure to solve a particular LP problem. The first challenge in developing a GA-based optimization technique is to form the chromosome structure that accordingly affects the entire GA procedure. Based on the size and nature of the problem, dissimilar LP problems require different chromosome representations, and therefore the existing GA procedures cannot be used to solve different problems from various complexity levels. The second difficulty is the construction of customized genetic operators to perform the mating process on the chromosomes. Lastly, the design of a constraint-handling mechanism is generally a complicated task that ensures the effective implementation of the model constraints.

18.6 A Case Study

In Section 18.4 attention has been paid to the mathematical modeling of an integrated LP problem using MINLF. Based on this formulation, this section presents a medium-sized case study and uses GAs to find the optimal logistics plan for a company producing and distributing home gym equipments. The following subsections will discuss the case problem, the optimization procedure, and the results achieved.

18.6.1 Case Problem

Body shape group (BSG) has four manufacturing plants, four warehouses, and five customer zones (end users) in major cities in Australia. BSG manufactures and distributes six types of home gyms ($i_1 - i_6$), including two types of benches (i_1 and i_2), two types of stack machines (i_3 and i_4), and two types of plate-loaded machines (i_5 and i_6). Plants 1 and 2 (m_1 and m_2) are located in Sydney and Melbourne and are able to produce all six types of products. Plant 3 (m_3) in Adelaide can only produce benches and stack machines ($i_1 - i_4$), and m_4 in Perth is able to produce stack machines and plate-loaded machines ($i_3 - i_6$). The planning horizon for the proposed LP problem is 1 year, which comprises 12 equal periods of 1 month ($t_1 - t_{12}$). The aggregated demands for the six types of products at four end users are known at every period of the planning horizon. Also known are the BSG manufacturing and distribution data, including production and outsourcing costs for each product type, inventory holding costs, transportation costs, and backlogging costs.

Production capacity constraints as well as holding and distribution capacity at stack buffers and warehouses are known for this case study. BSG has certain production and storage capacity constraints at each manufacturing plant, backlogging limitations enforced from the customers, and holding and distribution capacity constraints at its stack buffers and warehouses. Partial delivery is not allowed in this case problem, which implies that the customer demand for each product at each end user is to be delivered in one lot; otherwise, the associated demand will be backordered in order to be satisfied in subsequent periods.

18.6.2 Optimization Procedure

With the presence of numerous variables in a complex LP problem, the optimal solution can only be achieved by the delicate trade-off in allocating the production of products to the manufacturing plants in the way to minimize the overall logistics costs. For an LP problem of a realistic size, the size of the search space and the number of feasible solutions could be extremely large. The principle of GA optimization differs from most traditional optimization methods in that GA is capable of handling large search spaces easily [59,68,69].

In a typical type of GA (refer to Figure 18.2), the possible solutions of the problem are represented by chromosomes. Every chromosome is coded in the way to contain the potential solution to the problem. This process is called *chromosome representation* and forms the main difference between various GA models. In the first step of the GA process, a population of chromosomes is created with a group of randomly generated solutions (chromosomes). Then, all individuals in the population are evaluated by scoring them based on how fit they are (i.e., a fitness value for each chromosome). Following this, multiple individuals are randomly selected from the population to produce offspring through the application of genetic operators (generally includes crossover operator and mutation operators). The generated offspring form the new generation. Fitness values are evaluated again for the new generation. Based on the concept of natural evolution, this process evolves toward a better solution in consecutive generations. The reproduction process (i.e., scoring, crossover, and mutation) is repeated until a suitable solution is found or the termination criteria are satisfied. This process is illustrated in Figure 18.2.

For the proposed case problem in this chapter, the developed cost function in Section 18.4 (i.e., objective function in Eqn (18.1)) acts as the fitness function in

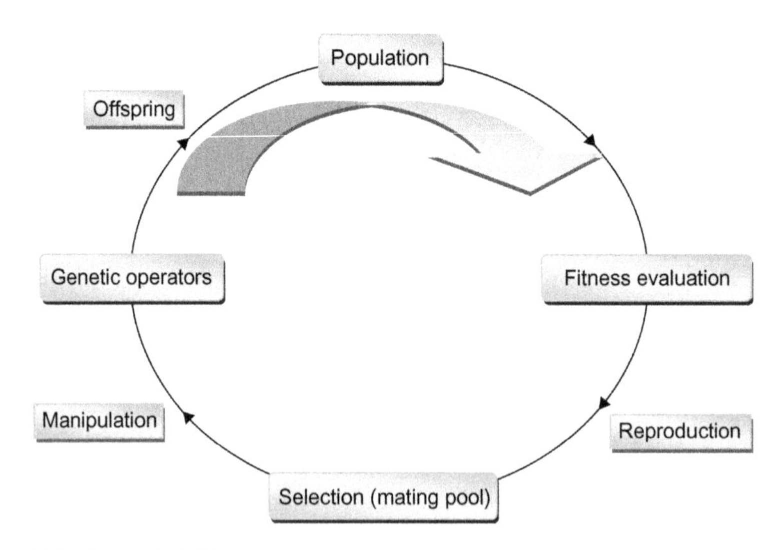

Figure 18.2 The typical GA process.

the GA process. It has been a common approach in solving LP problems to define the fitness function equal to the objective function, which in many cases is also the cost function [49,74,75,78]. In this case, the chromosome with the lowest fitness value signifies the fittest chromosome. To simplify the coding process, a linear chromosome representation is adopted for the presented case study. In this way, all of the decision variables (see Section 18.4) are located along a straight chromosome in which each gene of the chromosome accommodates the real value of a variable.

A multipoint crossover operator is employed for the proposed GA model. In a single-point crossover operator, a crossover point is randomly selected between the first and last bits of the parents' chromosomes. Parents 1 and 2 pass their binary codes from the left of the crossover point to offspring 1 and 2, respectively. Then the binary codes in the right of the crossover point of parent 1 and parent 2 go to offspring 2 and offspring 1, respectively. The number of randomly selected crossover points determines the type of crossover used: single point, two point, or multiple point. In terms of chromosome maintenance, the destructive effects of a multipoint crossover may lead to more exploration rather than exploitation in chromosomes, whereas having a small number of crossover points contributes to more exploitation than exploration [12].

A simple flip nonuniform mutation is adopted as the background operator in the sense that it helps GA find the solutions that crossover alone might not encounter. A mutation operator alters a certain percentage of the bits (genes) in a chromosome. This can stop GA from early convergence and ensure the feasibility of the developed offspring by maintaining diversity in the population [12]. Crossover rate and mutation rate (e.g., the proportion of chromosomes in the mating pool on which the crossover and mutation operator will apply) is to be determined experimentally (i.e., trial and error practice).

It is always difficult to prove the convergence to the optimal solution in the optimization of complex problems. Therefore, a termination point is generally set to stop the process. A multiple termination condition is adopted in the proposed GA procedure in this chapter: (1) stop by limiting the number of iterations or generations and (2) stop when minimal change in fitness value is observed in consecutive generations.

18.6.3 Results Achieved

The GA model presented in Section 18.6.2 was coded in Microsoft Visual Basic 6.0 using MS Excel as the program interface. The machine used to run the program was a PC with an Intel Core2 Duo Processor, with 4 GB DDR RAM at 1066 MHz. The program was run for a population size of 100 chromosomes, crossover rate of 0.6, and mutation rate 0.2. To track the evolution of fitness function and the rate of convergence, the overall logistics costs (the value of objective function) and its cost components (including production costs, distribution costs, and shortage or backordering costs) were recorded at every generation for 1500 iterations. Table 18.1 shows the achieved results (i.e., the average

Table 18.1 Numerical Results of the GA Model Recorded at Every 50 Generations (Average Values)

Generation	Objective Function Value	Objective Function Cost Components		
		Production Costs	Distribution Costs	Backlogging Costs
1	3362735.46	1535734.35	1785087.53	41913.59
50	3518056.29	1585712.71	1885604.83	46738.76
100	3432194.00	1570540.31	1816000.42	45653.27
150	3322349.64	1342030.76	1933676.20	46642.68
200	3332406.08	1395911.51	1890112.39	46382.18
250	3269328.90	1339857.54	1884197.35	45274.01
300	3300088.48	1279846.07	1972214.67	48027.74
350	3413263.23	1353061.36	2010694.26	49507.61
400	3237457.66	1182422.41	2003883.03	51152.22
450	3438569.33	1337718.85	2050300.23	50550.26
500	3335353.62	1317070.94	1964474.69	53807.99
550	3232867.64	1279079.26	1904696.90	49091.49
600	3113441.80	1143276.59	1921706.87	48458.34
650	3325795.10	1266882.61	2010020.71	48891.78
700	3239138.49	1188170.25	2002909.63	48058.61
750	3356759.78	1271412.43	2031592.60	53754.75
800	3383898.14	1301476.02	2031889.50	50532.62
850	3310228.58	1262840.23	1998644.31	48744.05
900	3231536.32	1236590.42	1945626.04	49319.87
950	3175533.05	1215303.27	1912822.28	47407.50
1000	3224699.33	1157885.90	2017513.73	49299.71
1050	3259193.36	1224197.47	1984848.84	50147.06
1100	3309834.29	1200427.46	2057050.37	52356.47
1150	3204893.68	1222135.88	1931309.80	51448.01
1200	3219992.61	1227578.66	1945790.49	46623.47
1250	3092748.56	1056950.94	1987857.14	47940.48
1262	3017958.97	1040282.48	1931422.03	46254.46
1300	3118698.65	1049077.61	2022842.91	46778.13
1350	3200044.66	1129250.06	2022542.90	48251.70
1400	3207720.89	1139170.30	2018833.82	49716.77
1450	3063926.54	994753.07	2023145.36	46028.12
1500	3185100.62	1097770.99	2040909.97	46419.66

fitness function values and their cost components) recorded at the intervals of 50 iterations.

The evolution of the average fitness function values is graphically illustrated in Figure 18.3, demonstrating the reasonable convergence speed of the GA model. The proposed model converges to optimality consistently with a typical reduction in overall logistics costs of more than 10% (optimal cost of $3,017,958.97 at

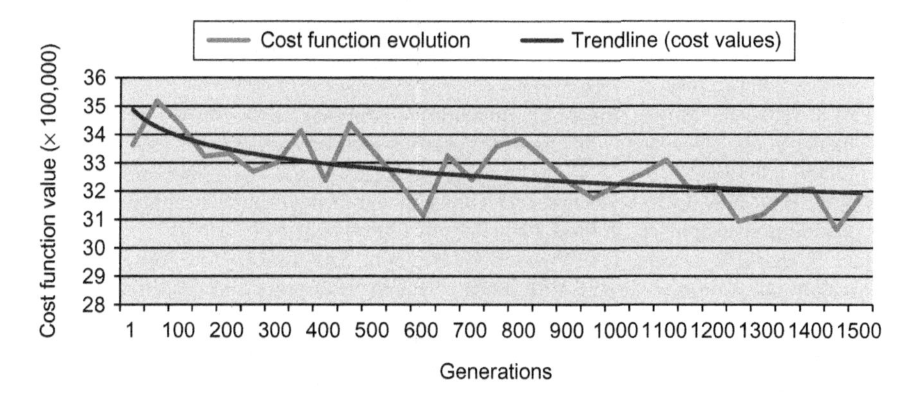

Figure 18.3 The evolution of fitness function (average values) in 1500 iterations.

generation 1262 compared to the original cost of $3,362,735.46 at generation 1). In another attempt, the model was run with the stopping condition of the model set to terminate the process when the difference between the overall SC costs in 10 consecutive generations becomes less than $1. The results of this experiment indicate that optimal results are typically achieved after 3200 iterations with only minimal difference in fitness function value compared to what we achieved in generation 1262 from the first experiment (i.e., about 11% reduction in overall logistics costs at generations 3224–3234 compared to just more than 10% in the first experiment at generation 1262).

In a nutshell, the results achieved from these experiments reveal that the proposed optimization model in this chapter yields significant cost-reduction benefits within acceptable model runtime through the global integration of activities across a logistics network.

References

[1] Y. Chang, H. Makatsoris, Supply chain modelling using simulation, Int. J. Simul. Syst. Sci. Technol. 2(1) (2001) 24–30.

[2] H. Mohammadi Bidhandi, R.M. Yusuff, M.M.H. Megat Ahmed, M.R. Abu Bakar, Development of a new approach for deterministic supply chain network design, Eur. J. Oper. Res. 198(1) (2009) 121–128.

[3] A.M. Sarmiento, R. Nagi, A review of integrated analysis of production-distribution systems, IIE Trans. 31(11) (1999) 1061–1074.

[4] F. Persson, J. Olhager, Performance simulation of supply chain designs, Int. J. Prod. Econ. 77(3) (2002) 231–245.

[5] A. Muriel, D. Simchi-Levi, Supply chain design and planning—applications of optimization techniques for strategic and tactical models, in: A.G. de Kok, S.C. Graves (Eds.), Handbooks in Operations Research and Management Science, (Volume 11), Elsevier, 2003, pp. 15–93.

[6] S.S. Pitty, W. Li, A. Adhitya, R. Srinivasan, I.A. Karimi, Decision support for integrated refinery supply chains: Part 1. Dynamic simulation, Comput. Chem. Eng. 32(11) (2008) 2767–2786.
[7] M.S. Sajadieh, M.R.A. Jokar, M. Modarres, Developing a coordinated vendor-buyer model in two-stage supply chains with stochastic lead-times, Comput. Oper. Res. 36(8) (2009) 2484–2489.
[8] S.W. Ryu, K.K. Lee, A stochastic inventory model of dual sourced supply chain with lead-time reduction, Int. J. Prod. Econ. 81–82 (2003) 513–524.
[9] E. Eskigun, R. Uzsoy, P.V. Preckel, G. Beaujon, S. Krishnan, J.D. Tew, Outbound supply chain network design with mode selection, lead times and capacitated vehicle distribution centers, Eur. J. Oper. Res. 165(1) (2005) 182–206.
[10] M. Leng, M. Parlar, Lead-time reduction in a two-level supply chain: non-cooperative equilibria vs. coordination with a profit-sharing contract, Int. J. Prod. Econ. 118(2) (2009) 521–544.
[11] D.J. Thomas, P.M. Griffin, Coordinated supply chain management, Eur. J. Oper. Res. 94(1) (1996) 1–15.
[12] A.M.N. Lair, An Integrated Model for Optimising Manufacturing and Distribution Network Scheduling, Ph.D. Thesis, School of Advanced Manufacturing and Mechanical Engineering, University of South Australia, Adelaide, 2008, p. 266.
[13] C. Sürie, M. Wagner, Supply chain analysis, in: H. Stadtler, C. Kilger (Eds.), Supply Chain Management and Advanced Planning: Concepts, Models, Software, and Case Studies, Springer, Heidelberg, Berlin, 2005.
[14] J. Banks, Getting Started with AutoMod, second ed., Brooks Automation, Inc., Chelmsford, MA, 2004 (permission of John Wiley & Sons, Inc.).
[15] D.W. Jordan, P. Smith, Mathematical Techniques: An Introduction for the Engineering, Physical, and Mathematical Sciences, third ed., OUP, Oxford, 2002.
[16] J.W. Chinneck, Binary and mixed-integer programming, in: Practical Optimization: A Gentle Introduction, Carleton University: Ottawa, Ontario, Canada, 2004 (Chapter 13).
[17] J.W. Chinneck, Introduction, in: Practical Optimization: A Gentle Introduction, Carleton University: Ottawa, Ontario, Canada, 2004 (Chapter 1).
[18] M. Chen, W. Wang, A linear programming model for integrated steel production and distribution planning, Int. J. Oper. Prod. Manage. 17(6) (1997) 592–610.
[19] A. Kanyalkar, G. Adil, An integrated aggregate and detailed planning in a multi-site production environment using linear programming, Int. J. Prod. Res. 43(20) (2005) 4431–4454.
[20] A. Kanyalkar, G. Adil, Aggregate and detailed production planning integrating procurement and distribution plans in a multi-site environment, Int. J. Prod. Res. 45(22) (2007) 5329–5353.
[21] A. Kanyalkar, G. Adil, A robust optimisation model for aggregate and detailed planning of a multi-site procurement-production-distribution system, Int. J. Prod. Res. 48(3) (2010) 635–656.
[22] T.-F. Liang, Fuzzy multi-objective production/distribution planning decisions with multi-product and multi-time period in a supply chain, Comput. Ind. Eng. 55(3) (2008) 676–694.
[23] K.S. Bhutta, F. Huq, G. Frazier, Z. Mohamed, An integrated location, production, distribution and investment model for a multinational corporation, Int. J. Prod. Econ. 86 (3) (2003) 201–216.
[24] J. Ferrio, J. Wassick, Chemical supply chain network optimization, Comput. Chem. Eng. 32(11) (2008) 2481–2504.

[25] H. Gunnarsson, M. Rönnqvist, D. Carlsson, Integrated production and distribution planning for Södra Cell AB, J. Math. Model. Algorithms 6(1) (2007) 25–45.
[26] M. Hamedi, R. Farahani Zanjirani, M.M. Husseini, G.R. Esmaeilian, A distribution planning model for natural gas supply chain: a case study, Energy Policy 37(3) (2009) 799–812.
[27] A.N. Haq, P. Vrat, A. Kanda, An integrated production-inventory-distribution model for manufacture of urea: a case, Int. J. Prod. Econ. 25(1–3) (1991) 39–49.
[28] Y. Kim, C. Yun, S.B. Park, S. Park, L.T. Fan, An integrated model of supply network and production planning for multiple fuel products of multi-site refineries, Comput. Chem. Eng. 32(11) (2008) 2529–2535.
[29] Z.M. Mohamed, An integrated production-distribution model for a multi-national company operating under varying exchange rates, Int. J. Prod. Econ. 58(1) (1999) 81–92.
[30] T. Paksoy, H. Kürsat Gules, D. Bayraktar, Design and optimization of a strategic production-distribution model for supply chain management: case study of a plastic profile manufacturer in Turkey, Selçuk J. Appl. Math. 8(2) (2007) 83–99.
[31] P. Tsiakis, L.G. Papageorgiou, Optimal production allocation and distribution supply chain networks, Int. J. Prod. Econ. 111(2) (2008) 468–483.
[32] H. Yan, Z. Yu, T.C.E. Cheng, A strategic model for supply chain design with logical constraints: formulation and solution, Comput. Oper. Res. 30(14) (2003) 2135–2155.
[33] H. Selim, C. Araz, I. Ozkarahan, Collaborative production-distribution planning in supply chain: a fuzzy goal programming approach, Transp. Res. Part E Logistics Transp. Rev. 44(3) (2008) 396–419.
[34] K. Demirli and A.D. Yimer, Production-distribution planning with fuzzy costs, in: NAFIPS 2006—Annual Meeting of the North American Fuzzy Information Processing Society, 2006.
[35] C.-L. Chen, W.-C. Lee, Multi-objective optimization of multi-echelon supply chain networks with uncertain product demands and prices, Comput. Chem. Eng. 28(6–7) (2004) 1131–1144.
[36] C. Dhaenens-Flipo, G. Finke, An integrated model for an industrial production-distribution problem, IIE Trans. 33 (2001) 705–715.
[37] B. Bilgen, Application of fuzzy mathematical programming approach to the production allocation and distribution supply chain network problem, Expert Syst. Appl. 37(6) (2010) 4488–4495.
[38] K. Das, S. Sengupta, A hierarchical process industry production-distribution planning model, Int. J. Prod. Econ. 117(2) (2009) 402–419.
[39] N. Rizk, A. Martel, S. D'Amours, Multi-item dynamic production-distribution planning in process industries with divergent finishing stages, Comput. Oper. Res. 33(12) (2006) 3600–3623.
[40] G. Barbarosoglu, D. Ozgur, Hierarchical design of an integrated production and 2-echelon distribution system, Eur. J. Oper. Res. 118(3) (1999) 464–484.
[41] V. Jayaraman, H. Pirkul, Planning and coordination of production and distribution facilities for multiple commodities, Eur. J. Oper. Res. 133(2) (2001) 394–408.
[42] T. Nishi, M. Konishi, M. Ago, A distributed decision making system for integrated optimization of production scheduling and distribution for aluminum production line, Comput. Chem. Eng. 31(10) (2007) 1205–1221.
[43] S.S. Syam, A model and methodologies for the location problem with logistical components, Comput. Oper. Res. 29(9) (2002) 1173–1193.
[44] J. Xu, Y. He, M. Gen, A class of random fuzzy programming and its application to supply chain design, Comput. Ind. Eng. 56(3) (2009) 937–950.

[45] A.M. Geoffrion, A guided tour of recent practical advances in integer linear programming, Omega 4(1) (1976) 49–57.
[46] G.B. Dantzig, Linear programming, Oper. Res. 50(1) (2002) 42–47.
[47] M.L. Fisher, An applications oriented guide to Lagrangian relaxation, Interfaces 15(2) (1985) 10–21.
[48] C.A. Méndez, J. Cerdá, I.E. Grossmann, I. Harjunkoski, M. Fahl, State-of-the-art review of optimization methods for short-term scheduling of batch processes, Comput. Chem. Eng. 30(6–7) (2006) 913–946.
[49] B. Park, H. Choi, M. Kang, Integration of production and distribution planning using a genetic algorithm in supply chain management, in: Analysis and Design of Intelligent Systems Using Soft Computing Techniques, 2007, pp. 416–426.
[50] J.W. Chinneck, Heuristics for discrete search: genetic algorithms and simulated annealing, in: Practical Optimization: A Gentle Introduction, Carleton University, Ottawa, Ontario, Canada, 2004 (Chapter 14).
[51] V. Jayaraman, A. Ross, A simulated annealing methodology to distribution network design and management, Eur. J. Oper. Res. 144(3) (2003) 629–645.
[52] M.A. Cohen, H.L. Lee, Strategic analysis of integrated production-distribution systems: models and methods, Oper. Res. Soc. Am. 36(2) (1988) 216–228.
[53] J.L. Coronado, An Optimization Model for Strategic Supply Chain Design Under Stochastic Capacity Disruptions, Doctoral Thesis, College of Engineering, Texas A&M University, Texas, 2008, pp. 1–110.
[54] L. Ozdamar, T. Yazgac, A hierarchical planning approach for a production-distribution system, Int. J. Prod. Res. 37(16) (1999) 3759.
[55] D.F. Pyke, M.A. Cohen, Performance characteristics of stochastic integrated production-distribution systems, Eur. J. Oper. Res. 68(1) (1993) 23–48.
[56] D.F. Pyke, M.A. Cohen, Multiproduct integrated production-distribution systems, Eur. J. Oper. Res. 74(1) (1994) 18–49.
[57] P. Yilmaz, B. Çatay, Strategic level three-stage production distribution planning with capacity expansion, Comput. Ind. Eng. 51(4) (2006) 609–620.
[58] W.-C. Yeh, A hybrid heuristic algorithm for the multistage supply chain network problem, Int. J. Adv. Manuf. Technol. 26 (2005) 675–685.
[59] K. Xing, Design for Upgradability: Modelling and Optimisation, Ph.D. Thesis, School of Advanced Manufacturing and Mechanical Engineering, University of South Australia, Adelaide, 2006.
[60] M.J. Tarokh, M. Golkar, Supply chain simulation methods, in: 2006 IEEE International Conference on Service Operations and Logistics, and Informatics, SOLI '06, 2006.
[61] E.J. Williams, A. Gunal, Supply chain simulation and analysis with SIMFLEX, in: 2003 Winter Simulation Conference, USA, 2003.
[62] C. Chung, Simulation Modeling Handbook: A Practical Approach (Industrial and Manufacturing Engineering Series), CRC Press, Boca Raton, FL, 2004.
[63] H. Pierreval, R. Bruniaux, C. Caux, A continuous simulation approach for supply chains in the automotive industry, Simul. Model. Pract. Theory 15(2) (2007) 185–198.
[64] H.S. Hwang, G.S. Cho, A performance evaluation model for order picking warehouse design, Comput. Ind. Eng. 51(2) (2006) 335–342.
[65] D. Burnett, T. LeBaron, Efficiency modeling warehouse systems, in: Proceedings of the 2001 Winter Simulation Conference, 2001.

[66] J.H. Holland, Adaptation in Natural and Artificial Systems: An Introductory Analysis with Applications to Biology, Control, and Artificial Intelligence, University of Michigan Press, Ann Arbor, 1975.
[67] K. Ganesh, M. Punniyamoorthy, Optimization of continuous-time production planning using hybrid genetic algorithms-simulated annealing, Int. J. Adv. Manuf. Technol. 26 (2005) 148–154.
[68] B. Fahimnia, Optimisation of Manufacturing Lead-Time, Using Genetic Algorithm, Master's Thesis, School of Advanced Manufacturing and Mechanical Engineering, University of South Australia, Adelaide, 2006, p. 151.
[69] R.A. Aliev, B. Fazlollahi, B.G. Guirimov, R.R. Aliev, Fuzzy-genetic approach to aggregate production-distribution planning in supply chain management, Inform. Sci. 177 (20) (2007) 4241–4255.
[70] F. Altiparmak, M. Gen, L. Lin, A genetic algorithm for supply chain network design, in: 35th International Conference on Computers and Industrial Engineering, 2005.
[71] F. Altiparmak, M. Gen, L. Lin, I. Karaoglan, A steady-state genetic algorithm for multi-product supply chain network design, Comput. Ind. Eng. 56(2) (2007) 531–537.
[72] F. Altiparmak, M. Gen, L. Lin, T. Paksoy, A genetic algorithm approach for multi-objective optimization of supply chain networks, Comput. Ind. Eng. 51(1) (2006) 196–215.
[73] F.T.S. Chan, S.H. Chung, S. Wadhwa, A hybrid genetic algorithm for production and distribution, Omega 33(4) (2005) 345–355.
[74] R.Z. Farahani, M. Elahipanah, A genetic algorithm to optimize the total cost and service level for just-in-time distribution in a supply chain, Int. J. Prod. Econ. 111(2) (2008) 229–243.
[75] M. Gen, A. Syarif, Hybrid genetic algorithm for multi-time period production/distribution planning, Comput. Ind. Eng. 48(4) (2005) 799–809.
[76] A. Syarif, Y. Yun, M. Gen, Study on multi-stage logistic chain network: a spanning tree-based genetic algorithm approach, Comput. Ind. Eng. 43(1–2) (2002) 299–314.
[77] A. Tasan, A two step approach for the integrated production and distribution planning of a supply chain, Intell. Comput. 4113 (2006) 883–888.
[78] W.-C. Yeh, An efficient memetic algorithm for the multi-stage supply chain network problem, Int. J. Adv. Manuf. Technol. 29(7) (2006) 803–813.
[79] A.D. Yimer, K. Demirli, A genetic approach to two-phase optimization of dynamic supply chain scheduling, Comput. Ind. Eng. 58(3) (2009) 411–422.
[80] A. Kazemi, M. Fazel Zarandi, S. Moattar Husseini, A multi-agent system to solve the production–distribution planning problem for a supply chain: a genetic algorithm approach, Int. J. Adv. Manuf. Technol. 44(1) (2009) 180–193.
[81] R. Marian, Optimisation of Assembly Sequences Using Genetic Algorithms, Ph.D. Thesis, School of Advanced Manufacturing and Mechanical Engineering, University of South Australia, Adelaide, 2003, p. 287.
[82] B. Sobhi-Najafabadi, Optimal Design of Permanent Magnet Generators, Ph.D. Thesis, School of Electrical and Information Engineering, University of South Australia, Adelaide, 2002, p. 331.
[83] R. Haupt, S. Haupt, Practical genetic algorithms, Discrete Applied Mathematics, John Wiley & Sons, Inc., New Jersey, 2004, pp. 1–261.

19 Optimization in Natural Gas Network Planning

Maryam Hamedi[1], Reza Zanjirani Farahani[2] and Gholamreza Esmaeilian[1,3]

[1]Department of Mechanical and Manufacturing Engineering, University Putra Malaysia, Serdang, Selangor, Malaysia
[2]Department of Informatics and Operations Management, Kingston Business School, Kingston University, Kingston Hill, Kingston Upon Thames, Surrey KT2 7LB
[3]Department of Industrial Engineering, Payam Noor Universiti, Iran

19.1 Introduction

19.1.1 Natural Gas Network Modeling

A vast number of real-world problems in various types of systems are presented using network modeling. A network model represents a powerful visual that helps to present connections among the system's components used in different fields of science. Optimization of network design, network flow, and network operation has been considered as a fundamental issue in different research fields from technical and financial perspectives. Usually, network problems are known as complex structural problems that, in real cases, seek optimal solutions. Considering the characteristics of networks optimization problems, a large number of algorithms have been developed in the literature, and many tried to solve them in reliable times to find the most suitable and optimal solutions.

As one of the most important sources of energy, natural gas is used to satisfy the needs of many commercial and residential users throughout the world through a huge and complex network. Each day, a large amount of money is spent on the different stages and main processes of this network, such as exploration, extraction, processing, production, transportation, storage, and distribution [1]. Since the 1960s, many different problems have been defined for the planning of design, flow, operation, and development of natural gas transmission and distribution networks. The natural gas network planning problems are multidisciplinary, and have been tackled by researchers in different fields from around the world.

Logistics Operations and Management. DOI: 10.1016/B978-0-12-385202-1.00019-0

19.1.2 Natural Gas Network Introducing

The aim of the natural gas network is to satisfying consumers' demands efficiently and at minimal cost. Therefore, some components should be used properly through the network, and some processes should be planned exactly. In this section, the main components of the natural gas network and its main processes are explained.

Natural Gas Network Components

Like other networks, natural gas network components can be divided into two physical categories: fixed entities and current entities.

Fixed physical entities include arcs, which correspond with pipelines, compressor stations, and valves and nodes that present physical interconnection points.

- Arc components
  - *Pipelines*: The two types of pipelines of concern to researchers are passive pipelines, which correspond to regular pipelines, and active pipelines, which are regular pipelines with compressors [2].
  - *Compressor stations*: The transmission capacity of a gas pipeline is limited but can be arranged based on the supply-and-demand nodes by setting differences among input and output pressures of the pipeline. Compressors are located at suitable locations through the network to enlarge the pressure differences between two nodes of pipelines to increase the network's transmission capacity [2].
  - *Valves*: To make the flow of natural gas stop for a certain section of pipelines in situations such as maintenance or replacement, valves are used along the entire length of interstate pipelines.
- Node components. As shown in Figure 19.1, components belonging to nodes in the natural gas network include the following:
  - Supply nodes, which have only output flow
  - Demand nodes, which have only input flow
  - Intermediate nodes, which have both input and output flows

Current physical entities can be classified as financial, informational, and physical flows. Usually, flows in the natural gas network are controlled by a dispatcher or dispatching organizations. These organizations obtain information about natural gas pressures and flows over pipeline systems and check warning signals from companies via simulation systems.

Natural Gas Network Processes

The natural gas is supplied through gas and oil wells and produced in refineries. In the natural gas network, some methods exist to move gas from producers to consumers in order to satisfy customers' demands, but the pipeline system is the most cost-effective way to transport gas over long distances. Consumers of natural gas are divided into three main groups: domestic and commercial subscribers, industrial consumers, and exports. Usually, the priority of natural gas networks is to serve domestic and commercial consumers.

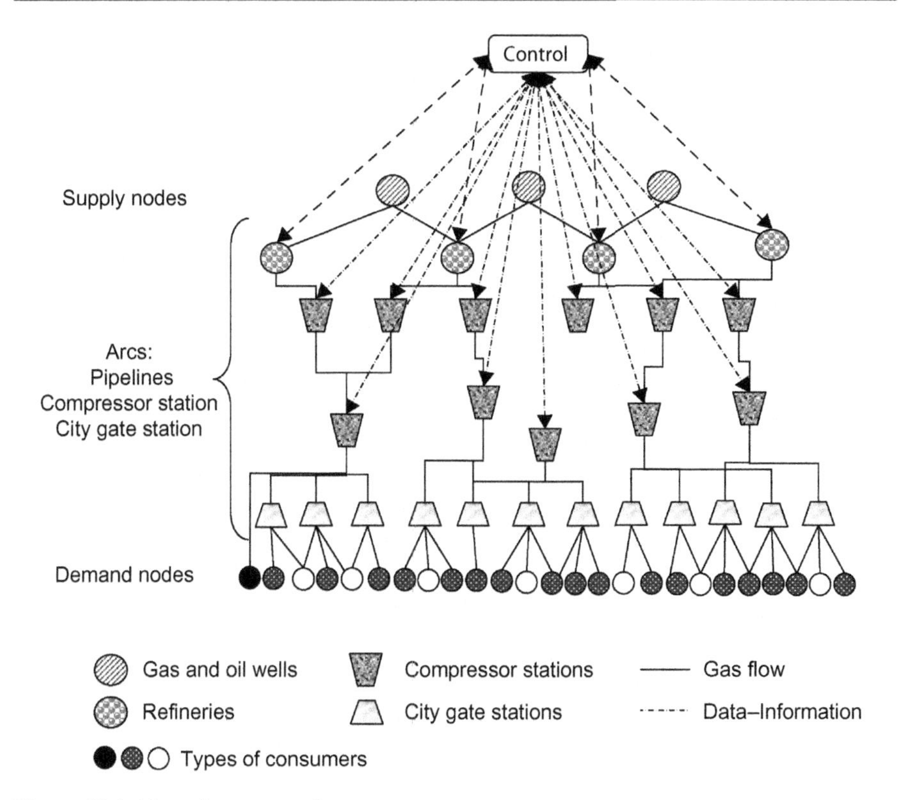

Figure 19.1 Natural gas network components.

Natural gas suppliers and natural gas consumers are connected through a complex and huge network in such a way that there is a long distance between them, and natural gas must flow by the use of suitable pressure. During long transportation, pressures are lost because of friction between the natural gas and the inner walls of pipelines. In addition, the natural gas volume is reduced because of heat transmitted from the environment. Therefore, compressor stations are installed to set and hold pressure continuity along the network and to periodically determine the capacity of the transmitted gas.

Generally, compressor stations are one of the most complex entities in a natural gas network because they consist of several compressor units (typically 15–20) that have been connected in different configurations such as series, parallel, or a combination of both, and they have different types [3]. Two of the main types of compressor units are centrifugal and reciprocating units. Centrifugal units are more common in the industry and consequently in related research assumptions [4]. Without considering which type of the compressor unit is used in the model, both types of unit have two options turning on and turning off—which makes their behavior nonlinear. When the demand of the customer increases, the pressure of the pipeline drops. Therefore, at least one compressor should be opened until the gas pressure resumes an acceptable level [5].

The whole process of the natural gas network can be concluded in four main parts: supply, transportation, storage, and sale of the natural gas in the market places.

Supply

Supply usually starts with development and exploration, extracting the gas reserves, and processing the extracted gas. In practice, the same company performs all three functions. Exploration is concerned with locating natural gas and petroleum deposits. After a team of exploration geologists and geophysicists has located a potential natural gas deposit, in the extraction phase a team of drilling experts digs down to where the natural gas is assumed to be [30].

Transportation

Natural gas transportation is the most important process in the natural gas industry. It consists of a complex pipeline network that moves natural gas from various origins to consumers in order to satisfy their demands. The network is divided into two main networks, namely, transmission and distribution. Moving a large volume of gas at high pressure over long distances from a gas source to distribution centers is done in transmission networks. Routing gas to individual consumers is done through distribution networks [6]. The efficiency of transportation is a suitable criterion to estimate the whole of a natural gas system's performance. Considering the aim of the network and consequently the physical characteristics of its components, the natural gas network can be split into the transmission network and distribution networks. The scope of both networks is presented in Figure 19.2.

Transmission network: Ríos-Mercado et al. [6] have shown how a gas transmission network, including pipelines, junction nodes, and compressor stations, is different from conventional networks by two special characteristics. (1) Beside the flow variables, which present the mass flow rate for each arc, a pressure variable is

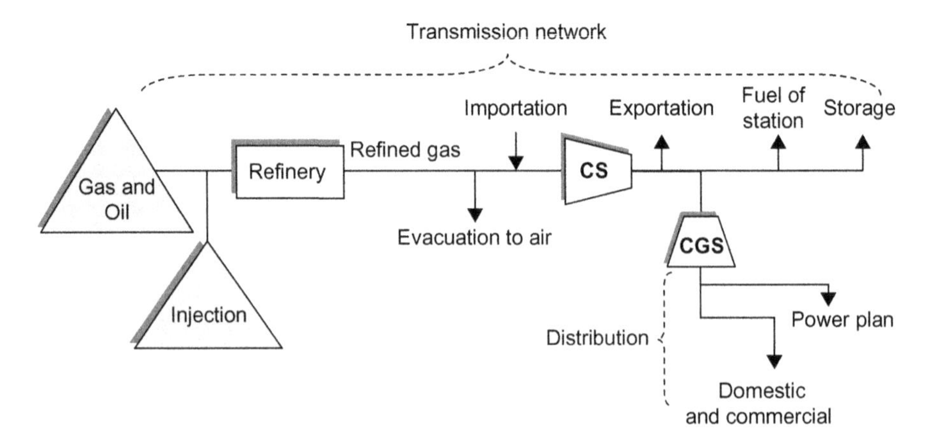

Figure 19.2 The process of transportation during the natural gas network.

defined at every node. (2) Unlike most networks, which consider only mass flow balances, transmission networks take into account two other types of constraints:

- A nonlinear equality constraint on each pipe caused by the connection that exists between flow and pressure drop.
- The feasible limits that are available to operate pressures and flows inside each compressor, which are represented by a nonlinear and nonconvex set.

Usually, the operating expenses of the natural gas transmission network are estimated through the operating costs of the compressor stations, which can be determined based on the fuel consumed at each compressor station. For example, Borraz-Sánchez and Ríos-Mercado [7] estimated that 3–5% of transmitted gas is consumed by compressor stations. Because a huge volume of natural gas is being transported through the network, about 25–50% of the total operating budget of companies is spent on running the compressor stations. Therefore, minimizing the total fuel consumption of the compressor stations along the network is one of the main objectives for transmission networking because of its effects on overall gas operation costs [6].

Distribution network: Distribution networks are different from transmission networks in several perspectives. They do not have valves, compressors, or nozzles, and pipes act under fewer pressures. Therefore, pipelines are smaller, and networks are simpler [8]. Most natural gas users, which are domestic and commercial consumers, receive the gas from local distribution companies. A smaller number of natural gas users such as power-generation companies receive the natural gas directly from high-capacity interstate and intrastate pipelines. In large municipal areas, local gas companies usually deliver gas to users through stations called *city gates*. For the design of distribution networks, first the network topology should be defined by technical teams; second, required features must be determined so pipelines and pumps can meet the flow of their nodes and pressure requirements [8].

Storage

Because natural gas processes, including exploration, production, and transportation, are time consuming and because all of the produced natural gas is not always needed at various destinations, a part of the extra gas is injected into storage units, which usually are located near market centers and are usable for unlimited periods. Gas storage is one of the new and critical steps of the natural gas network process that must respond to the demands of different periods of the year. Traditionally, during summer months, natural gas was stored to respond to increased demands during the coldest months, but nowadays natural gas demand in summer has increased because special users such as power-generation companies must produce electricity for air conditioners during summer. In addition, natural gas storage plays a critical role in unexpected events such as natural disasters, which may affect production and transportation. In general, some of the main reasons behind using storage along the natural gas network are its capability to respond to cyclic fluctuations when temperatures vary and consumption is high, improving services to all customers, keeping market shares competitive with other sources of energy, and achieving operations with high load factors. Natural gas storage can be done in different ways, but underground reservoirs are the most important method. The storage deals with pipelines, local distribution companies, producers, and pipeline shippers (US Department of Energy, US Energy Information Agency, March 1995).

Sales and Marketing

Marketing natural gas means selling natural gas or organizing its business from well to end users at various levels. At this stage, all of the required intermediate steps are considered: transportation arrangements, storage, accounting, and especially sales. Marketers in the natural gas industry play a complex role and may be joint endeavors with producers, pipelines, or local utilities or may be an independent group concerned with selling natural gas to retailers or end users. Natural gas usually has three to four owners before reaching customers. Marketers utilize their skills to reduce their exposure to risks and increase throughput by forecasting the behavior of the natural gas market, finding buyers, and securing ways to deliver the natural gas to end users.

19.2 Natural Gas Network Problems

19.2.1 *Formulating*

A number of useful notations and concepts are common in most of the developed models dealing with network optimization in the natural gas industry and can be useful to researchers who want to start modeling. As mentioned previously, the natural gas network is composed of pipelines and compressor stations as arcs and nodes. Several general indices and parameters will be presented here and are depicted in Figure 19.3. For more details, see Ríos-Mercado [3].

Indices:

i,j: index of nodes, $i,j \in N = \{1, \ldots, |N|\}$
k: index of compressor stations, $k \in C = \{1, \ldots, |C|\}$
m: index of units (turbocompressor) of compressor station, $m \in U = \{1, \ldots, |U|\}$
l: pipelines, $l \in L = \{1, \ldots, |L|\}$
n: nodes
c: compressor stations

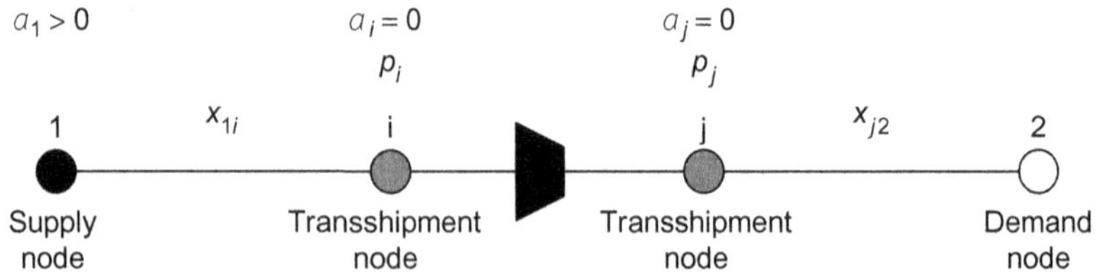

Figure 19.3 Basic notations for planning models in natural gas networks.

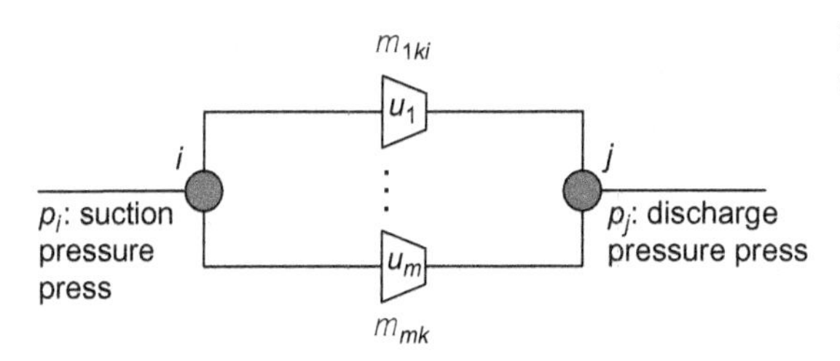

Figure 19.4 The behavior of compressor station units.

l: pipelines
u: units
S_a: set of arcs

Parameters:

P_{ij}: associate capacity between ith and jth nodes
C_{ij}: transshipment cost for each unit of natural gas between ith and jth nodes

Variables:

P_i: the gas pressure at node i
X_{ij}: the mass flow rate between nodes i and j
α_i: the net flow through the node i

$$\mu_{mk} = \begin{cases} 1, & \text{if unit } m \text{ of compressor station } k \text{ must be opened} \\ 0, & \text{if unit } m \text{ of compressor station } k \text{ must be closed} \end{cases}$$

Considering the above notations, $S_a = L \cup C$ in a manner that $L \cap C = \varnothing$.

In some problem areas, the nonlinear behaviors of compressor stations units (turbocompressors) are considered. Figure 19.4 presents the positions of turbocompressors in a compressor station. Turning these turbocompressors on and off is one of the main decisions that have to be taken in the natural gas network [3].

19.2.2 Optimization

In optimization problems, the search for the optimal solution is done by iteratively transferring the current solution to a newer and hopefully better solution. Optimization methods can usually overcome numerical simulation approaches because of two main limitations. First, there is no guarantee from simulation approaches that the achieved result is optimal (or cost is minimal). Second, determining pipe diameters depends only on the experience of users. Therefore, for the same problem, different users always take different decisions, which is not of interest [8].

A typical pipeline network for delivering natural gas requires a vast number of facilities and limitations, which should be considered. Because of the complex nature of the natural gas pipeline network, problems defined in this scope seek different aims and methods that certain requirements have to be considered in their optimization methodologies to achieve satisfactory and robust enough solutions to cover the most important aspects of the network. In such complex and huge networks, proper planning for transmission and distribution networks has a special importance because even a small reduction in operation expenses and investment costs can include considerable amounts of money and improvements in the system utilization, which is more valuable in gas-rich countries. Growing natural gas networks, make them more complex, and from the optimization perspective, developing effective algorithms becomes more important.

Network Optimization

According to Osiadacz [19], network optimization means finding a certain objective function in such a way that design parameters, development structures, and parameters of the network operation are optimum. In the last two decades, so many researchers in the natural gas area have paid attention to optimization methods to find the optimal solution in various fields of the natural gas industry. Depending on which decisions are going to be made and what are the variables that are sought to make optimum objective function, all optimization problems defined in this field can be decomposed into four groups: optimal design, optimal flow, optimal operation, and optimal expansion.

Network Design

Network design decisions are key strategic decisions, and the consequences of making these decisions poorly are often severe [10]. The network design problem occurs in many diverse application areas, including facility location, material-handling systems, natural gas or electric power transportation, and telecommunications.

In the optimal design of a natural gas network, the main design parameters of basic components of the network including pipelines and compressor stations are provided over a planning horizon in such a way that considering the network constraints the customers are satisfied with a minimum annualized cost [11,12]. Outputs of the system will be the design characteristics of pipelines, including diameters, pressures, and flow rates, and such design parameters of compressor stations as location, suction pressure, pressure ratio, station throughput, fuel consumption, and station power consumption. Each parameter and characteristic influences the overall construction and operating cost to some degree [12].

Mohitpour et al. [13] defined and explained the major influencing factors on pipeline system design: properties of fluid, design conditions, magnitude or locations of demand and supply nodes, codes and standards, route, topography, access, environmental impacts, financial matters, hydrological impacts, seismic and volcanic impacts, material, construction, operation, protection, and long-term integrity.

Network Flow

The main objective of network flow optimization in the natural gas and other industries is minimizing costs and providing sufficient services to customers, which is close to operational decisions. In this type of problem, decision variables are defined to determine the volume of gas flowing through the network. Many of the network flow's problems such as minimum cost flow problems, shortest path problems, maximum flow problems, and transmission network planning can be modeled as different forms of mathematical programming with linear or nonlinear functions and integer or mixed-integer variables.

To date, many models have been developed to describe the gas flow though the network as well, but there are several difficulties to find the suitable solution for the developed models because of their nonlinear and nonconvex nature [3].

By making use of the introduced notations, the general form of network flow model, taken from Ahuja et al. [14], can be presented as follows:

Minimize

$$\sum_{(i,j)\in N} C_{ij}X_{ij} \tag{19.1}$$

subject to

$$\sum_{\{i:(i,j)\in N\}} X_{ij} - \sum_{\{j:(j,i)\in N\}} X_{ji} = \alpha_i \tag{19.2}$$

$$0 \le X_{ij} \le P_{ij} \quad \forall i,j \in N \tag{19.3}$$

Set of the first constraints (Eqn (19.2)) is mass balance constraints, and set of the second constraints (Eqn (19.3)) presents the capacity boundaries for gas flowing between the ith and jth nodes.

Network Operation

Some operational decisions should be taken into account for the network to ensure that the demand for natural gas is met. At high pressures of natural gas, the operation cost of the network is determined based on the operation of compressors because of the significant percentage of running costs of compressor stations in the total budget of companies. In low and medium pressures, an optimal operational cost is achieved through leakage reduction by optimizing the nodal pressures [9]. In general, the operating cost belonging to the natural gas network normally takes up more than 60% of the total cost of the pipeline [5]. Therefore, operational decisions have a significant effect on the network performances. Given the fact that the amount of natural gas in the pipeline system is set by compressor stations and that the cost associated with the operation of compressor stations, including turning them on and off, the most critical operational decision in a natural gas network is selecting compressors. This important decision, which is influenced by the compressors' capacity and the energy required to turn the compressor units on and off, significantly affects total natural gas operation cost. Another critical factor on the performance of the natural gas network is starting or stopping compressors because of their different outputs [5]. Therefore, efficient operation of the complex networks of natural gas can substantially reduce airborne emissions, increase safety, and decrease the daily operating cost [3].

Network Expansion

In today's competitive markets, natural gas companies are interested in expanding the network and consequently serving potential customers because their market shares will be larger and the achieved profits will increase. In network expansion, generally the objective is scheduling the investments to supply an economic and reliable energy with minimal cost, which is not easy to achieve [15]. To make an optimal capacity expansion of natural gas network, several decisions regarding the time, size, and location of expansion should be made [11]. The projects dealing with networks expansion have various steps that are different from country to country and from company to company based on rules and governmental economic policies [16].

Referring to the literature, researchers mentioned different aspects to current difficulties of network development and expansion. Davidson et al. [10] have indicated these difficulties from some angles: the many existing options for expanding and generating a prespecified layout, existing uncertainties in absorbing the customers and profits, difficulty in estimating construction costs because of difficulties in calculating the length and unit cost per length for new pipes to expand, and finally the dynamic nature of the problem. In this matter, Kabirian and Hemmati [16] have paid attention to the presentation of an integrated strategic plan, which considers different aspects of the network development on a long-time horizon. They introduced the difficulties of this subject in various points, including covering development and strategic planning in both short and long run, identifying the locations and schedules of new compressor stations and pipelines in the network, determining the best type and routing of the pipelines, selecting the best combination of natural gas procurement from available sources, and providing the best operating conditions for compressor stations in long-run horizons [16]. By accuracy of data input for new substation availability, substation reinforcements, local generation, and future load location, which should be sent under dynamic or nondetermined status, a robust decision making is possible. These uncertainties can be presented in mathematical models as well, but the nature of the problem, which is nonconvex and multiobjective, makes it difficult to solve. However, it can be simplified through linearization of the objective functions and simplify the problem description [15].

Chung et al. [17] focused on transmission networks planning through a mathematical model with three objectives, including investment cost, reliability, and environmental impacts. The model was formulated using the approach of goal programming and solved by a genetic algorithm (GA). For analyzing the decisions, a fuzzy decision method was used to select the best scheme. In distribution networks, Carvalho and Ferreira [15] proposed an evolutionary algorithm for the stochastic planning of the large-scale networks under uncertain conditions and introduced the difficulties of optimizing networks expansion, including multistage investment decisions, the large-scale distribution network, and a huge variety of operation policies, variable demands, investment costs, equipment variabilities, and locations that make the decision very insightful.

The reviewed papers in the scope of optimization in the natural gas industry based on the decisions tried to make have been classified in Table 19.1.

19.2.3 Model Characteristics

Each defined problem on natural gas networks follows several assumption. The assumptions are presented in the form of constraints and affect the problem complexity and formulation. In addition, sometimes researchers focus on part of a network that has special characteristics. Usually, in this area some properties are determined as the problem statement first. Some of the most usual attributes of natural gas network problems have been illustrated in this section.

Steady State or Nonsteady State

The state of natural gas pipeline networks in different models is presented with two main categories: steady state and transient state. These states are determined through considering or not considering a partial differential equation involving derivation with respect to time [4]. In other words, this classification is dependent on how the gas flow changes in relation to time.

Steady state: In a large number of previous researches with optimization problems in the field of natural gas network, the operation of systems is assumed in steady states because in the previous decades there was no need to quick responses to variability of demands and conditions and problems were simplified by converting to subproblems in steady states [18].

In a steady-state system, the flow of gas is determined with some values which are independent from the time and constraints of the system, especially the ones describing the pipelines gas flow are described by algebraic nonlinear equations [6]. In the steady-state assumptions, it is possible to work out the partial differential equation and reduce to a nonlinear equation with no derivatives, which from the optimization view makes the problem more tractable [4].

Because loads and supplies are not a function of time in steady-state problems, the structure of the network—including the number of sources, compressor stations, valves and regulators, and the optimal parameters of operations including pressures and flows—are determined once [9]. General equations for steady-state flows in natural gas networks have been collected in Coelho and Pinho [19].

Nonsteady (transient) state: When load variations in a system are high, steady-state operations of that system are not desirable or even possible to consider such as when factors like deregulation and peak shaving are being considered. Therefore, efficient and responsive operations in dynamic statuses are essentially required to respond rapid variations in demands and conditions [18]. In a transient state system, the system variables such as mass flow rates through the pipelines and gas pressure levels at each node are defined as the functions of the time

Table 19.1 Classified Papers Based on Their Problem's Objectives

Author	Type of Network		Network Design			Network Flow and Operation				Network Expansion
	Transmission	Distribution	Investment Cost Minimization	Maximizing NPV	Maximize Customer Satisfaction	Minimize Transportation Cost	Compressor Selection to Minimize Costs	Fuel Cost Minimization	Maximize Customer Satisfaction	Optimize Scheduling of Investments
Carvalho and Ferreira [15]	—	√	—	—	—	—	—	—	—	√
Wu et al. [22]	√	—	—	—	—	—	—	√	—	—
Ríos-Mercado [3]	√	—	—	—	—	—	—	√	—	—
Uraikul et al. [5]	√	—	—	—	—	—	√	—	√	—
Chung et al. [17]	√	—	—	—	—	—	—	—	—	√
Borraz-Sánchez and Ríos-Mercado [4]	√	—	—	—	—	—	—	√	—	—
Mora and Ulieru [24]	√	—	—	—	—	—	—	√	—	—
Davidson et al. [10]	—	√	—	√	—	—	—	—	—	—
Ríos-Mercado et al. [6]	√	—	—	—	—	—	—	√	—	—
Kabirian and Hemmati [16]	√	—	√	—	—	—	—	—	—	—
Abbaspour et al. [18]	√	—	—	—	—	—	—	√	—	—
Wu et al. [8]	—	√	√	—	—	—	—	—	—	—
Sadegheih and Drake [28]	√	—	—	—	—	—	—	—	—	√
Borraz-Sánchez and Ríos-Mercado [7]	—	—	—	—	—	—	—	√	—	—
Hamedi et al. [1]	√	√	—	—	—	√	—	—	—	—

dynamically. Usually, descriptive models are used to analyze transient states because of their intractable from the optimization point of view [7].

Cyclic Topology or Noncyclic Topology

Two fundamental types of network topologies are cyclic topology and noncyclic topology.

Cyclic: A cyclic topology is concerned with a network in which at least one cycle is present, including two or more compressor station arcs such as in Figure 19.5. In practice, effective algorithms for cyclic topologies do not exist [7].

Noncyclic: Most of the pipeline systems have noncyclic structures. A serial (or gun-barrel) structure is a special type of a noncyclic network where the associated reduced network is a simple path [3]. Tree structures are another type of the noncyclic topology. A tree structure involves multiple converging and diverging branches in such a way that all nodes have in-degree equal to one, except one node, which has in-degree equal to zero [3]. Figure 19.6 is a sample for a serial topology, and Figure 19.7 presents a tree topology in natural gas networks.

To recognize the natural gas network topologies, Borraz-Sánchez and Ríos-Mercado [7] explained a usual methodology. First, remove the compressor arcs from a given network temporarily. Second, merge the remaining connected components and eventually put the compressor arcs back in place. The obtained network is a reduced network. Three cases will occur from the reduced network. If it has a single path, the given network has a serial (gun-barrel) topology. If in the reduced network the compressors are arranged in branches, then the topology is a tree. If in the reduced network compressor stations are arranged to form cycle, the topology is cyclic.

19.2.4 Types of Methods

After introducing the natural gas problems and their main characteristics to distinguish them from each other, a basic classification is done regarding methods of solving the natural gas pipeline networks. To find the best solution for the network problems, estimating the problem complexity is very important. It is quite clear to scholars in this area that the problems with cyclic structure are more difficult to

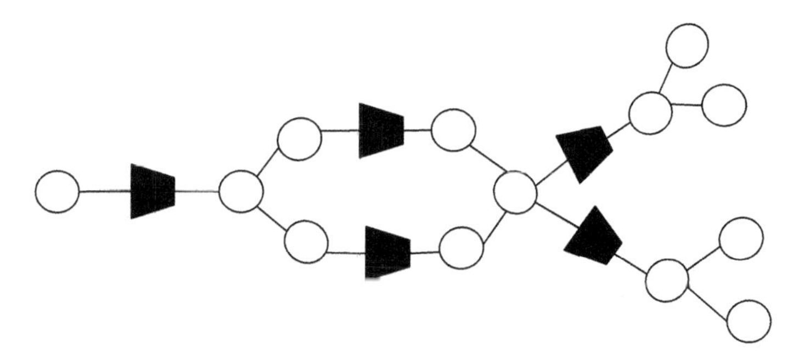

Figure 19.5 A cyclic topology.

Figure 19.6 A serial (gun-barrel) topology with three compressor stations.

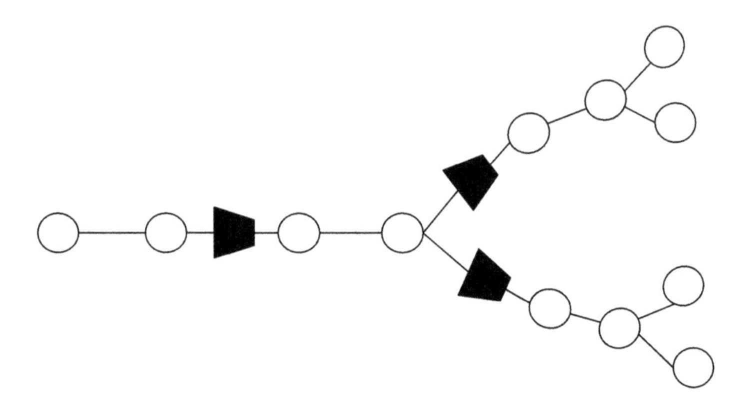

Figure 19.7 A tree topology.

solve than problems with noncyclic topology. In other words, the dimension of problems with cyclic topology is usually large and cannot be reduced by removing or fixing variables as happens in some noncyclic topology problems. The majority of the noncyclic gas network topologies have been developed based on dynamic programming and there are a large number of optimization algorithms for this type of topologies. Before we explain the suitability of methods for solving the planning problems of the natural gas network, we present Table 19.2.

Dynamic Programming

For the last few decades, dynamic programming (DP) has been utilized to optimally solve very large noncyclic networks such as gun-barrel and diverging branch tree systems, and some subclasses of cyclic networks. In general, to solve network problems with noncyclic systems by DP, flow variables are determined in advance and pressure variables are kept. Therefore, by converting a multidimensional problem into one dimension, the problem is simplified and solved easily. In a diverging branch, the problem is decomposed into a sequence of several one-dimensional DP problems in such a way that each deals with a single branch [6].

Since a DP simply satisfies constraints of any natural gas network and overcomes to nonconvexity and nonlinearity difficulties of feasible solutions, it can be used for noncyclic topologies but its computation difficulties increase with problems dimensions exponentially [7]. Unlike noncyclic topology, the applicability of DP for cyclic topologies is limited because the cycles break the linear structure of the network and the flow variables must be explicitly managed. In other words, the

Table 19.2 Priority of Methods to Solve Natural Gas Network Problems

Method	Topology	
	Cyclic	Noncyclic
DP	4	1
Gradient search	2	2
Hierarchical programming	3	3
Mathematical programming	1	4

DP for cyclic networks will be multidimensional. The main limitation of DP regarding the cyclic topology is that to solve this type of the problem the flow variables must be fixed. Therefore, the achieved solution is optimal only with respect to a prespecified set of flow variables [4]. As it would appear from the literature, by increasing the consideration of cyclic topologies in the defined problems belonging to the natural gas system, the success of DP has been reduced.

Gradient Search

In 1987, the generalized reduced gradient (GRG) was introduced for the first time. GRG is based on a nonlinear optimization technique for noncyclic structures. In comparison to DP approaches, in the dimensionality issue for cyclic topologies GRG acts well, but it does not guarantee a global optimal solution, especially in cases where decision variables are discrete [4,7].

Hierarchical Control Mechanisms

In some transmission and distribution network problems, which are difficult to solve in an integrated way, other techniques such as hierarchical structures can be used in the process of solution to decompose the solution space to several levels. In the case of natural gas network hierarchical approaches, Ríos-Mercado et al. [6] illustrated that the overall network is decomposed into two levels: the network state level as the highest level, and the compressor station level as the lowest level.

Mathematical Programming

Since DP can not avoid trapping into the local optimum solutions, DP-based approaches and gradient searches have not had a valuable success rate to overcome difficulties of cyclic topologies in natural gas network problems. Therefore, these methods are more useful for the problems, which have fixed the flow variables, and consequently the optimality of the solution is only with respect to a prespecified set of flow variables. For more than half a century, mathematical programming approaches have been used in various sections of the natural gas industry. Because of nonconvexity of feasible solutions and nonlinearity and nonconvexity of objective functions of natural gas optimization problems formulated by mathematical models, a large number of local optimum solutions exist where metaheuristic methods help to escape from the local optimality. Overall, a rapid improving in optimization algorithms is seen for solving complex mathematical models of natural gas networks, which has had a significant growth especially for cyclic topologies because of difficulties in solving the problems.

19.3 Survey on Application of Optimization

Many optimization approaches have been developed to make significant improvements in different fields of the natural gas networks with a number of general assumptions, but still a tremendous potential exists in this field. By increasing the complexity of natural gas problems, more algorithms are being defined from the optimization perspective. Therefore, analyzing the previous researches can be helpful to scholars for future research. Ríos-Mercado [3] focused on reviewing fuel cost minimization, which is only one field of optimization in the natural gas industry. Considering the application of optimization methods in different fields of this critical industry and its importance in all fields, this section focuses on comprehensively surveying the most important optimization problems in the natural gas network and organizing the latest papers on this topic.

19.3.1 Subnetworks

Transmission

Some of the problems of natural gas, which have been formulated mathematically, obey the general frame of mathematical models. This means each mathematical model includes an objective function, some constraints, and a number of variables in such a manner that the differences among these components separate developed models.

Chung et al. [17] developed a fuzzy mathematical model that considered more than one objective to solve the planning problem of transmission networks, which tries to optimize investment cost, reliability, and environmental impacts. The developed model was solved through a GA, and efficient results were achieved. To minimize the operational cost of natural gas networks, one of the most critical operations studied by some researchers is the compressor selection. This problem is important because it is associated with the cost of turning compressors on and off, which is a considerable part of total operation cost. Uraikul et al. [5] presented the compressor selection problem in the form of a mixed-integer linear programming (MILP) and by considering three types of the cost including operating cost, start, and stop penalties. The mentioned penalties refer to the cost of turning the compressors on and off in such a way that the energy used for starting a compressor is more than the required energy for stopping it. To escape from being trapped in this model into nonlinearity, some types of cost, which are based on time or have uncertainties such as maintenance cost, have not been considered in the developed model. Among researchers who have focused on the natural gas network optimization, only a few have adopted the difficulties of cyclic topologies and have not simplified the problem to the linear or tree structure. The research work of Borraz-Sánchez and Ríos-Mercado [4] is in this group. They presented the problem of optimal operation of a natural gas pipeline system in cyclic topologies through combining a nonsequential DP approach within a Tabu search (TS) technique through four main phases: preprocessing, finding an initial feasible flow, finding an optimal set of pressure values, and flow modifications. What makes this work different from noncyclic research is flow modification, which was done in noncyclic approaches by determining a unique set of optimal flow values in the preprocessing phase. What makes this work a little far from reality are its steady-state assumptions.

Kabirian and Hemmati [16] developed a nonlinear optimization mathematical model for formulating a strategic planning model to find the best feasible development plan for natural gas transmission networks. The objective of the developed model was determining the type, location, and installation schedule of pipelines and compressor stations over a long planning horizon with the goal of least cost and with a consideration of network constraints. To achieve optimal or near optimal plans, an algorithm based on random searches was applied. Mahlke et al. [20] formulated the problem of transient technical optimization in the form of a mixed-integer nonlinear problem with the aim of minimizing the fuel gas consumption. Because of difficulties with the time-dependent natural gas transmission network, they limited their work to achieve a good feasible solution in a suitable run time through a simulated annealing (SA).

Distribution

In spite of the simplicity of distribution networks and the relatively low importance of the transmission network from the perspective of design cost, it is very critical to satisfy costumers. A considerable number of optimization methods in the form of mathematical models have been developed for distribution networking to find the best design and optimal operation along the pipelines. Among reviewed papers, Carvalho and Ferreira [15] tried to develop a mathematical model for large-scale distribution networks to make robust expansion on variable conditions and under deregulation. Wu et al. [8] considered a nonlinear network and proposed the problem of minimizing the investment cost through the distribution network under steady-state assumptions. They developed a model by introducing new variable and converting the primal problem to a nonconvex constrained problem. Therefore, by escaping from the available difficulties in solving the primal nonsmooth and non convex problem of designing a distributed layout, a global optimization approach was achieved. Moreover, Davidson et al. [10] also focused on investment planning in the natural gas distribution networks.

19.3.2 Main Problems

Fuel Cost Minimization

Because of the long distances among the supply and consumption nodes in the natural gas network, many compressor stations are used along the route to set the natural gas pressure throughout the pipeline systems. By considering the tremendous amount of transported gas in pipelines per day, minimizing the gas consumed by compressors is critically important. Global optimization can be lead to a 20% saving in fuels consumed by compressor stations [21]. To date, a great deal of research has been performed to develop new techniques to decrease the consumed fuel of consumption in compressor stations.

In the problem of fuel cost minimization, the decision variables are pressure dropped at each node of the network, flow rate at each pipeline, and the number of units operating within each compressor station [22]. In general, defined problems for the fuel cost minimization differ from each other because of some assumptions and methodologies applied by researchers to determine the value of variables in the optimal case. In a number of previous works, to avoid or decrease the nonlinearity of the model, the number of compressor units in each compressor station has been considered as fixed. In addition, some of the developed models have been simplified by considering only one unit for each compressor station, whereas compressor stations usually have multiple units. Balancing or not balancing the network is another matter. If the network is assumed balanced, then in each node of the network the sum of all net flows will equal zero. This means there are no differences between the total output flows of supply nodes and input flows to demand nodes [23]. Other assumptions may be related to a steady state or a transient state of the model or topology of the networks, which are referred to the problem statement. In addition, regarding the methodology, in some research if there is more than one variable, the values of variables are achieved simultaneously. In contrast, some of the researchers have proposed methodologies based on multistage iterative procedures. Ríos-Mercado et al. [6] developed a two-stage procedure to optimize the fuel cost minimization in such a way that gas flow variables were fixed at the first stage and optimal pressure variables were found via DP. Then the pressure variables were considered fixed at the second stage, and a set of flow variables was achieved, taking the network topology into consideration to improve the objective function. Some authors relax the nonconvex and nonlinear models by relaxation techniques because generally such problems are very difficult to solve. For example, for fuel cost minimization, Wu et al. [8] developed a mathematical model with steady-state assumptions and a nonconvex feasible domain; a nonlinear, nonconvex, and discontinuous fuel function; and a nonconvex set of pipeline flow equations. To solve the developed model, it was relaxed in two ways. First, the fuel cost objective function is relaxed; second, nonconvex and nonlinear compressor domains are relaxed. In their procedure solution, the optimal solution of the original problems involves upper bound, and the optimum solution of the relaxed problems is lower bound. The general formation of fuel cost minimization in the natural gas network, considered to be the most applied variables including flow rate and pressure, has been presented by Ríos-Mercado [3]. Another research that investigated the fuel cost minimization of compressor stations belongs to Mora and Ulieru [24]. It focused on developing a new method to achieve a near optimal feasible solution in a shorter reasonable time for minimizing the amount of natural gas consumed by the compressor station units.

Some of the latest papers on minimizing the fuel costs of compressor stations and variables that have been used to achieve optimal values are cited in Table 19.3.

Investment Cost Optimization

Carvalho and Ferreira [15] presented the general form of optimizing investment policies, which have been adapted to the information structure of scenarios, based on the minimum cost as follows:

Minimize $F(U)$subject to

$$\begin{aligned} &U \in Y \\ &U(s) \in \Omega_s \\ &s \in S \end{aligned} \tag{19.4}$$

where F is a function of operational and investment costs, U presents a policy, and Y is universe of not opposing decision policies; the universe of admissible policies for the sth scenario has been presented in Ω_s; and, finally, S illustrates a set of available scenarios.

To make an investment strategy for minimizing the risk and increasing profits, Davidson et al. [10] developed a dynamic model integrated with a geographical

Table 19.3 Common Decision Variables in Fuel Cost Minimization Problems

Author	Mass Flow Rates Through Each Arc	Suction Pressure and Discharge Pressure	Number of Compressor Units
Wu et al. [22]	√	√	√
Cobos-Zaleta and Ríos-Mercado [23]	√	√	√
Borraz-Sánchez and Ríos-Mercado [4]	√	√	—
Mora and Ulieru [24]	√	√	—
Ríos-Mercado et al. [6]	√	√	—
Abbaspour et al. [18]	√	√	—
Borraz-Sánchez and Ríos-Mercado [7]	√	√	
Chebouba et al. [29]	—	√	√

information system (GIS). Among investment projects with sequencing, budget, and timing limitations, the model made a trade-off to maximize the expected net present value (NPV) and minimize the variance among NPVs. This model considers both the revenues and cost while selecting the best expansion project with the aim of taking a decision support system to present the revenue of serving new customers and related costs to constructions as well as considering the uncertainties. It was solved by rollout heuristic algorithms to improve the solution quality. In this case, GIS helps to identify opportunities in potential network expansion, data collection, and perception made because of the developed model. Kabirian and Hemmati [16] developed a model with the aim of least discounted operating and capital cost to plan for the natural gas transmission network.

Minimizing the Cash-Out Penalties of the Shipper

In drawing a contract between the shipper and a pipeline company to deliver a certain volume of gas among several points, a problem may occur in marketing natural gas because of the differences among the promised amount of gas and the real amount actually delivered along a transmission network. In those cases where imbalances occur, pipeline companies penalize shippers by imposing a cash-out penalty policy, which is a function of daily imbalance. Therefore, the problem, which should be solved optimally, is making decisions for shippers to minimize their incurred penalty by carrying out their daily imbalances [25].

Table 19.4 presents a summary for some of the reviewed papers that have focused on optimization problems in the natural gas industry.

19.3.3 Mathematical Models Classifications

To the best of our knowledge, a vast number of works have been done on the optimization approaches and developing suitable algorithms to find the optimal solutions for the natural gas distribution and transmission networks. In this part, we classify the earliest papers that focused on the mathematical modeling in Table 19.5. Some general findings are obviously seen in this table. For example, the optimization has a more effective role in transmission network in comparison to distribution network because the natural gas spends more time under high pressures in transmission networks because of the long distances among producers and city gate stations and because more instruments are used in the transmission network. Therefore, more problems have been defined in this segment. Moreover, as it would appear from the table, although some data in the natural gas industry are not deterministic, to simplify the problem researchers have not considered uncertainties in the form of fuzzy or statistical data such as what Carvalho and Ferreira [15] and Davidon et al. [10] did. Other issues such as type of problems, number of objectives, and more useful solution methods regarding the optimization of mathematical models developed for natural gas network planning are presented in the following subsections.

Table 19.4 A Summary of Optimization Problems in the Natural Gas Industry

Author	Transmission (T) or Distribution (D)	Type of Optimization			System		Topology			Solution Method			
								Noncyclic					
		Fuel Cost Minimization	Investment Cost Minimization	Cash-Out Penalties of the Shipper	Steady State	Transient	Cyclic	Gun Barrel	Tree	DP	Gradient Research	Hierarchical	Mathematical Programming
Wu et al. [22]	T	√	—	—	√	—	—	—	—	—	—	—	√
Carvalho and Ferreira [15]	D	—	√	—	√	—	—	—	—	—	—	—	√
Chung et al. [17]	T	—	—	—	—	—	—	—	—	—	—	—	√
Uraikul et al. [5]	T	—	√	—	√	—	—	—	—	—	—	—	√
Borraz-Sánchez and Ríos-Mercado [4]	T	√	—	—	—	—	√	—	—	—	—	—	√
Davidson et al. [10]	D	—	√	—	—	√	—	—	—	—	—	—	√
Ríos-Mercado et al. [6]	T	√	—	—	—		√	—	—	√	—	—	—
Abbaspour et al. [18]	T	√	—	—	—	√	—	—	—	—	—	—	—
Wu et al. [8]	D	—	√	—	√	—	—	—	√	—	—	—	√
Mahlke et al [20]	T	√	—	—	—	√	—	—	—	—	—	—	√
Chebouba et al. [29]	T	√	—	—	√	—	—	—	√	—	—	—	√
Borraz-Sánchez and Ríos-Mercado [7]	T	√	—	—	√	—	√	—	—	—	—	—	√

Table 19.5 Summary of Developed Mathematical Models For Natural Gas

Author	Network		Number of Objectives		Type of Model	Variables	Type of Data			Solution Method		
	Distribution	Trans-mission	Single	Multi			Deter-ministic	Fuzzy	Stochastic	Exact	Heuristic	Metaheuristic
Carvalho and Ferreira [15]	√	—	—	√	MILP	Expected value of the scenario	—	—	√	—	√	—
Chung et al. [17]	—	√	—	√	NLP	Number of possible circuit additions in each path	—	—	—	—	—	GA
Uraikul et al. [5]	—	√	√	—	MILP	On–off compressor units (compressor selection)	√	—	—	√	—	—
Borraz-Sánchez and Ríos-Mercado [4]	—	√	√	—	NLP	Mass flow rates through each arc, gas pressure level at each node	√	—	—	—	—	TS
Davidson et al. [10]	√	—	—	√	INLP	Each phase of each project in each year is done or not	—	—	√	—	√	—
Wu et al. [8]	√	—	√	—	NLP	Length of the pipes' diameters, pressure drops at each node of the network, and mass flow rate at each pipeline	√	—	—	√	—	—
Kabirian and Hemmati [16]	—	√	√	—	NLP	Type, location, and installation schedule of pipeline and compressors	√	—	—	—	√	—
Mahlke et al. [20]	—	√	√	—	MINLP	Gas flow in valves and compressors, gas pressure at beginning and end of pipeline, fuel gas consumption, on–off compressor or valves	√	—	—	—	—	SA
Borraz-Sánchez and Ríos-Mercado [7]	—	√	√	—	NLP	Mass flow rate in each arc, gas pressure in each node	√	—	—	—	—	TS
Hamedi et al. [1]	—	√	—	√	NLP	Transported gas volume, shortage volume, on–off compressor units	√	—	—	√	√	—

Types of Problem

In general, optimization problems can be classified based on the type of variables (continuous, integer, or mixed) and nature of functions used as the objective function and constraints. Considering these two factors, six types of problems are defined as presented in Table 19.6. Because of the nonlinearity behavior of the compressor station units and other factors and existence of mixed continuous and integer variables, most of the optimization problems of the natural gas network planning are categorized in nonlinear problems. Dependent on the defined variables, they can be Non-Linear Programming (NLP), Integer Non-Linear Programming (INLP), and Mixed Integer Non-Linear Programming(MINLP).

Number of Objectives

In practice to plan for a natural gas network optimally, more than one objective, generally conflict objectives, should be considered. For example, minimizing network flows or investment cost versus the maximum satisfaction of the customers involve two conflict objectives, which should be achieved simultaneously.

In theory, if all objectives are seen in the solution methodology, problems become more difficult, especially when a large number of objectives are considered. In a few cases, all objectives are transformed into a single objective, but in most situations, it is not possible. Therefore, a number of researchers focus only on one objective and do not pay attention to others or relax them, and another group of researchers keeps the nature of problems and applies a multiobjective optimization method based on an approach. For example, if goal values of objective functions are known, the goal programming approach can be a suitable option.

Solution Methods

The nature of natural gas network problems that are categorized in the group of NP-hard problems is nonconvex and nonlinear. Therefore, the design and selection of proper solution methods are very critical in this field. Mallinson et al. [26] described two general methods to optimize natural gas network problems. In the first method, numbers of optimization problem variables are reduced by eliminating the flow variables; in the second one, to achieve a better behavior, the optimization problem is solved without removing any variable. By reduction techniques, the solution of the problem becomes easier, but finding the suitable algorithm has

Table 19.6 Classification of Optimization Problems in Mathematical Models

Functions	**Variables**		
	Continuous	**Integer**	**Mixed**
Linear	LP	ILP	MILP
Nonlinear	NLP	INLP	MINLP

troubles because the selected algorithm could perhaps not provide a feasible solution or in several cases it shows error messages. In addition, in some cases, however, a valid solution seems to be achieved, but after inspecting the results, it was detected that some constraints have not been satisfied [26].

In general, three options exist to solve mathematical models. Researchers try to choose the best one based on the model's complexity and solution time limitations. These options are exact methods, heuristic methods, and metaheuristic methods.

Exact Techniques

The problem featuring in the natural gas transmission and distribution networks because of its nonlinear and nonconvex nature cannot be solved using classical techniques like exact methods from mathematical programming because these methods are usually time consuming and unable to solve NP-hard problems even on a small scale. A number of researchers have tried to solve the developed models by exact techniques, but they had to oversimplify their approaches and compressor station models, which in practice may be inaccurate.

Heuristic Technique

The heuristic methods give the final solution in shorter time in comparison with exact methods, but there is a risk of trapping in the first local optimality. Therefore, achieving a global optimal solution is not guaranteed.

Metaheuristic Techniques

The best choice to solve NP-hard problems, that their solution time is dependent on the problem size exponentially, is the metaheuristic method, which guarantees finding the global optimum solution through decreasing the problem complexity without any limitations regarding the problem size. Some of the common effective methods, which researchers in different fields are interested in and in natural gas network planning also achieved many successes, are GA, SA, and TS. Chung et al. [17] used a GA for the problem of transmission networks planning to avoid arriving at local optimality and utilized a fuzzy decision analysis to select the best possible planning scheme. Mahlke et al. [20] exploited an SA to find a feasible solution in a reliable short time because of its simplicity to apply. TS allows designers to take advantages of the previous information in the selection of algorithms and subalgorithms. In optimization problems dealing with natural gas networks, the high nonconvexity of objective functions and the capability of TS to escape from local optimality have made it very efficient with an appropriate discrete solution space. Borraz-Sánchez and Ríos-Mercado [4] combined TS with nonsequential DP for the fuel cost minimization in the natural gas transmission network.

19.4 Case Studies

Some researchers applied developed optimization models to the natural gas industry in a number of special cases and achieved significant results. Two case studies are described in this section to illustrate the real application of optimization models dealing with the natural gas industry.

19.4.1 Case 1: Optimization of Planning in the Natural Gas Supply Chain

Hamedi et al. [1] developed a mathematical model to optimize the flowing gas through the network along a six-level supply chain with the aim of minimizing direct or indirect distribution costs. The mathematical model is a mixed-integer nonlinear programming (MINLP) model converted to linear programming to solve and is limited to six groups of constraints, including capacity, input and output balancing, demand satisfaction, network flow continuity, and relative constraints to the required binary variables. To reduce the model's complexities for the large-size problems, it has been divided into two parts based on the relations among pipelines and solved hierarchically. In such a manner that in each step, one section of the problem is solved exactly through Lingo software and its outputs are passed to the next part as inputs. Therefore, by decreasing the computational complexity a nearly optimized solution is achieved.

The result of applying the developed model and hierarchical algorithm on the natural gas network of a gas-rich country presents the 19.839% improvements (near to 452,754.222 cost unit with the given parameters) in the defined objective function in contrast with implemented plans in the reality. Even assuming that a part of this improvement is due to simplifier assumptions, the huge cost of transmission and distribution of natural gas to the consumers makes this improvement valuable for the natural gas industry.

19.4.2 Case 2: Optimization of a Multiobjective Natural Gas Production Planning

Barton and Selot [27] formulated a nonconvex MINLP model for the upstream natural gas production system, which has been considered from the wells to the liquefied natural gas plants (excluding the plants). The upstream production-planning model involves two important components, including the model of actual production facilities and networks (the infrastructure model) and the customer requirements (the contractual rule model). In this model, the natural gas network has been presented as a multiproduct network with nonlinear pressure-flow rate relations in the wells and the trunk line network. Moreover, production-sharing contracts (PSCs) and operational rules have been considered. The developed model, which comprises three objective functions, is a relatively large nonconvex MINLP (several hundred continuous variables and tens of binary variables). Maximizing dry gas production to satisfy contractual demands, maximizing Natural Gas Liquid (NGL) production to increase revenue for the upstream operator, and prioritizing production from certain fields are the objectives that the model seeks to achieve them. Because it was not possible to obtain global optimization approaches directly, the model has been reduced through a reduction heuristic and solved by a global branch-and-cut algorithm.

The result of applying the developed model in a real-world case study in Malaysia indicates its efficiency in increasing the secondary products production, achieving optimal long-term asset management beside satisfying contractual requirements of gas supply and customers in short term.

19.5 Conclusions and Directions for Further Research

Because using many instruments, including pipelines, compressor stations, valves, and regulators over long distances and using a variety of network topologies and technologies, natural gas networks have been known as a complex and difficult problem to solve. Therefore, when this problem is mathematically modeled, the problem will be NP-hard and cannot be solved easily. On the other hand, gradient search and DP approaches have had limitations to consider real characteristics of network models because of their limitations in avoiding trapping a local optimality. From the optimization point of view, to solve planning models of natural gas networks, mathematical models and consequently metaheuristic algorithms seem the most desirable solution methods. This is more valuable when the problem is formulated based on transient assumptions and the cyclic topologies are considered. As it would appear from reviewing papers implemented in real cases, by optimal design of natural gas networks, which is possible using mathematical models and solving with suitable algorithms to find the closest optimum solution, considerable improvements can be achieved. Because of the enormity of the problem, even a small improvement in a natural gas network could save a huge amount of money per day, and the need to develop more models and algorithms is strongly felt among planners. Within optimization problems in natural gas networks, minimizing the fuel cost consumed by compressor stations has received more attention among researchers although there are not many developed models to optimize expansion or investment costs. To date, many optimization algorithms in different fields of natural gas networks planning have been introduced for all the steady state, the transient state, and different topologies, cyclic or noncyclic. Although the proposed optimization algorithms have been really successful, comprehensive models are needed to consider all constraints simultaneously and solve the problem aggregately. Moreover, in the developed models, transient systems have not been of interest to the researchers during the last decades to optimize because of increasing difficulties. In addition, cyclic topologies have had a few successes in researches and implementations. Because of difficulties available in natural gas networks planning, researchers usually avoid considering varieties in demand data, production data, and other fuzzy or statistical data. Therefore, it can be a suitable point for future researches to develop new models. Furthermore, a number of technical perspectives can make scientific gaps for new researches in the planning of natural gas networks, including considering the temperature as a new variable, considering various types of compressor station units, and presenting the network in low or medium pressures instead of high pressures only.

References

[1] M. Hamedi, R.Z. Farahani, M.M. Husseini, G.R. Esmaeilian, A distribution planning model for natural gas supply chain: a case study, Energ. Policy 37 (2009) 799–812.
[2] J. Muno, N. Jimenez-Redondo, J. Perez-Re, J. Barquin, Natural gas network modeling for power systems reliability studies, in: Bologna PowerTech Conference ed., Bologna, Italy, 2003, pp. 23–26.
[3] R.Z. Ríos-Mercado, Natural gas pipeline optimization, in: P.M. Pardalos, M.G.C. Resende (Eds.), Handbook of Applied Optimization, Oxford University Press, New York, 2002, pp. 813–825.
[4] C. Borraz-Sánchez, R.Z. Ríos-Mercado, A Hybrid Meta-heuristic Approach for Natural Gas Pipeline Network Optimization, Springer-Verlag, Berlin Heidelberg, 2005, pp. 54–65.
[5] V. Uraikul, C.W. Chan, P. Tontiwachwuthikul, MILP model for compressor selection in natural gas pipeline operations, Environ. Informat. Arch. 1 (2003) 138–145.
[6] R.Z. Ríos-Mercado, S. Kim, E.A. Boyd, Efficient operation of natural gas transmission systems: a network-based heuristic for cyclic structures, Comput. Oper. Res. 33 (2006) 2323–2351.
[7] C. Borraz-Sánchez, R.Z. Ríos-Mercado, Improving the operation of pipeline systems on cyclic structures by tabu search, Comput. Chem. Eng. 33 (2009) 58–64.
[8] Y. Wu, K.K. Lai, Y. Liu, Deterministic global optimization approach to steady-state distribution gas pipeline networks, Optim. Eng. 8 (2007) 259–275.
[9] A.J. Osiadacz, Dynamic optimization of high pressure gas networks using hierarchical systems theory, in: 26th Annual Meeting PSIG (Pipeline simulation Interest Group) ed., San Diego, California, 1994.
[10] R.A. Davidson, J.L. Arthur Jr., J. Ma, L.K. Nozick, T.D. O'Rourke, Optimization of investments in natural gas distribution networks, J. Energ. Eng. (2006) 52–60.
[11] Z. Huang, A. Seireg, Optimization in oil and gas pipeline engineering, J. Energy Resour. Technol. 107 (1985) 264–267.
[12] F. Tabkhi, L. Pibouleau, C. Azzaro-Pantel, S. Domenech, Total cost minimization of a high-pressure natural gas network, J. Energy Resour. Technol. 131 (2009) 043002.
[13] M. Mohitpour, H. Golshan, A. Murray, Pipeline Design and Construction: A Practical Approach, ASME Press, New York, NY, 2003.
[14] R.K. Ahuja, T.L. Magnanti, J.B. Orlin, Some recent advances in network flows, Soc. Ind. Appl. Math. 33 (1991) 175–219.
[15] P.M.S. Carvalho, L.A.F.M. Ferreira, Planning large-scale distribution network for robust expansion under deregulation, in: Power Engineering Society Transmission and Distribution Conference (IEEE, ed.), 2000, pp. 1305–1310.
[16] A. Kabirian, M.R. Hemmati, A strategic planning model for natural gas transmission networks, Energ. Policy 35 (2007) 5656–5670.
[17] T.S. Chung, K.K. Li, G.J. Chen, J.D. Xie, G.Q. Tang, Multi-objective transmission network planning by a hybrid GA approach with fuzzy decision analysis, Int. J. Electr. Power Energy Syst. 25 (2003) 187–192.

[18] M. Abbaspour, P. Krishnaswami, K.S. Chapman, Transient optimization in natural gas compressor stations for linepack operation, J. Energy Resour. Technol. 129 (2007) 314–324.
[19] P.M. Coelho, C. Pinho, Considerations about equations for steady state flow in natural gas pipelines, J. Brazil. Soc. Mech. Sci. Eng. (2007) 262–273.
[20] D. Mahlke, A. Martin, S. Moritz, A simulated annealing algorithm for transient optimization in gas networks, Math. Meth. Oper. Res. 66 (2007) 99–115.
[21] D.W. Schroeder, Hydraulic analysis in the natural gas industry, in advances in industrial engineering applications and practice, Int. J. Ind. Eng. (1996) 960–965.
[22] S. Wu, R.Z. Rios-Mercado, E.A. Boyd, L.R. Scott, Model relaxations for the fuel cost minimization of steady-state gas pipeline networks, Math. Comput. Model. 31 (2000) 197–220.
[23] D. Cobos-Zaleta, R.Z. Ríos-Mercado, A MINLP model for a minimizing fuel consumption on natural gas pipeline networks, in: Congreso Latino Iberoamericano de Investigación de Operaciones (CLAIO) ed., Chile, 2002.
[24] T. Mora, M. Ulieru, Minimization of energy use in pipeline operations – an application to natural gas transmission systems, Industrial Electronics Society, IECON, 31st annual conference of IEEE, 2005. ISBN:0-7803-9252-3.
[25] V.V. Kalashnikov, R.Z. Ríos-Mercado, A natural gas cash-out problem: a bilevel programming framework and a penalty function method, Optim. Eng. 7 (2006) 403–420.
[26] J. Mallinson, A.E. Fincham, S.P. Bull, J.S. Rollett, M.L. Wong, Methods for optimizing gas transmission networks, Ann. Oper. Res. 43 (1993) 443–454.
[27] P.I. Barton, A. Selot, A production allocation framework for natural gas production systems, in: V. Plesu, P.S. Agachi (Eds.), 17th European Symposium on Computer Aided Process Engineering—ESCAPE17, Elsevier, Amsterdam, 2007.
[28] A. Sadegheih, P.R. Drake, System network planning expansion using mathematical programming, genetic algorithms and tabu search, Energ. Conver. Manag. 49 (2008) 1557–1566.
[29] A. Chebouba, F. Yalaoui, A. Smati, L. Amodeo, K. Younsi, A. Tairi, Optimization of natural gas pipeline transportation using ant colony optimization, Comput. Oper. Res. 36 (2009) 1916–1923.
[30] E. Dept, What Exactly "IS" Natural Gas?, from http://oilprice.com/Energy/Natural-Gas/What-Exactly-IS-Natural-Gas.html., 2009.

20 Risk Management in Gas Networks: A Survey

Reza Zanjirani Farahani[1], *Mohammad Bakhshayeshi Baygi*[2] *and Seyyed Mostafa Mousavi*[3]

[1]Department of Informatics and Operations Management, Kingston Business School, Kingston University, Kingston Hill, Kingston Upon Thames, Surrey KT2 7LB
[2]Mechanical and Industrial Engineering Department, University of Concordia, Montreal, Canada
[3]Centre for Complexity Science, University of Warwick, Coventry, UK

20.1 Structure of Gas Networks

Natural gas networks are complicated, starting from the wells to end users. Natural gas is found in some places underground, and there are many exploration methods to determine whether or not natural gas exists in a particular place. After investigation, the act of drilling starts and a well is made, a process called *extraction*. After technically and economically ensuring that recovering the existing gas is feasible, the gas is lifted up. This gas cannot be used in its raw state and needs to be processed. Refining stations are usually close to the wells to separate parts of the gas and prepare it for customer use. For natural gas, pipes are the usual mode of transportation. The pipes that deliver natural gas from the wells to refinery stations are called *gathering pipes*; they are usually low pressure and low diameter. If the gas extracted from a well has more than the standard amounts or levels of sulfur and carbon dioxide, special types of gathering pipes are required; the so-called sour gas is dangerous, and care should be taken in its transportation.

Because gas wells are usually located in places far from customers, a complex system is needed to deliver the gas. A transportation system called the *mainline system* of gas networks is used to deliver the gas through transmission pipes and compressor stations to customers. Transmission pipes are usually of high pressure, large diameter, and long distance. The task of compressor stations is to balance the gas pressure in the pipes. The mainline system needs large amount of investment, approximately 80% of total investment. The amount of investment depends on the parameters of the system, including pipe diameter, thickness, pressure, length, and compression ratio. A large number of articles have tried to optimize this system from various aspects. Ruan et al. [1] presented a mathematical model that took into

Logistics Operations and Management. DOI: 10.1016/B978-0-12-385202-1.00020-7

account all of the parameters important to the amount of investment. Kabiriana and Hemmati [2] presented a strategic planning model to determine the type, location, and installation schedule with a cost-minimization objective function. Cheboubaa et al. [3] proposed a metaheuristic algorithm called *ant colony optimization* (ACO) to determine the number of compressor stations and the discharge pressure for each.

In the next step in gas networks, which is called *distribution*, gas is delivered to the end user. Local distribution companies (LDCs) receive the gas in city gates, transfer points from transmission pipes to LDCs, and deliver it to individual customers. This delivery is done with the help of an extensive network of small-diameter distribution pipes throughout municipal areas. End users of natural gas from LDCs are residential, commercial, and industrial sectors and power-generation customers. Note, however, that some large commercial and industrial customers receive natural gas directly from the high-pressure pipelines. Literature is extensive on different aspects of gas distribution. Hamedi et al. [4] presented a six-level supply chain to minimize the cost of gas transmission and distribution. Generally, there are many articles related to the transmission or distribution of gas, among which Herran-Gonzalez et al. [5]; Martin et al. [6]; Ríos-Mercado et al. [7]; Wong and Larson [8]; and Wu et al. [9] can be mentioned.

Natural gas is not always used when delivered, so it is usually stored underground. This storage capability can be very helpful, especially when shortages occur in the network. Gas networks also use physically manipulated and automated controls such as the supervisory control and data acquisition (SCADA) system to ensure appropriate communications between equipment and control center.

Figure 20.1 demonstrates the schematic view of gas networks with all key parts shown.

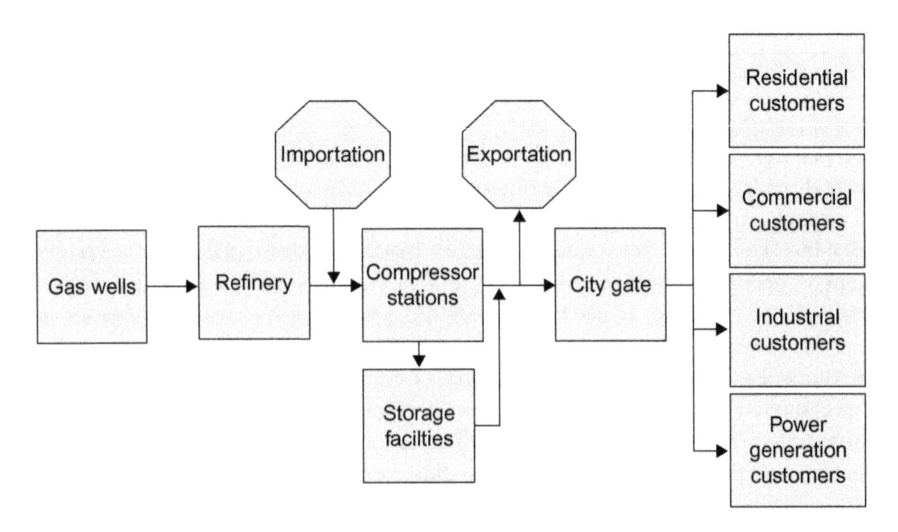

Figure 20.1 Schematic view of a gas network.

20.2 The Vulnerabilities and Risks of Gas Networks

Why is risk important in gas networks? Is it important to think about the risks? In the first part of this section we will discuss why we consider risk in gas networks. The remainder of this section thoroughly explains all existing vulnerabilities and risks in gas networks.

20.2.1 Why Is Risk Investigation Important?

The material presented in this section has been mostly excerpted from a US federal commission [10]. Gas has a very complicated network and is the subject of much attention nowadays in the energy sector, which is why it is really important to consider the associated risks. These can be considered an important challenge for countries, especially in the future. The reliability of gas networks has been affected by many factors, including the following:

- Gas has become very popular, giving rise to a competitive market in which equipment has been extensively used.
- The age of pipelines is an important matter. Pipes are usually exposed to failure risks because of leakage. In many countries, pipes are very old, creating an increasing risk of failure.
- Environmental considerations have become more crucial lately, and gas networks will be financially affected as they consider these issues. In fact, gas networks have already caused considerable harm to the environment.
- Many countries are dependent on imported gas from other countries. This creates great risk when they do not know how reliable their supply may be for many reasons, including politics and the environment.
- The use of information technology (IT) has led to more efficient operations, but on the other hand it has caused major problems. The networks are increasingly exposed to cyber attacks. In addition, the commitment of employees has declined in recent years.

With the aforementioned reasons in mind, we now consider certain challenges that exist in all countries that need to be addressed and solved, including the following:

- Reliable systems of delivery that can manage change must be developed.
- The increasing complexity of the network must be managed well because it is interdependent with other systems such as electrical systems.
- Mainline systems must be protected against disturbances that can cause long-term problems in delivery.
- Having the conditions in which the network can be seen as a whole, so the public and private sectors can experience helpful interactions.
- Appropriate regulations and policies must be developed and applied to ensure reliable delivery of natural gas.
- Communication systems must be upgraded so that information can be shared rapidly by system components at appropriate times.
- Security systems must be able to protect the network from cyber disturbances.

The preceding points show the importance of studying risk. The next sections investigate current gas network vulnerabilities and risks.

20.2.2 What Are the Vulnerabilities and Risks of Gas Networks?

Before discussing the available vulnerabilities and risks, it is important to know the relationship between vulnerability and risk. Vulnerability is related to a weakness that exists in the system. This weakness can lead to different losses, according to its environment, so risk is the probability that a weakness is misused to attack the system. To manage the risk, it is important to identify the vulnerabilities of gas networks and the risks related to those vulnerabilities, which can be different in different environments. In fact, the first step in recognizing risk is identifying the vulnerabilities that exist in a system.

The Vulnerabilities of Gas Networks

Gas networks have always had various types of vulnerabilities. In the past, physical security of the networks could mitigate the risk of physical attacks. Natural disasters can also cause loss to gas networks. Today, with the advent of modern technologies, these vulnerabilities are managed efficiently, but new types of vulnerabilities have appeared that did not exist before. The security of various information systems used in networks is an important challenge nowadays. The Internet has given many attackers the opportunity to try and attack systems, so the number of attacks has increased dramatically. Cyber attacks can lead to significant losses for the system, so they are considered to be a significant threat to the gas networks. Globalization and downsizing are other phenomena that have seriously affected the systems because the management of these risks at the international level is difficult, especially when a great many employees are not as committed to corporations because of downsizing, among other reasons. For more information, the National Petroleum Council has classified the vulnerabilities of gas networks [11]. In this section, we use the same categorization of vulnerabilities and discuss them more briefly while acknowledging the latest works done in each area. Vulnerabilities that can affect the strength of the system have been classified into the following seven categories.

Vulnerability 1: Information Systems

Vulnerabilities in information systems is considered to be critical today. The information revolution has made big changes in many different aspects of our lives, and IT has made many changes in businesses. Nowadays, the world is considered a village in which even faraway places can communicate with each other within seconds. Information systems are also developing rapidly and new advances occur every day. These technologies have increased the efficiency of many processes, and most businesses are so dependent on these processes that they cannot function properly without them. In fact, businesses have made themselves so reliant on these technologies that they can be harmed by various threats. Gas networks, like other businesses, now face significant security challenges. The wide range of vulnerabilities include the following:

- Manual systems have been widely replaced by the information systems on which the gas companies depend. There are no manual backups for automated processes, so there is no possibility of returning a system to a manual status.
- Because of the competitive market, gas industries rapidly welcome new technologies to reduce their cost and increase their efficiency. However, the security of these systems is an important concern.
- The use of joint systems can be considered a type of vulnerability. For example, many companies are interested in having joint systems for their e-commerce, so a problem in one system can be transferred to other systems as well causing huge losses.
- The information systems have increased access from local to national and international levels; as a result of this wider access, systems are exposed to more electronic vulnerabilities.
- The IT advances have allowed attackers to attack from almost anywhere. They can attack systems even from home, so it becomes difficult to determine the origins of an attack.
- There is a great competition in software industry market, so most vendors try to offer their products to the market as soon as possible. As a result, many softwares do not have adequate security features. These insecure systems are exposed to different professional attacks and cannot provide enough security. Updating software systems with frequent security patches is critical.
- The gas industries widely use e-commerce, so it is exposed to different virus attacks. Some antivirus programs are available, but they are reactive and there are more viruses in the IT environment than these programs can handle.
- In the gas industries, most of today's equipment has become automated in order to increase their efficiency. The gas networks are highly reliant on the Internet, intranets, and extranets, or they depend on satellites, fiber-optic cables, microwave, phones, and so on. A disruption in one of these systems can even cause gas networks to fail to respond appropriately to customers because most operations are done automatically and need these communication systems.

This vulnerability has become one of the most important threats to the gas industries and its risks are rapidly increasing. Currently, the widespread hacking tools that are available are leading to more people, even amateurs, using these tools. Hackers have become more professional and have gained the ability to better exploit vulnerabilities and attack the systems.

Vulnerability 2: Globalization of Economies

According to the Merriam-Webster dictionary, *globalization* is the "development of an increasingly integrated global economy marked especially by free trade, free flow of capital, and the tapping of cheaper foreign labor markets." As a result, each country sees itself not only as a nation but also as part of the world. Gas industries have experienced globalization by foreign ownership and consolidations of multinational corporations, among other events. Most companies try to develop their services to an international level. This issue has made dramatic changes in the industry, especially by changing business models and the mix of various stakeholders. Globalization has increased the complexity of the industry because there are many differences between nations, from cultural to regulatory, and this has made new vulnerabilities for the network. Some of the important vulnerabilities are as follows:

- Globalization has made businesses highly dependent on each other. Gas industries have become multinational. In many countries, such as the United States, foreigners can own the companies. This has made the economies and problems of countries dependent on each other. A problem in one country may have many consequences for others in which they have no control. For example, OPEC decisions can have worldwide effects on the prices.
- Because there are no international standards for security, it is not possible to have worldwide protection. Many countries do not have strong systems, and this inconsistency can affect other countries negatively, even those that have good measures to protect themselves.
- Increasing interdependency is part of globalization. Many sectors, such as information systems, finance and banking, and transportation, should work efficiently to support globalization, and the management of this added complexity is a real concern.
- One important vulnerability of globalization is related to cultural differences. To some extent, each country is in cultural transition, which can lead to instability. Different work ethics in different countries affect productivity. This instability can have large effects on the industry because investment would be hard to attract because of its high risk.

Vulnerability 3: Business Restructuring

Business restructuring, according to the *American Heritage Dictionary of Business Terms*, is "a significant rearrangement of a firm's assets or liabilities. A firm's restructuring may include discontinuing a line of business, closing several plants, and making extensive employee cutbacks. Restructuring generally entails a one-time charge against earnings."

The competitive market has made most of the gas companies reduce their costs and downsize their workforces with increasing automation. In the past, employees usually have had strong commitments to a company, but nowadays this commitment has lessened because of downsizing, outsourcing, and different social contracts, among other trends. Many employees who have left companies are angry about the terms and conditions, and employees who are currently working are not satisfied with their heavy workloads. All of these influences have led employees to pay much more attention to their own welfare rather than the interests of the company and also exploit the vulnerabilities of the system. The following, which is a type of business restructure, has caused various vulnerabilities which are discussed in more detail below:

- Outsourcing is a transaction in which one company sells some of its processes to another company but is also responsible for final services. Outsourcing leads to considerable vulnerability. Many employees detach themselves as outsourcing occurs, and new employees are usually hired by a contracting company to do their former tasks, sometimes even for their former company. These employees are not as committed as full-time employees. It is not uncommon that the contracting company has less-reliable procedures for doing the function. When the critical processes are outsourced, the employees of the contracting company find good information about the procedures. This also creates vulnerability, because the corporation would have serious problems if the contracting company fails for any reason.

- A strategic alliance is an agreement between firms of different countries to cooperate on any activity or joint venture that is created from at least two different firms. This matter has shown some vulnerability, especially as related to intellectual property (IP). IP is difficult to protect in a venture or alliance, and it would be more at risk if the companies separate because employees of the joint venture may continue their personal relationships with each other.
- Just-in-time logistics is related to the concept that material and equipment should be in place at the time they are needed in the process. Just-in-time logistics creates great vulnerability when the equipment is not delivered in the required time. In fact, this matter makes the corporate very dependent on vendors and transportation infrastructures for timely delivery of the equipment.

Vulnerability 4: Political and Regulatory Concerns

In most developed countries, individual companies take responsibility for investing in infrastructure. These companies naturally seek their own profit. The amount of capital in gas infrastructures is significant and has a long-term payoff period, so investing in infrastructure is really a strategic but difficult decision for these firms to make. Political and regulatory concerns are among the factors that can increase the uncertainty of making these decisions and make decision making more difficult. These regulations should guarantee the profit of all stakeholders, including consumers, governments, and companies. There exist many examples in which regulations did not consider the profit of the companies very well and caused serious problems for these firms. The increase in the risk of investment in critical infrastructure diminishes the robustness of the system, which is never acceptable. In fact, according to the International Energy Agency [12], governments should play a more important role in reliable delivery of gas to customers by setting clear policies for the whole system instead of just managing part of the system.

Pelletier and Wortmann [13] presented a multistage model to evaluate the risk of investment in infrastructures in Western Europe under the uncertainty of transport tariff changes. This risk is important to calculate to determine whether or not there is enough motivation for investment, especially after the restructuring of the gas market because of liberalization in Western Europe, which led to the shortening of gas transport contracts. Jepma [14] and Jepma et al. [15] conducted comprehensive studies on the consequences of the tariff differences in gas distribution while considering only the profit of shippers as the objective. According to works on tariffs, differences have made some illogical rerouting in the network, causing a change from the route with the shortest path to the one with the least expenses. As a result, congestion may appear in the cheapest route. Apart from congestion, this problem can lead to the false expansion of the cheapest grid, leading to a suboptimal network. This phenomenon—congestion and false investment motivation—is known as *Jepma effect*.

Vulnerability 5: Interdependencies of Businesses

Today, most businesses are dependent on others. In other words, they rely on other systems for their operations. In fact, the advent of IT made many businesses highly dependent on each other, such as banking, gas, power, and transportation. Gas networks are highly dependent on other systems, especially electricity and transportation.

The interdependency of gas on electricity can be seen from two points of view. First, IT is playing an important role in gas industries, and it cannot operate unless electricity is provided. In fact, the gas industry would stop operating without electricity. Second, many combined-cycle gas and electricity plants are appearing because of the efficient working of these plants, so the gas company can easily choose, according to the prices of gas and electricity, whether it wants to sell gas as fuel or use it to generate electricity and sell this as the final product. The combined-cycle power plants have increased the interdependency of gas and electricity and have largely added to the complexity of the system. There are many researchers in this field, especially in recent years as these plants have become more popular. Unsihuay-Vila et al. [16] present a model for the expansion of the integrated gas and electricity systems. Whiteford et al. [17] assess the risks of the interlinked gas and electricity systems in the United Kingdom, because these plants using gas for the electricity generation are producing about a third of the total demand of electricity in the United Kingdom. Arnold et al. [18] introduced a system for controlling the combined electric and gas systems to work more efficiently and to mitigate the risks of interdependency of gas and electricity. Munoz et al. [19] designed a model for the combined gas and electricity systems in order to investigate their reliability. Fedora [20] studied the reliability analysis of the linked gas and electricity systems in North America.

The gas industry is also highly dependent on transportation systems. A failure in one of these systems can lead to many problems in the gas industry because just-in-time logistics have become critical in gas delivery.

These interdependencies can cause problems. A failure in one system may lead to failures in all other systems. Outages in one system can lead to outages in other systems. These incidents can even be seen as a natural disaster.

Vulnerability 6: Physical and Human Matters

Gas networks are comprised of complicated and capital-intensive equipment. This equipment and other assets are exposed to many threats and risks. Disruption in each asset has a different effect on the network, so they are ranked according to their potential impact. Some assets have low vulnerability because their damage would only have local disruptions of short duration. Others have medium vulnerability because their damage would lead to regional effects with economic losses and even losses in human life. The other assets have high vulnerability because their damage would even have national or international effects and might cause high economic losses and extreme hardship for consumers. Transmission pipelines, compressor stations, and storage and distribution facilities are usually considered among the assets with high vulnerability. Some of these types vulnerabilities are as follows:

- Underground pipelines are very vulnerable to accidental damage. Although these pipelines are marked for easy identification, they are also easily damaged, especially by construction equipment.
- The use of equipment has increased considerably, especially because of the advent of IT. As a result, the loss of equipment would have a severe impact.
- Companies try to decrease their inventories of spare parts, so that it can be used to increase in outage durations in the case of any problem.
- Nowadays, the gas industry widely uses automated facilities, and reaching and repairing them when needed can also be time consuming.

There are many risks and threats that use the physical and human matter-vulnerability, (some of which are human error, pipe breaks, contamination, and employee dissatisfaction) as justification.

Vulnerability 7: Natural Disasters

There is a great number of natural disasters which can cause much damage to the gas infrastructures. Among them, earthquakes, storms, hurricanes, tornadoes, blizzards, floods, and volcanic eruptions can be mentioned. In these cases, emergency acts are really needed to mitigate the losses. There are many countries, like United States, which have saved great record in this regard. However, it seems that it is more difficult now, in comparison with the past, to act appropriately when these disasters strike because business restructuring and downsizing led to the resignation of many skilled workers who could be of great help in times of these disasters. Also, the increase of interdependency of the industry on other industries has increased the degree of difficulty in dealing with these problems.

The Risks of Gas Networks

In the previous section, all of the vulnerabilities of gas networks were investigated and some of their probable threats or risks were mentioned. Those vulnerabilities can have different impacts in different environments. In fact, recognizing the vulnerabilities is a major step in recognizing the risks because the vulnerabilities are the cause and effect of the risks and it is important to reduce the risks to an acceptable level. Natural gas risks have been classified as falling into the following five categories [12]:

- Technical risks are problems related to such externals as weather conditions.
- Political risks are limitations in or interruption of external supplies.
- Regulatory risks create inabilities to deliver because of wrong regulation.
- Economic risks create problems in gas delivery because of such events as price increases.
- Environmental risks include contamination of deliveries.

The discussed vulnerabilities can each lead to many risks in different conditions. The risks can also be divided into two categories along a time horizon. Short-term risks are those that have minimal effects. Long-term risks, however, are considered the ones that have the longest lasting effects.

20.3 How to Manage Risks in Gas Networks?

The previous sections introduced gas networks and a concise review of all possible vulnerabilities and risks. This section provides a fairly deep insight into developing plans to evaluate and lower risks in the new economy, and it also tries to help the reader understand key components of risk-management steps.

We defined risk as the possibility of exploitation by an attacker using existing vulnerabilities. "Risk management can therefore be considered the identification, assessment, and prioritization of risks followed by coordinated and economical application of resources to minimize, monitor, and control the probability or impact of unfortunate events" [21]. There are many risks associated with the gas industry that are common in other industries, too. Therefore, the need for complete risk-management guidelines is required. A high degree of interconnections between businesses has made identification of the risk of a company a much more difficult task. With the emergence of thermal plants, security of power supply has become highly dependent on the security of the gas supply, which was discussed in depth in the previous section.

Inspired by the National Petroleum Council [11], a six-step risk-management process that is applicable to the gas industry has been proposed. They are (1) valuating key assets and estimating losses, (2) identifying and describing vulnerabilities and threats, (3) doing risk assessment, (4) developing usable options for risk abatement, (5) analyzing options to select the cost-effective ones, and (5) implementing chosen activities. A schematic view of the steps is depicted in Figure 20.2.

Risk management is a dynamic and continuous process that should be integrated into the culture of organization with an effective policy and promising program to give us what we expect. This dynamic process should be repeated periodically (e.g., annually or every 3 years). It stands to reason that the risk environment,

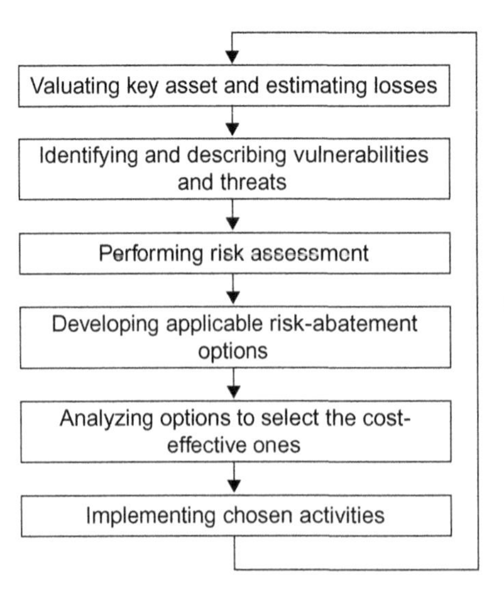

Figure 20.2 Risk-management steps.

vulnerabilities, laws, and so on are changing constantly. In the remainder of this chapter we will look carefully at all of the steps.

20.3.1 Valuating Key Asset and Estimating Losses

The first step is to valuate assets and estimate the losses. In this part, the main focus is on determining the value of all assets that exist in the company and whether they are vulnerable. By having the value of each asset, the incurred loss if the asset that was damaged or destroyed can be calculated. Because we want to protect assets that are more vulnerable and would create greater financial losses for a company, a simple rating system that uses qualitative criteria can be used. Furthermore, Vaidya and Kumar [22] reviewed the application of the analytic hierarchy process (AHP) method in order to prioritize and rank different options that can be used as an advanced rating system. According to the National Petroleum Council [11], asset valuation and loss estimation in more sophisticated systems is measured in monetary units. "These values may be based on such parameters as the original cost to create the asset, the cost to obtain a temporary replacement for the asset, the permanent replacement cost for the asset, costs associated with the loss of revenue, an assigned cost for the loss of human life or degradation of environmental resources, costs to public/stakeholder relations, legal and liability costs, and the costs of increased regulatory oversight" [11]. With the appearance of cyber threats and their incredible losses to assets, the estimation of cost has become more difficult. Methods for assessing this value rely heavily on principles of asset management and available data.

Knowing the value of each asset will determine the level of effort that should be made to protect various assets. For example, trade secrets and control systems are vital to all companies, and no insurance company can pay the incurred loss if something bad happens.

For more information on methodologies that can be used to put a value on each asset, readers are referred to asset-management books.

20.3.2 Identifying and Describing Vulnerabilities and Threats

Risk identification tries to find ways in which a company is exposed to risk. This requires an analyst to have at least a primary knowledge of the organization; the legal, social, and political structure of the environment; and the objectives of the organization on an operational and strategic level [23]. In Section 20.2, we discussed many threats and vulnerabilities that exist in gas networks. The problem here is to identify them. Although every organization can hire outside consultants to perform the job, an in-house worker trained in appropriate techniques (shown in Figure 20.3) can be used to identify various risks.

Choosing an appropriate technique to identify risk in a company depends on many factors such as the assigned budget. According to Dunjó et al. [24], hazard and operability (HAZOP) methodology is a process hazard analysis technique used worldwide to consider and analyze problems related to hazards and system

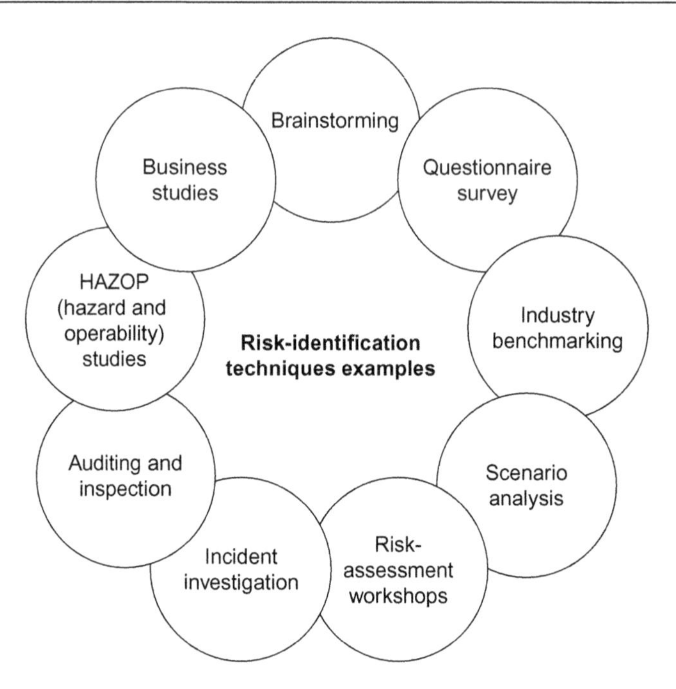

Figure 20.3 Examples of risk-identification techniques [23].

operability by evaluating the impacts of any deviations from design specifications. They present a concise review on all papers published in this area. In business studies, many aspects of a company such as its accountancy, money matters, marketing, and organizational behavior are studied. Questionnaire surveys are methods for acquiring information (usually statistical) about the attributes, attitudes, or actions related to a population with the help of a structurally defined set of questions. Preston [25] provided very good guidelines for the appropriate time to use a questionnaire and efficient ways that can be used to design, construct, administer, and present a questionnaire.

Brainstorming is an individual or group methodology for generating ideas, increasing creative efficiency, and finding solutions to various problems. Accident and incident investigation is a tool for uncovering not only earlier missed hazards but also hazards where loss of controls occurred. "Effective [accident and incident] investigations should be capable of identifying a broad range of factors that may have contributed to an [accident or incident], from an operator's action moments before an [accident or incident] to a senior-level executive decision made years earlier" [26]. As previously noted, vulnerability and threat identification is a dynamic process, especially in the case of cyber risk. After identifying all the risks, we should also describe them in the best formatted manner (may be a standard table is the most useful tool). The first column of each row can be the risk items, and subsequent columns can represent consequence of each risk, its effect on the system, degree of damage that occurred, and class of business activity that was affected such as tactical, operational, or strategic activities. For example, Lee and Shu [27] used an intuitionalistic fuzzy fault tree analysis (FTA) to find out the most important components of the system for the problem of failure analysis of a liquefied natural gas (LNG) terminal emergency shutdown (ESD) system to understand weak trail in the ESD and show the areas in which important improvements could be done.

20.3.3 Performing Risk Assessments

The goal of risk assessment is to evaluate the risks to each asset using information gathered from previous steps. Risk assessment is the main part of the risk-management process. Here we need to integrate the cost incurred and the probability that an asset will be destroyed or damaged. We know the vulnerabilities of the understudied system, but integrating the probability and impact of these vulnerabilities on the system is the outcome of this process. The probability of occurrence is a matter of contemplation because it has many components. Just because there is a vulnerability does not mean an attacker will absolutely exploit it, so this is associated with a probability. Furthermore, this attempt can be successful or not. Finally, the degree of success (or cost incurred in each state) itself has a random distribution that makes probability assessment much more difficult. On the other hand, the impact of each vulnerability dictates how much the company should invest to mitigate possible losses [11]. For instance, damage resulting from an earthquake in a department store is a risk that can strongly affect its operation, but the impact of this vulnerability may be more severe on a plant that produces petrochemical materials. This company should make more efforts to control and reduce the consequences of this phenomenon should it occur.

Although managers tend to use prescriptive measures to assess the risk, a great number of quantitative risk assessment (QRA) methods have been widely presented: Cagno et al. [28]; Krueger and Smith [29]; Metropolo and Brown [30]; Jo and Ahn [31]; Sklavounos and Rigas [32]; Jo and Crowl [33]; Brito and De Almeida [34]; Suardin et al. [35]; Brito et al. [36]; and Han and Weng [37], among others.

Jo and Ahn [31] presented a QRA approach applicable in the planning and building phases of new pipelines or modifying existing ones. By using the information of pipeline geometry and population density from geographic information systems, they estimated the parameters of fatal length and cumulative fatal length. The former is used to determine individual risk (the probability of loss of life at any special location because of all unwanted events) and the pipeline failure rate. The latter plus the failure rate can be used to estimate risk (the relationship between the frequency of an event and the number of its casualties).

Han and Weng [37] present an integrated QRA method that is composed of the probability assessment of accidents, consequence analysis, and risk evaluation. This method analyzes consequences, including those outside and inside gas pipelines. Individual and societal risk caused by different accidents are related to the outside risk of pipelines and economic risk derived from pressure redistribution related to outside risk of pipelines. In their method, using the FTA, event tree analysis, and historical data or modified empirical formula, the expected failure rate per unit pipeline is calculated as follows:

$$\rho = \sum_{k} \rho_k K_k(a_1, a_2, \ldots) \tag{20.1}$$

In the above formula, ρ shows the expected failure rate per unit pipeline (1/year km), ρ_k demonstrates the basic failure rate per unit length of pipeline (1/year km), and K_k shows the correction function for any failure cause, a_1,a_2, ... These are the arguments of the correction function, and the subscript k indicates each failure cause such as corrosion, construction defects, external interference, or ground movement.

The main sources of harm to gas pipelines from outside are toxic gas diffusion, jet flames, fireballs, and unconfined vapor cloud explosions. They measure all of these adverse effects by quantitative criteria and finally calculate the fatal probability for destruction.

> According to the dose—effect relationship between the dose of the concrete harmful load as toxicity, heat or pressure and such recipient categories as death or injuries, the function of fatality probability unit P_T is defined to quantitatively describe the harmful load. Fatality probability unit can be used for the measurement of the damage from an accident and that is the critical basis of the calculation of death probability percentage, which is the final result of the accident consequence. [37]

The main source of harm from outside gas pipelines is economic loss. For example, it is possible to estimate P_T from an accident using the following formula:

$$P_T = a + b \ \ln I_f \tag{20.2}$$

where empirical constant a represents the hazard only related to a studied harmful load and b (also an empirical constant) represents the vulnerability of recipients to the load. I_f, for a given exposure time, is a dose of the load. Also note that a relationship exists between the death probability percentage and the fatality probability unit, by which we can calculate death probabilities.

They finally do the risk evaluation process in which risk is defined as a function of the probability of an accident and its consequences. For example, they calculate economic risk as follows (considering the assumption that economic production is in direct proportion to the gas supply pressure):

$$E(R) = *(K' * (P_{\text{node}} - P_{\text{node, now}})) \tag{20.3}$$

where $E(R)$ is the financial risk, $K'*$ is the expected failure rate of the nodes (calculated in the first step), P_{node} is gas supply pressure of the nodes in a normal situation, and $P_{\text{node, now}}$ is the gas supply pressure after a disruption. Note that the term in second parenthesis represents financial loss.

Brito et al. [36] tried to design a multiattribute model for investigating risk in natural gas pipelines. An evolutionary version of Brito and Almeida [34], this paper used the ELECTRE TRI method integrated with utility theory to do so. Identifying the hazard scenarios, they divided the pipeline into a definite number of sections (each section has specific threats and vulnerabilities), and then the impact of each scenario on each section is calculated. After estimation of payoff sets (H, E, F) in which H stands for human, E for environment, and F for financial, the utility function of each is elicited (using the utility theory). Data from consequence probabilities on human environment and financial and utility functions are used to calculate human, environment, and financial losses, and these data are combined with hazard scenario probabilities to calculate human, environmental, and financial risks for each section. Finally, a number of risk categories are defined and all of the identified risks are put into these categories (using ELECTRE TRI).

For the consequence analysis of the outside pipelines, heat and overpressure are considered to calculate the individual risk and societal risk. The economic risk of the gas pipeline network is used to study the results of the inside gas pipelines. A sample of an urban gas pipeline network is used to show the presented integrated quantitative risk-analysis method.

For more information on guidelines for QRA, readers are referred to the TNO Purple Book [38].

20.3.4 Developing Applicable Risk-Abatement Options

In this step, different options to mitigate risk are introduced and categorized. The outputs of the previous step are much like the causes of pain, and output of this step resemble remedies that are prescribed to alleviate these pains. Different options can be employed to fight the risks. Some are preventive and should be done before a loss event, and some are just applicable when the event occurs. Preventive measures include deterring the sources of threats and eliminating vulnerability events. Event-dependent activities include reductions of potential loss, managing the crisis, and rapid restoration to get back to normal. "Risk abatement is achieved through policies and procedures, technology, and insurance" [12].

Setting preventive governmental and international laws to deter the sources of threats are part of a common strategy. For example, severe international laws can be set to act against terrorism. Furthermore, some policies can be made to eliminate vulnerabilities, such as forcing all companies in a country to use computerized systems in which more reliable information systems are provided. Establishing institutions such as the North American Electricity Council (NERC) in power sectors can be helpful in establishing operating policies and planning standards to ensure the reliability of gas systems.

With the usage of the high tech systems, risk can be abated. For example, some sophisticated control systems can be used to determine the defects of gas pipelines. Using new technologies, gas can be transported in a liquefied state (LNG). Recent advances in the area of operations research and mathematical modeling brought about more accurate solutions for problems in the real world. For example, in the area of gas networks, Unsihuay-Vila et al. [16] presented a long-term, multiage, and multistage model for the supply and interconnections expansion planning of integrated electricity and natural gas. The model of this work considered the value chain of both natural gas and electricity. This robust model, which minimizes investment and operational cost, dictates that to reduce the risk of power supply we are recommended to have natural gas storage when hydropower is considered. Shi et al. [39] proposed to use natural gas hydrate (NGH) instead of LNG and compressed natural gas (CNG) because NGH has particular advantages such as moderate conditions of production and storage, simple technology, and high security.

Insurance companies are the best example for managing risk and rapidly restoring a situation to normal.

20.3.5 Analyzing Options to Select Cost-Effective Ones

Although the previous step tries to clarify the risk environment, threats, and vulnerabilities, this step tries to find strategies and activities that should be used to act against this situation and abate the risk. Already-known and different strategies that can be taken should be recognized, and the efficient ones should be elected through a cost–benefit analysis process. Having the information of the previous steps is crucial: If we do not know the risk environment, then there is no guarantee an activity can reduce the risk. Employing the AHP tool, decision trees, the use of a plus/minus/interesting (PMI) tool which weighs pros and cons of each option is also helpful. A review of decision-making techniques is given in Huitt [40]. The journal *Risk Analysis* [41] presents a 0–1 knapsack formulation that tries to perform a valid cost–benefit analysis when a number of mitigation options and a limited amount are available. Although monetary value is incurred by owners and operators of the gas network, there are benefits that the aforementioned article tried to maximize subjected to total some of cost which can be paid for. Knowing this, the following formula is used:

$$\max \sum_{i=1}^{N} b_i x_i \tag{20.4}$$

subject to

$$\sum_{i=1}^{N} b_i w_i \leq W \tag{20.5}$$

$$x_i \in 0, 1 \quad \forall i \tag{20.6}$$

where

b_i is the benefit derived from implementation of countermeasure i
w_i is the cost of countermeasure i
W is the available budget
N is the total number of available countermeasures

In this formula, the total benefit (Eqn (20.4)) derived from implementing cost-effective countermeasures is trying to be maximized, while in Eqn (20.5) the total cost spent is trying to not be more than W. The problem is solved using a dynamic programming method. It may be good to say that in the same paper a concise risk-management procedure is presented that is partly similar to what we have presented here.

A range of different options may be used. In fact, a combination of these options should be chosen based on a cost–benefit analysis. For example, to have a robust system in a gas supply network, we can either purchase high-quality pipes with no control system or low-quality pipes with high-tech control systems, or even a combination of both.

20.3.6 Implementing Chosen Activities

Implementing risk-abatement activities is the sixth step that should be undertaken. If a program is not executed effectively and efficiently, then the time used to plan the previous steps has been wasted. To do a good job, we must define effective plans and procedures, assign and train promising staff, monitor the work to guarantee that the plans are implemented carefully and accordingly, and look dynamically at new situations that may have different threats, vulnerabilities, and risk environments.

References

[1] Y. Ruan, Q. Liu, W. Zhou, B. Batty, W. Gao, J. Ren, T. Watanabe, A procedure to design the mainline system in natural gas networks, Appl. Math. Model. 33 (2009) 3040–3051.
[2] A. Kabiriana, M.R. Hemmati, A strategic planning model for natural gas transmission networks, Energy Pol. 35 (2007) 5656–5670.
[3] A. Cheboubaa, F. Yalaoui, A. Smati, L. Amodeoa, K. Younsi, A. Tairi, Optimization of natural gas pipeline transportation using ant colony optimization, Comput. Oper. Res. 36 (2009) 1916–1923.
[4] M. Hamedi, R.Z. Farahani, M. Moattar Husseini, G.R. Esmaeilian, A distribution planning model for natural gas supply chain: a case study, Energy Pol. 37 (2009) 799–812.
[5] A. Herran-Gonzalez, J.M. De La Cruz, B. De Andres-Toro, J.L. Risco-Martın, Modeling and simulation of a gas distribution pipeline network, Appl. Math. Model. 33 (2009) 1584–1600.
[6] A. Martin, M. Möller, S. Moritz, Mixed integer models for the stationary case of gas network optimization, Math. Program. 105 (2006) 563–582.
[7] R.Z. Ríos-Mercado, S. Kim, E.A. Boyd, Efficient operation of natural gas transmission systems: a network-based heuristic for cyclic structures, Comput. Oper. Res. 33 (2006) 2323–2351.
[8] P.J. Wong, R.E. Larson, Optimization of tree structured natural gas transmission networks, J. Math. Anal. Appl. 24(3) (1968) 613–626.

[9] S. Wu, R.Z. Rios-Mercado, E.A. Byod, L.R. Scott, Model relaxation for the fuel cost minimization of steady state gas pipeline networks, Math. Comput. Model. 31(2–3) (2000) 197–220.
[10] Critical Infrastructure Assurance Office, Tab B Preliminary Research and Development Roadmap for Protecting and Assuring the Energy Infrastructure. Preliminary Research and Development Roadmap for Protecting and Assuring Critical National Infrastructures (Transition Office of the President's Commission on Critical Infrastructure Protection and the Critical Infrastructure Assurance Office, Washington, DC), 1998. Available from: http://www.oe.energy.gov/DocumentsandMedia/Preliminary_RandD_Roadmap_for_Prot_and_Assur_Critical_Natl_Infr.pdf (accessed April 2010).
[11] National Petroleum Council, Securing oil and natural gas infrastructures in the new economy. A report by the National Petroleum Council: Committee on Critical Infrastructure Protection, 2001.
[12] International Energy Agency, Security of gas supply in open markets. International Energy Agency (IEA), 2004.
[13] C. Pelletier, J.C. Wortmann, A risk analysis for gas transport network planning expansion under regulatory uncertainty in Western Europe, Energy Pol. 37 (2009) 721–732.
[14] C. Jepma, Gaslevering onder druk, Stichting JIN. Available from: www.jiqweb.org, 2001 (52pp.) (in Dutch).
[15] C. Jepma, M. Broekhof, W. van der Gaast, Hydra, Aantasting van leveringszekerheid stichting JIN, 2004 (106pp.) (in Dutch).
[16] C. Unsihuay-Vila, J.W. Marangon-Lima, A.C. Zambroni de Souza, I.J. Perez-Arriaga, P.P. Balestrassi, A model to long-term, multiarea, multistage, and integrated expansion planning of electricity and natural gas systems, IEEE Trans. Power Syst. 25(2) (2010) 1154–1168.
[17] J.R.G. Whiteford, G.P. Harrison, J.W. Bialek, Electricity and gas interaction: a UK perspective and risk assessment, IEEE, 2009, 978-1-4244-4241-6/09.
[18] M. Arnold, R.R. Negenborn, G. Andersson, B.D. Schutter, Model-based predictive control applied to multi-carrier energy systems, IEEE, 2009, 978-1-4244-4241-6/09.
[19] J. Munoz, N. Jimenez-Redondo, J. Perez-Re, J. Barquin Natural gas network modeling for power systems reliability studies. 2003 IEEE Bologna PowerTech Conference, June 23–26, 2003, Bologna, Italy.
[20] Fedora P.A., Reliability review of North American gas/electric system interdependency, in: Proceedings of the 37th Hawaii International Conference on System Sciences, 2004.
[21] D.W. Hubbard, The Failure of Risk Management: Why It's Broken and How to Fix It, John Wiley & Sons, Hoboken, NJ, 2009.
[22] O.S. Vaidya, S. Kumar, Analytic hierarchy process: an overview of applications, Eur. J. Oper. Res. 169(1) (2006) 1–29.
[23] The Institute of Risk Management, A risk management standard, The Institute of Risk Management, 2002. Available from: www.theirm.org.
[24] J. Dunjó, V. Fthenakis, J.A. Vílchea, J. Arnaldos, Hazard and operability (HAZOP) analysis: a literature review, J. Hazard. Mater. 173(1–3) (2010) 19–32.
[25] V. Preston, Questionnaire survey, Int. Encycl. Hum. Geog. (2009) 46–52.
[26] S. Reinach, A. Viale, Application of a human error framework to conduct train accident/incident investigations, Accident Anal. Prevent. 38(2) (2006) 396–406.
[27] C. Lee, M. Su Fault-tree analysis of intuitionistic fuzzy sets for liquefied natural gas terminal emergency shut-down system. IIH-MSP, vol. 2, pp. 574–577, Third

International Conference on International Information Hiding and Multimedia Signal Processing (IIH-MSP), 2007.
[28] E. Cagno, F. Caron, M. Mancini, F. Ruggeri, Using AHP in determining the prior distributions on gas pipeline failures in a robust Bayesian approach, Reliability Eng. Sys. Safety 67 (2000) 275–284.
[29] J. Krueger, D. Smith, A practical approach to fire hazard analysis for offshore structures, J. Hazard. Mater. 104 (2003) 107–122.
[30] P.L. Metropolo, A.E.P. Brown, Natural gas pipeline accident consequence analysis, in: 3rd International Conference on Computer Simulation in Risk Analysis & Hazard Mitigation, Sintra, Portugal, 2004, pp. 307–310.
[31] Y. Jo, B.J. Ahn, A method of quantitative risk assessment for transmission pipeline carrying natural gas, J. Hazard. Mater. A123 (2005) 1–12.
[32] S. Sklavounos, F. Rigas, Estimation of safety distances in the vicinity of fuel gas pipelines, J. Loss Prevent. Process Ind. 19 (2006) 24–31.
[33] Y.D. Jo, D.A. Crowl, Individual risk analysis of high-pressure natural gas pipelines, J. Loss Prevent. Process Ind. 21 (2008) 589–595.
[34] A.J. Brito, A.J. de. Almeida, Multi-attribute risk assessment for risk ranking of natural gas pipelines, Reliability Eng. Sys. Safety J. 94(2) (2009) 187–198.
[35] J.A. Suardin, A.J. McPhate, A. Sipkema, Fire and explosion assessment on oil and gas floating production storage offloading (FPSO): an effective screening and comparison tool, Process Safety Environ. Protect. 87 (2009) 147–160.
[36] A.J. Brito, A.J. de Almeida, C.M.M. Mota, A multicriteria model for risk sorting of natural gas pipelines based on ELECTRE TRI integrating utility theory, Eur. J. Oper. Res. 200 (2010) 812–821.
[37] Z.Y. Han, W.G. Weng, An integrated quantitative risk analysis method for natural gas pipeline network, J. Loss Prevent. Process Ind. 23(3) (2010) 428–436.
[38] TNO Purple Book, Guidelines for quantitative risk assessment. CPR18E, first ed., Committee for the Prevention of Disasters, The Netherlands, 1999.
[39] G. Shi, Y. Jing, X. Zhang, G. Zheng Prospects of natural gas storage and transportation using hydrate technology in China, IEEE, 2009, 978-1-4244-2800-7/09.
[40] W. Huitt, Problem solving and decision making: consideration of individual differences using the Myers-Briggs type indicator, J. Psychol. Type 24 (1992) 33–44.
[41] Risk management for natural gas pipeline distribution networks. Risk Anal. J. (2009).

21 Modeling the Energy Freight-Transportation Network

Mohsen Rajabi

Department of Industrial Management, Faculty of Management,
Tehran University, Tehran, Iran

21.1 Introduction

21.1.1 Energy in the World

Societies all over the world are entirely dependent on energy for achieving and sustaining their development. Energy, in any form, plays a vital role in the industrial environment. Whenever sufficient energy is available, economic development occurs. A wide range of energy forms are being consumed, including natural gas (NG) for industrial uses and heating, electricity for light and to power electronic devices, petrochemicals to produce plastics, and petroleum derivatives such as gasoline and diesel for transportation.

Because modern life is based on energy, individuals, corporations, and firms should have a reliable access to energy in all its forms. Problems involving the continuous transportation and delivery of energy can lead to the disruption and breakdown of economic, industrial, and social infrastructures. An energy resource cannot be reliable unless, first of all, it is delivered to right place at the right time, and secondly, it must be available in the consumer desired quantity. Maintaining the efficient and reliable generation, transportation, and distribution of energy is the most important in a world which is so dependent on energy.

Different kinds of energy are used in today's industries and in people's lives. Some are fossil fuel resources such as oil and gas, but electricity, nuclear energy, solar energy resources, wind energy, and wave energy are also being used. All of these types should be delivered to consumers. The two most useful forms are fossil fuels and electricity. Electrical and gas systems are quite similar. Both are designed to carry energy from suppliers to consumers. The process of generating the energy and delivering it to the final consumer can be structured into:

- Suppliers (electrical power plants or gas fields)
- Transmission (high-voltage or high-pressure networks)
- Distribution (medium-low voltage network or medium- or low-pressure network)
- Customers (electricity customers or gas customers) [1].

Logistics Operations and Management. DOI: 10.1016/B978-0-12-385202-1.00021-9

Nevertheless, there are differences between these two types of energy. NG is directly obtained from gas fields, whereas electrical energy is a secondary form of energy that is produced by the transformation of a primary form in a power plant. Moreover, gas can be stored for future use in peak load periods, whereas electrical energy cannot be efficiently stored.

As societies develop, they inevitably need more and newer sources of energy. Planning for power-generation expansion is necessary to meet the rising need for energy. Planning determines which types, where, and when new generation or transmission installations should be constructed [2]. Clearly, in any expansion plan, there is an interaction between the energy supply and transport and power plants. For example, gas is carried from gas fields (suppliers) to consumers in liquefied natural gas (LNG) ships or flows through pipelines. Gas consumers can be classified as domestic or industrial customers [1]. Combined-cycle power plants are NG industrial customers that use gas to generate electricity. The expansions of pipeline networks, which are the part of the energy-transportation network, are complex and extensive. It is also possible to increase the capacity of the pipelines. However, the transmission capacity of a gas pipeline is not unbounded and depends on the pressure difference between the two ends of the pipeline. To increase the capacity of the network, compressors can be added at certain locations. Compressors enlarge the pressure difference between the two nodes of pipeline network. By expanding the network or increasing the capacity, the amount of gas that can be supplied to consumers is always limited.

21.1.2 The Importance of Energy Around the World

The organization and structure of the power industry has undergone important change in many countries over the last few years. Reasons for change are political, economic, and technical. The development of combined-cycle power plants is the best example of a technical change. These plants use NG as the primary fuel to generate electricity. They have efficient procedures and several advantages compared to traditional thermal and nuclear power plants. In addition to that, they require less startup investment cost and shorter depreciation periods. The traditional economies of scale that were the reason for the existence of big regulated utilities have either disappeared or greatly reduced. This new insight compels the restructuring of the power industry into a free electricity market. The general trend is to move toward a greater competition, which means larger risks for private companies [1].

NG is considered one of the most reliable sources for supplying the world's growing energy demands. The industrial heating sector is the largest consumer of NG. Electrical power generation comes after that. It has grown strongly after the introduction and development of combined-cycle generation technology (CC-NG) in the 1980s. One reason for this growth has been abundant NG resources around the world: the United States, Russia, Europe, Latin America, and the Middle East. These resources have favored gas-fired generation as the key factor in the total growth of NG consumption. As there are still NG reserves that have not been yet discovered, this consumption is certain to keep growing. As a result, investment in infrastructure such as terminals, pipeline networks, and compressors will increase in the future.

Countries around the world have understood the importance of energy and its role in the development of their societies. The following examples show how countries pay attention to and invest in the energy resources and on-time transportation of energy to the customer.

In recent years, Latin America has been one of the most intensive development regions for NG and electricity [3]. The region is highly dependent on hydropower (about 57% of the region's installed capacity is hydro), and the need to diversify away from heavy investments in hydropower and oil is driving many countries to promote the use of NG, especially for power generation. Examples of these developments are in Brazil, Chile, and Colombia. Countries in the region are diverse in size, electrical installed capacity, electrical power demand, and electrical transmission and NG network characteristics (level of meshing and geographical extension) [4].

Economic reforms let private sectors invest in energy-generation and energy-freight sectors that were previously reserved to national governments. These reforms boosted the development of an energy infrastructure in Latin America. Electricity and NG pipelines developed in the region, separately in each country and in cross-border electricity–gas interconnections. The introduction of NG in the energy matrix of these countries was more aggressive at the end of the 1990s with the construction of cross-border gas pipelines (Bolivia–Brazil, Argentina–Chile, etc.) and the development of local gas production fields [3].

NG is mostly used in industries and automotive sections. As these two sections have experienced significant growth, their consumption rates of NG have increased. Another reason for the increase in the use of NG is the growing numbers of gas-fired thermal generation plants producing electricity.

In addition, Chile and Brazil decided to implement regasification plants in order to start importing LNG beginning in 2009 [4]. The motivation for both countries is similar: (1) to diversify the gas supply for the country (in the case of Chile, to diversify from Argentina; and in the case of Brazil, to diversify from Bolivia) and (2) to create a flexible supply able to accommodate the use of gas to power generation.

In China, NG is becoming one of the main resources of energy, in addition to coal and oil, and its consumption is rapidly increasing [5].

In Spain, gas systems have been restructured from a regulated market to a free and competitive market in the last few years. Gas companies are building combined-cycle power plants to get into the electricity market [1]. Such companies have the opportunity to act in both gas and electricity markets. They can either sell their gas as fuel in the gas market or generate electricity and sell it in the electricity market. Thus, these two markets are related to each other with respect to market price, and the coupling between them is much stronger.

In the European Union, only a few countries have the advantage of having substantial energy resources that not only meet their domestic needs but also can be exported to other countries in the region. Norway, the Netherlands, Denmark, and the United Kingdom are among these countries. There have been considerable investments in pipeline networks in order to transport energy to other European terminals.

The growth in energy-transportation networks then led to the establishment of organizations to manage and coordinate the transmission system. They were also in charge of negotiating import and export transactions between countries that possessed energy and those that did not.

In the United Kingdom, one of the energy exporter and importer countries in the European Union, NG will play a great role in the energy sector in the future. With the closure of coal-fired power plants and the retirement of nuclear plants, the need for NG is inevitable. However, the country's gas supply has decreased as the country has become more dependent on gas to generate energy in recent years. This fact made the country rely more on imported energy sources than domestic resources. Ensuring on-time supply and delivery of energy to the country is necessary in order to increase the energy sector's security.

Over the past few years, more and more gas has been drawn from indigenous North Sea supplies to meet significantly rising domestic consumption. Rather than government looking to protect and prolong supplies (the practice in Norway, for example), policy changes in the 1980s led to the fastest possible promotion of NG production in the area known as the *UK Continental Shelf* [6].

21.1.3 Energy Freight Transportation

Generally speaking, freight transportation is a vital component of the economy. It supports production, trade, and consumption activities by ensuring the efficient movement of raw materials and finished goods and their on-time delivery. Transportation accounts for a significant part of the final cost of products and represents an important component of the national expenditures of any country [7].

Freight transportation has always been an integral component of economic development. It has now emerged as one of the most critical and dynamic aspects of the transport sector, where change has become the norm. Freight transportation is the main element supporting global commodities and, more generally, supply chains, complex and functionally integrated networks of production, trade, and service activities that cover all stages of production from the transformation of raw materials to market distribution and after-market services [8].

The cost of a transportation system directly influences the cost of finished goods. The rising cost and complexity of shipping and delivering goods is adding to pressures faced by manufacturers and producers across the globe. However, as a result of the surge in global activities over the past 10 years, this issue has taken on new dimensions and importance. Thus, the freight-transportation industry must perform at a high level in order to become economically efficient. With respect to quality standards, transportation has to offer high-quality services while being reliable.

The political evolution of the world also affects the transportation sector. This fact becomes more important as the significant political and economic role of energy is taken into consideration. The appearance and expansion of energy terminals in increasing free-trade zones, political changes that result in new markets, and growing economic globalization have tremendous consequences for the evolution of transportation systems. Not all of the consequences are studied or understood well, and they need to be evaluated to become apparent.

NG is a vital component of the world's supply of energy. NG has made a strong comeback in the global energy balance since the mid-1970s as a direct response to increasing crude oil prices that began then. This development was given further impetus in the late 1980s in light of new concerns about potential global warming and climate change. The low-carbon intensity of NG (lowest among the fossil fuels) has made it the fuel of choice from an environmental point of view [1].

The transportation economics of NG, as one major form of energy, depend greatly on the annual volume of gas fields and transport distances. However, consumers or existing gas pipelines are usually at long distances from NG fields where gas pipelines cannot be built and operated economically. In developing countries, economical transporting and storing of NG is an important issue.

Two technologies have been introduced to condense NG to make it easy to transport safely. LNG and compressed natural gas (CNG) technologies have been applied in the gas industry for several years. LNG technology can reduce its volume by about 620 times using the liquefaction of NG, whereas CNG technology reduces its volume by about 200 times using compression [5]. Unit transport costs and unit storage costs are greatly reduced by both technologies.

The production, processing, transportation, and consumption of LNG and associated products are an issue of great interest in the energy industry. An LNG supply chain consists of loading ports shipping LNG to one or more receiving ports [9]. A typical supply chain is depicted in Figure 21.1. The loading ports and receiving ports structures are depicted in more detail in Figures 21.2 and 21.3, respectively.

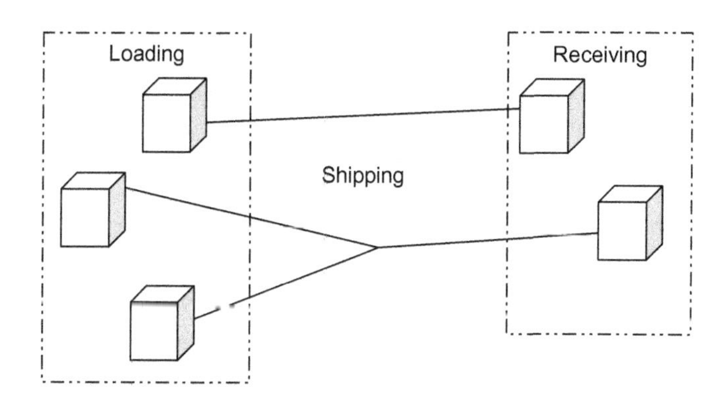

Figure 21.1 A typical supply chain [9].

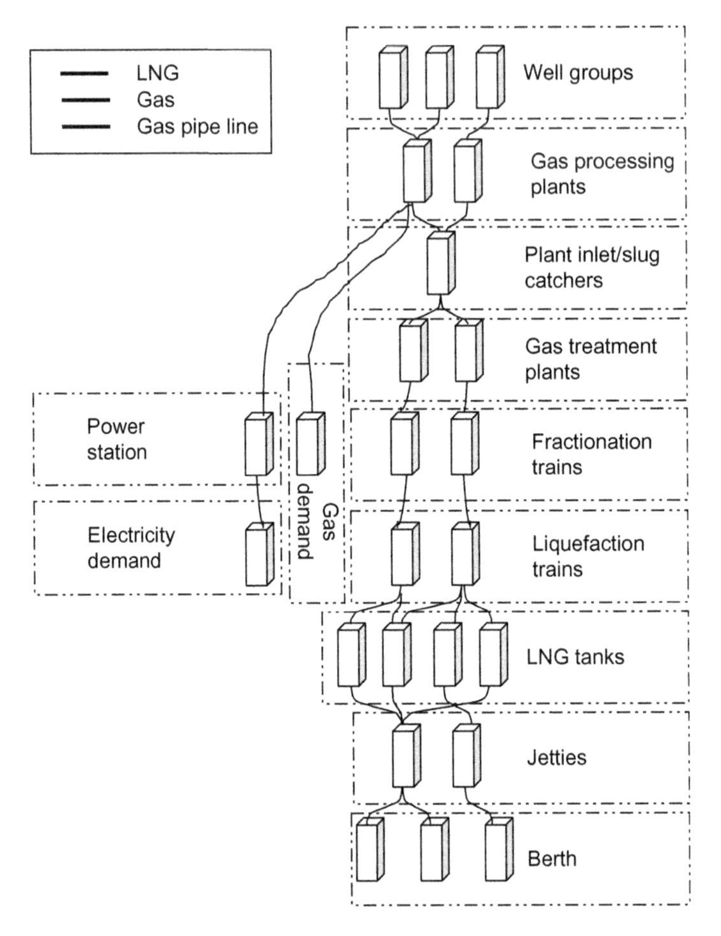

Figure 21.2 The structure of a loading port [9].

NG is one of the cleanest, most efficient, and most useful of all energy sources. It is basically formed of methane but also contains heavier hydrocarbons such as ethane, propane, and butane. After extraction, it is purified to make it easier to transport and store. Impurities such as water, sulfur, sand, and other compounds make NG harder to transport, so they must be separated and removed.

In its raw form, NG is unsuitable for delivery, so it is condensed into a liquid at almost atmospheric pressure by cooling it to approximately $-162°C$. LNG is about 1/614th the volume of NG at standard temperature and pressure. To transport it over long distances where pipelines do not exist, it is carried by specially designed cryogenic vessels and cryogenic tankers and stored in specially designed tanks [10].

Huge advances in maritime transportation technology have been made in recent decades. These advances now increase opportunities to transport energy, oil in particular, over long distances. Since 1980, the world's maritime fleet has grown in

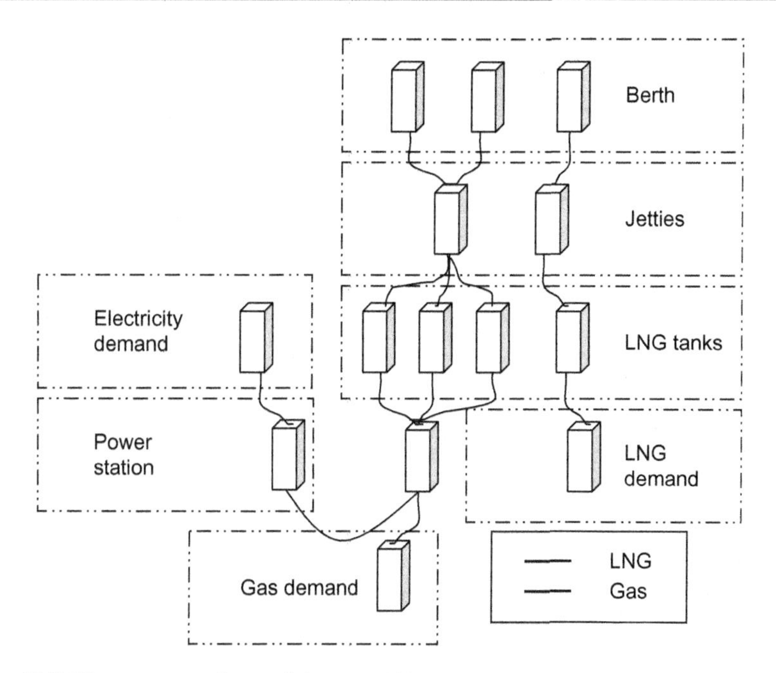

Figure 21.3 The structure of a receiving port [9].

Table 21.1 World Fleet by Vessel Type (Million DWT)

Year	Oil Tankers	Bulk Carriers	General Cargo	Container Ships	Other	Total
1980	339	186	116	11	31	683
1990	246	234	103	26	49	658
2000	286	281	103	69	69	808
2001	286	294	100	77	69	826
2002	304	300	97	83	60	844
2003	317	307	95	91	47	857

parallel with the seaborne trade. Table 21.1 shows the growth of world fleet in the period of 23 years from 1980 to 2003 [11].

The ocean shipping industry has a monopoly on the transportation of energy to faraway continents and countries. Pipelines are the only transportation mode that is cheaper than ships, but they are far from versatile because they can move only fluid types of energy over fixed routes, and they are feasible and economical only under specific conditions. Other modes of energy transportation (rail and truck) have their advantages, but ships are probably the least regulated mode of transportation because they usually operate in international waters, and few international treaties cover their operations.

Pipeline transportation is more economical over short distances, whereas LNG shipping is more attractive over greater distances [12]. Once NG is in the transmission network, it travels from suppliers to customers over long distances. The gas network can be described as having supply nodes, demand nodes, and intermediate nodes. The gas is injected into the system through the supply nodes and flows out of the system through the demand nodes, which are also known as consumers. Demand nodes are classified as electrical customers and nonelectrical customers [1]. Electrical customers are combined-cycle power plants that use the gas as fuel to produce electrical energy. Nonelectrical customers are the remainder of NG system customers.

Except for the two technologies, some technologies, especially natural gas hydrate (NGH) technology, are being developed to store and transport NG.

Energy in transit is exposed to unexpected dangers that may cause extensive damage to life, property, and environment. For NG, incidents such as leak or spill, irregular high or low temperatures, explosion, and flame can occur as it is transported or stored [10]. Industry analysts have analyzed the statistical likelihood of these events occurring simultaneously, and transient analysis has been used to derive the level of support necessary in pipe design to ensure there would be no system failure [6].

Safety is a high priority with companies that are in charge of moving and distributing NG. It is very important that they analyze the probabilities of failure in the system, assess the worst results of such incidents, and provide guidance in developing safety and security requirements.

In addition to establishing rules for a well-functioning internal gas market, the European Community wanted to provide measures that would provide an adequate level of security for gas supplies. The directive (2004–2004/67/EC) set out certain instruments that were to be used by each member state to enhance security [13]. The instruments are as follows:

- Provide pipeline capacity to enable diversion of supplies and system flexibility.
- Transmit system operator cooperation to coordinate dispatch.
- Invest in infrastructure for gas imports in the form of regasification terminals and pipelines.

The evolution of technology has also had a major influence on energy freight transportation. The problem is whether the transportation system can adapt to advances in new technologies and fuels and whether it will be well organized and operated in the new era. Freight transportation must perform within rapidly changing technological, political, and economic conditions and trends. Significant changes in technology are not about traditional hardware but about advances in information technology and software. The introduction of the Internet and the increasing use of it has dramatically changed the process of transportation, the way it is being controlled, and the interaction between carriers, shippers, and terminals.

Intelligent transportation systems, on the other hand, offer means to efficiently operate and raise new challenges, as illustrated by the evolution toward modifying planned routes to respond in real time to changes in traffic conditions or new demands. More complex planning and operating procedures are a direct result of these new policies, requirements, technologies, and challenges [14]. Freight transportation must adapt to and perform within these rapidly changing political, social, and economic conditions and trends. In addition, freight transportation is in itself a complex domain. Many different firms, organizations, and institutions, each with its own set of objectives and means, make up the industry. Infrastructure and even service modifications are capital intensive and usually require long implementation delays; important decision processes are often strongly interrelated. It is thus a domain in which accurate and efficient methods and tools are required to assist and enhance analysis, planning, operation, and control processes.

21.2 Energy Freight-Transportation Network

21.2.1 Application of Energy Freight-Transportation Models

Transportation network problems in the real world are studied through network models. A model is a representation of an object or a problem being studied. Models and methods of freight-transportation networks present the evolution of the network as well as the response to rapidly changing environments. Models must be capable of responding to various changes and modifications: modifications in existing infrastructure, the introduction of modern carriers, the construction of new facilities, the introduction of new technologies resulting in changes of volumes and patterns of supplying energy, the increasing rate of energy consumption resulting in faster transportation, variations in energy prices, changes to labor conditions, new regional or international policies and legislations, and so on. Network models help better explain the relationships between network components.

Transportation scientists try to explain spatial interactions that result in the movement of objects from place to place. It includes research in the fields of geography, economics, and location theory. Transportation science goes back several centuries. Its methodologies draw from physics, operations research, probability, and control theory. It is fundamentally a quantitative discipline, relying on mathematical models and optimization algorithms to explain the phenomena of transportation [14].

Scientists try to solve the problems in models in order to reach to an optimal solution. Different aspects of a network such as network design, network flow, and network operation are varieties of basic issues being considered in sciences such as operation research, management, accounting, and engineering. Therefore, algorithms of different types have been developed to solve the problems in reasonable time to find the optimal solution.

Transportation of energy for industrial or individual uses is of great importance in developing and underdeveloping countries. As a large amount of money is invested in energy-related issues such as supplying transportation and storage of energy, it is a priority to study freight-transportation network models. There are researchers of different fields working on the network models as the problems of transportation and distribution of energy are multidisciplinary. As the result, varieties of algorithm and solutions have been found to deal with the problems.

21.2.2 Energy Freight-Transportation Network

The need for energy freight transportation derives from the significant distances between energy's production and supply point and the consumption point. Whether it is petroleum or the downstream products of it such as gas and gasoline, a transportation system must move and deliver it to the final point to meet demands. Therefore, the transportation system requires carriers to use any appropriate means to facilitate the delivery of energy at such distances. Another reason for using varieties of carriers would be the need to move energy according to schedule, in reliable containers, with the lowest possibility of happening hazards, and in reasonable costs while meeting the quality standards to satisfy consumer demands. Achieving this aim requires that the components of transportation system operate properly.

A transportation system can be defined as a set of elements and the interactions between them that produces both the demand for travel within a given area and the provision of transportation services to satisfy this demand. Many elements operate in every transportation network, so identifying and controlling all of these interacting elements in order to design and implement the network is hardly possible. Therefore, it is inevitable to isolate the elements that are relevant more to the problem being studied and to keep the remaining ones as external factors.

Levels of Planning

In general, transportation systems are classified into three levels, according to the planning level of the system: strategic, tactical, and operational. Because a transportation system has close relationships with management decisions and policies and is a complex organization of components from human and material resources to facilities, infrastructures, carriers, and containers, it is necessary for such system to be planned in detail.

Strategic Level

Long-term planning—or, in other words, strategic planning—is directly concerned with the design of physical network and related models of transportation and its evolution, allocating the location of terminals, ports, and the same facilities; the expansion of transportation capacity; and tariff policies. It determines the general policies and the development trend of the system in the long-term horizon. Therefore, strategic planning involves the highest level of managers and may need large capital investments to be executed.

Knowing that transportation networks are not just national but in many cases international, strategic planning is also done at national, international, and even regional levels. As discussed previously and as elaborated on in "Route" section, models of transportation networks are designed in the strategic level of planning.

Tactical Level

Over a medium-term horizon, tactical planning determines resource allocation and utilization in order to make the system operate efficiently. Medium-term planning includes carrier scheduling and routing, and it is responsible for the design of network service. The decisions for this level of planning are taken by medium-level managers of the organization.

Operational Level

Performed by local and operational management, this level of planning is concerned with the issues happening in the short term. In today's rapidly changing environment, time has an important role, so scheduling and implementing carriers, services, crews, and maintenance activities and efficiently routing and allocating carriers and crews in the short term are the problems that operational managers have to deal with.

21.2.3 Classifications of Energy Freight-Transportation Networks

The energy-transportation process is usually divided into three parts, two of them being similar. The process begins at the production point, which can be a port, a terminal, a petroleum platform in the middle of the ocean, or an electricity production zone, and generally any production point for any kind of energy. The process next comes to containers of energy. Energy is mainly transported through railways, shipping lines, and energy container trucks. It can be said that pipelines and cable networks are other examples of energy transportation. However, these are the facilities mostly used to distribute the energy rather than transport it. Transporting energy usually includes long distances and large amounts that pipelines and cable networks are not capable of handling. The third and last part of the process is the receiving ports or terminals, where the energy goes through distribution to be delivered to the consumption points.

Freight transportation also can be categorized into shippers, carriers, and governments. Shippers originate the demand for transportation, whereas carriers are utilities such as railways, motor carriers, and shipping lines. Governments are responsible for providing transportation infrastructures such as railroads, ports, and platforms and to pass relevant laws.

Another classification of energy freight-transportation components, like other types of transportation, has two main elements: demand and supply [15].

With respect to distance, freight transportation can be divided into long-haul versus short-haul transportation. In long-haul freight transportation, the carrier moves over long distances and between national or international ports and terminals. Large vessels, railways, and, in rare cases, trucks are used in this type of transportation. However, in short-haul transportation, energy and related products are transported by large trucks. Although trucks are the main means of short-haul transportation, railways are also appropriate when there is a need to deliver large amounts of energy—for instance, gasoline—to short distances.

Clearly, government plays a large role in developing institutions responsible for facilitating the transportation process. Governments contribute the infrastructure: roads, highways, and a significant portion of ports, internal navigation, and rail facilities. Governments also regulate (e.g., dangerous and toxic goods transportation) and tax the industry [14].

The following sections are brief notes about the different components of the transportation process.

Energy Production and Receiving Point

The process of energy freight transportation begins at a production point such as an oil platform. Where the energy is produced and is ready to be transported, based on the type of energy, the production process and method of storage are varied. For instance, at an offshore oil platform, oil is extracted from wells and stored in industrial facilities called *oil depots*. Oil that is stored in depots is in the final step of refining and therefore is ready for customer use. Oil depots usually have particular reserve tanks that are used for discharging the product to transportation vehicles.

Energy Containers

Because the production points of energy freights are most often long distances from consumption points, transportation is inevitable, so selecting the suitable mode of transporting energy products is an important concern.

For transporting energy freight to customers, a number of carriers and methods can be used. The dominant ones are trucks, pipelines, marine lines, and railroads. Each method has its own advantages and disadvantages regarding issues such as reliability, cost, safety, security, and accessibility. Nowadays with the changes in environmental conditions, other elements should be taken into consideration, such as pollution problems, noise production, traffic jams, and energy consumption of a node [16]. Choosing the proper container depends on optimizing these various criteria, and sometimes it is more beneficial to use a combination of them to better serve customers. This is called *intermodal transportation*.

The following part of this section introduces some common carriers for transporting energy freight.

Railways

Rail is one common method of freight transportation. This is a cost-effective method, especially for carrying energy freights. Although this method has less speed and somehow lower reliability, it costs much less than other methods, thus making freight more affordable. Moreover, compared to truck transportation, it can transport bulkier and heavier commodities such as coal, chemicals, and petroleum in large volume to more distant areas [16]. In the United States, coal is the leading commodity of rail transportation. Another advantage of railroads is that service providers can use existing infrastructures; in most countries, governments provide the infrastructure and therefore it needs less investment [14]. However, in some countries, especially underdeveloped ones, not all of a region is covered by railways. As a result, there is less opportunity to use this mode to transport energy freight on a national scale.

Shipping Lines

Among the different modes of transporting energy freights, maritime transportation is most often used to transport critical and strategic energy commodities such as oil and related products between countries and continents. Today, more than 60% of all oil is transported by ships. Maritime lines are used to transport crude oil and its products, and their related costs are often less than other modes.

Special oil tankers are used to transport oil; these are the largest vessels in the world [17].

Trucks

Trucks are among the most popular methods of transporting commodities within and between countries. It is the leading way to transport commodities in most countries such as the United States. Like other modes of transportation, it has advantages and disadvantages. Its wide coverage, convenient accessibility, fast responsiveness, and flexibility make it popular. Furthermore, in some circumstances the application of trucks and road transportation is inevitable, because not all consumption points are connected by rail or water. Although in most cases it is more suitable for short distances and lightweight shipments, it is more costly than rail transportation.

In addition to that, this mode of transportation has disadvantages that must be considered in making decisions, such as its pollution and the amount of fossil fuels it consumes. It also causes traffic congestion and has low levels of safety. In most countries, the number of road accidents and tolls are very high, and it is not possible to get rid of it completely even with precautionary rules and standards [16].

Demand and Supply

Travel demand derives from the need for energy in other parts of the region or the world that is deprived of energy resources. As shown in Figure 21.4, the demand flows in transportation systems rise from the fact that there are discrepancies in each and every freight transportation—that is, the distance between production and receiving points varies from one place to another, the energy containers stretch from trucks to vessels, and the types of energy vary from petroleum to NG. Their movements make up freight travel's demand flows.

21.2.4 Introducing the Energy Freight-Transportation Network Models

The network of energy freight transportation is of different levels, from regional movements through highways and trucks, to national movements through railways, and to international movements through shipping lines and large vessels. Designing

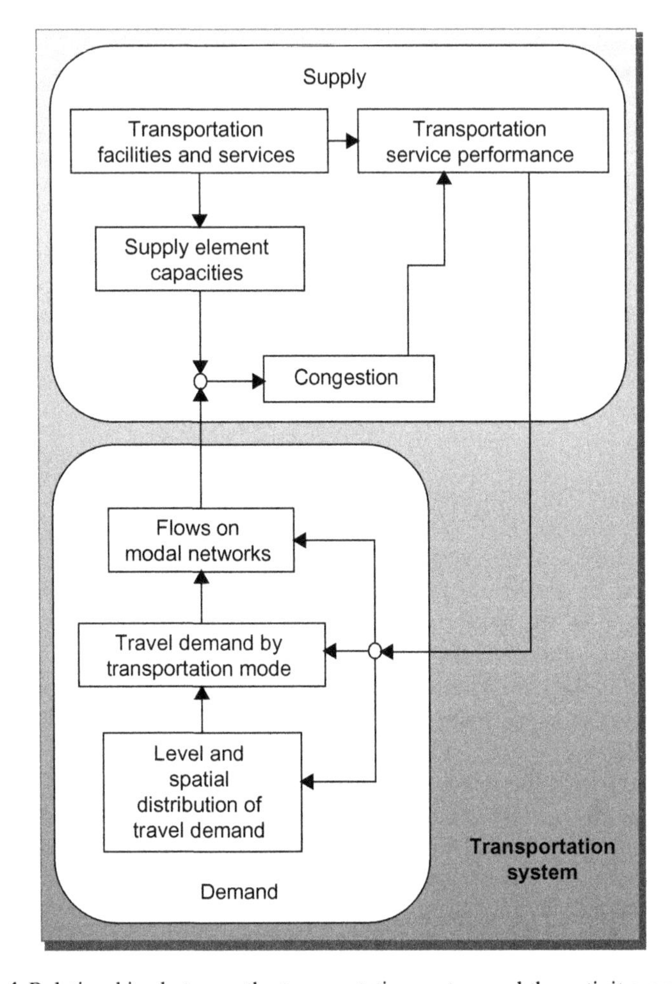

Figure 21.4 Relationships between the transportation system and the activity system [15].

and evaluating the models of such networks requires quantifying interactions among the elements of existing and future transportation systems [15].

Although every element cannot be identified or controlled in modeling, it still plays a central role in the design and evaluation of transportation systems. Factors such as a region's or country's transportation infrastructure, constraints of delivery points, and marine and road traffics all influence the behavior of a network model but can hardly be modeled or controlled. It is therefore necessary for scientists of energy-transportation networks to account for these hidden parameters when designing a network model.

Transportation planning, from goods to energy transportation, has been widely discussed in books and papers, but most of them are about road transportation by truck rather than other modes of transportation. It may be questioned why there is a lower level of attention, in spite of the large capital investments and operating costs associated with these other modes. Although research on rail planning problems has increased considerably over the last 15 years, it is not the same for maritime transportation.

Christiansen [18] has some explanations. First, there is low visibility; people mostly see trucks or trains rather than ships, and ships are not the major transportation mode worldwide. In addition, large organizations that sponsor research mostly operate fleets of trucks, not ships. Second, the planning problems of shipping networks are less structured than the other modes. This makes the planning more expensive because of the customization of decision-support systems. There is more uncertainty in maritime operations because of weather conditions, mechanical problems, and incidents such as strikes. Slacks in maritime transportation planning are few because they have high costs. Most quantitative models originated in vertically integrated organizations where ocean shipping is just one component of the business. This occurs because there are many small family-owned companies, because the ocean shipping industry has a long tradition and it is not open to new ideas.

Modeling the Transportation of Hazardous Materials

The US Department of Transportation (USDOT) defines a hazardous material as any substance or material capable of causing harm to people, property, or the environment [19]. It has categorized a list of hazardous materials into nine classes according to their physical, chemical, and nuclear properties. Gases and flammable and combustible liquids are among the classes.

It should be mentioned that most hazardous materials (hazmats) originate at locations other than their destination. Oil, for instance, is extracted from oil fields and shipped to a refinery (typically via pipeline); many oil products, such as heating oil and gasoline, are refined at a refinery and then shipped to storage tanks at different locations within a country or abroad.

The risks associated with the transportation of oil and gases and their consequences can be significant because of the nature of the cargo: fatalities, injuries, evacuation, property damage, environmental degradation, and traffic disruption. Reductions in hazmat transportation risks can be achieved in many different ways. Some of these ways are not related to modeling and planning the transportation network, such as driver training and regular vehicle maintenance. Others can be studied through operation research and modeling.

As mentioned in previous sections, energy can be moved over roads, rails, or water. In some cases, shipments are intermodal; they are switched from one mode to another during transit. Hazmat transportation incidents can occur at three points: the origin when loading, the destination when unloading, and en route. To identify the route that minimizes fuel costs and travel times between production and receiving points, operation research models are designed with the related constraints.

According to different routes, energy transportation as a kind of hazardous material is a typical multiobjective problem with multiple stakeholders that are difficult to solve. Transport by truck, for instance, has choices between selecting short routes while moving through heavily populated areas or selecting longer routes through less populated areas, which makes the transportation cost more and expose to risks.

Mathematical models that are described in the following sections allow representation and analysis of the interactions among the various elements of a transportation system.

Components of Energy Freight-Transportation Models

Modeling any transportation network requires identification of components that are acting reciprocally. Ghiani et al. [20] introduce cost as the major component of the transportation model and then classify the problems based on relevant costs.

As mentioned in the previous section, despite the fact that factors affect the transportation network model, they can hardly be identified, quantified, and modeled. Some of the main factors are categorized under the name of *external factors*, which is a subcategory of operational factors [21].

The transport infrastructure is of great importance. Lacking such a capability could affect the scheduling and delivery of the energy shipment. In some regions, there are not proper rail networks and essential facilities in the terminals to transfer energy to a location. Ports have to be well equipped for large vessels to berth and transfer the energy freight.

In addition to that, a transportation network is affected by trade barriers as well as laws and taxation policies. Variation in any of these parameters around the world may affect the decision concerning the most appropriate mode of transportation and routings for cost reasons. Legal requirements are likely to differ from one country to another. As a result, there would be problems in costs and planning while trying to adapt to the requirements.

Because parameters and problems of modeling ship fleets are different from those of other modes of transportation, ships operate under different conditions. Table 21.2 provides a comparison of the operational characteristics of the different freight-transportation modes. Shipping lines are mostly in international territories, which means they are crossing multiple national jurisdictions. In energy freight transportation with ships, each unit represents a large capital investment that translates into a high daily cost because they must pay port fees and operate in international routes.

In addition to that, other means of energy freight transportation generally come in a small number of sizes and similar models and designs, whereas among ships we find a large variety of designs that result in nonhomogeneous fleets.

More than that, ships have higher risks and lower certainty in their operations because of their higher dependence on weather conditions and on technology, and because they usually pass multiple jurisdictions. However, because ships operate around the clock, their schedules usually do not have buffers of planned idleness that can absorb delays. As far as trains are concerned, they have their own dedicated rights of way, they cannot pass each other except for specific locations, and their size and composition are flexible (both numbers of cars and numbers of power

Table 21.2 Comparison of Operational Characteristics of Freight Transportation Modes [18]

Operational Characteristics	Mode				
	Ship	**Aircraft**	**Truck**	**Train**	**Pipeline**
Barriers to entry	Small	Medium	Small	Large	Large
Industry concentration	Low	Medium	Low	High	High
Fleet variety (physical and economic)	Large	Small	Small	Small	NA
Power unit is an integral part of transportation unit	Yes	Yes	Often	No	NA
Transportation unit size	Fixed	Fixed	Usually fixed	Variable	NA
Operating around the clock	Usually	Seldom	Seldom	Usually	Usually
Trip (or voyage) length	Days–weeks	Hours–days	Hours–days	Hours–days	Days–weeks
Operational uncertainty	Larger	Larger	Smaller	Smaller	Smaller
Right of way	Shared	Shared	Shared	Dedicated	Dedicated
Pays port fees	Yes	Yes	No	No	No
Route tolls	Possible	None	Possible	Possible	Possible
Destination change while underway	Possible	No	No	No	Possible
Port period spans multiple operational time windows	Yes	No	No	Yes	NA
Vessel–port compatibility depends on load weight	Yes	Seldom	No	No	NA
Multiple products shipped together	Yes	No	Yes	Yes	NA
Returns to origin	No	No	Yes	No	NA

NA, not applicable.

units). As a result, the operational environment of ships is different from other modes of freight transportation, and they have different fleet-planning problems.

Energy Freight-Transportation Costs

There are different costs during a transportation network. They can be divided into transportation costs and handling costs [15]. Transportation costs include the cost of operating a fleet, the cost of transporting a shipment, the cost of hiring carrier if not owned, and the cost of a shipment when a public carrier is used. Handling costs are not discussed in energy freight transportation, because they are incurred when inserting individual items into a bin, loading the bin onto an outbound carrier, and reversing these operations at a destination.

The Cost of Operating a Fleet

The main costs are related to crews' wages, fuel consumption, container depreciation, maintenance, insurance, administration, and occupancy. It is obvious that wages and insurance are time dependent, fuel consumption and maintenance are distance dependent, and that depreciation depends on both time and distance whereas administration and occupancy costs are customarily allocated as a fixed annual charge.

The Cost for Transporting a Shipment

This type of cost is paid by a carrier for transporting a shipment. It is rather arbitrary because it would be difficult to assign a trip cost to each shipment, where several shipments are moved jointly by the same carrier—that is, a large vessel containing barrels of petroleum and other downstream products simultaneously.

The Cost of Hiring Carrier

Although hire charges are parts of a transportation total cost, they are still unidentified and hard to evaluate.

The Cost of a Shipment Using a Public Carrier

The cost for transporting a shipment when using a public carrier can be calculated on the basis of the rates published by the carrier. The size and equipment of a carrier as well as the origin, destination, and route of the movement are factors that are taken into account when calculating this cost.

Risk

As discussed in "Modeling the Transportation of Hazardous Materials" section, energy in the form of gas and oil is one type of hazardous material. As a result, possible incidents during loading, transporting, and unloading should be considered when making models. To estimate the probability and cost of a hazmat release incident, various consequences must be considered. The consequences can be categorized as injuries and fatalities (often referred to as population exposure) [22,23], cleanup costs, property damage, evacuation, product loss, traffic incident delays, and environmental damage. It is clear that all impacts must be converted to the same unit (e.g., dollars) while modeling in order to permit comparison and complication of the total impact cost.

Route

Some models presented in the field of energy-transportation networks seek to minimize travel distances between production and consumption points. It first occurs that the shortest possible route—roads and railways or marine lines—would be the answer. However, looking profoundly at all of the issues concerning routing problems shows that there are significant components that prevent the model from being designed and solved in such an easy way.

The previous sections contain explanations about the parameters dealing with routing problems. As mentioned, not all shortest distances have the lowest expense. Models of freight transportation seek to solve a multiobjective function in which more than two factors are optimized. A routing model should give decision makers the shortest route with the minimum cost simultaneously. Because it would be quite hard to achieve such a solution, the models show an appropriate solution that does not necessarily have the minimum distance or cost.

More than that, previous sections explained one important issue that has arisen in recent years. The security of the routes matters considerably as the rates of lost or attacked energy freight increase. There are routes with lower levels of security that have a minimum cost or distance. Meanwhile, secure roads or marine lines certainly cost more for longer distances. Routing model planners have to design models that can achieve a good solution while at the same time accounting for as many issues involved in the problem as possible.

Models of Energy Freight-Transportation Network

Modeling problems of energy freight-transportation networks contain assumptions, constraints, and one or more objective functions. Models usually focus on one attribute of the network—for instance, minimizing the cost of moving energy while ignoring other effective attributes or considering them as constant parameters.

As discussed in previous sections, particularly "Energy Containers" section, modes of energy freight transportation vary from trucks to trains to fleet. The tactical planning level perspective is missing in ship routing and scheduling studies reported in the literature. Fleet scheduling is often performed under tight constraints. Flexibility in cargo quantities and delivery time is often not permitted. So the shipping company tries to find an optimal fleet schedule based on such constraints while trying to meet the objective functions—that is, maximizing profit or minimizing costs. Brønmo et al. [24] and Fagerholt [25] have developed models that consider flexibility in shipment sizes and time windows. The models are not specified in energy but would be applicable in shipping energy problems as well. The results of their studies show that there might be a great potential in collaboration and integration along the factors of a transportation process—for instance, between shippers and shipping companies.

Christiansen et al. [18] introduce a planning problem in which a single product is transported and call it the single-product-inventory ship-routing problem (s-ISRP). The assumptions and constraints of the model are close to reality—that is, transporting energy using ships. The production and consumption rate of the transported product—energy, in this case—is constant during the planning horizon. The advantage of the model is that contrary to similar scheduling problems, neither the number of calls at a given port during the planning horizon nor the quantity to be loaded or unloaded in each port call are not predetermined. There needs to be some initial input in order to determine the number of possible calls at each port, the time windows for the start of loading, and the range of feasible loads for each port of call. The initial information would be the location of loading and unloading ports, supply and demand rates, and inventory information at each port. Eventually, the planning problem finds routes and schedules that minimize the transportation cost without interrupting the production or consumption processes.

Ghiani et al. [20] continue the problems based on transportation cost, discussing freight-traffic assignment problems and classifying them as static or dynamic. Static models are appropriate when decisions related to transportation are not affected explicitly by time. The graph $G = (V, A)$ is then applied, where the vertex set **V** often corresponds to a set of facilities as terminals, ports, and platforms in production and receiving points, and the arcs in the set A represent transportation carriers linking the facilities.

In addition to that, they take a time dimension into account in dynamic models, including a time-expanded directed graph. In a time-expanded directed graph, a given planning horizon is divided into a number of time periods, T1, T2..., and a physical network is replicated in each time period. Then temporal links are added. A temporal link connects two representations of the same terminal at two different periods of time. They may describe a transportation service or the energy freight waiting to be loaded onto an incoming carrier.

Some linear and nonlinear models based on cost parameter are as follows: minimum-cost flow formulation; linear single-commodity, minimum-cost flow problems; and linear multicommodity, minimum-cost flow problems.

As explained in "Modeling the Transportation of Hazardous Materials" section about oil and gases as types of hazardous materials, transporting them contains risks that have to be measured. Erkut et al. [26] talk about risk along an edge or route while transporting hazmats in what they call *linear risk*. They focus on hazmat transportation on both roads and railways. A road or rail network is defined as *nodes* and *edges*. The nodes stand for the production and consumption points, road or rail intersections, and population centers. The road segments connecting two nodes are called the *edges*. It is assumed that each point on an edge has the same incident probability and level of consequence. As a result, a long stretch of a highway or railway moving through a series of population centers and farmland should not be represented as a single edge but as a series of edges. This is the difference between a hazmat transportation network and other material networks. Erkut and Verter [27] discuss this difference as a limit to the portability of network databases between different transport applications. Also, along with Erkut and Verter [27], Jin et al. [28] and Jin and Batta [29] suggest a risk model that considers the dependency to the impedances of preceding road segments.

Transporting energy from place to place requires a detailed plan and a schedule in order to minimize the costs during the process and determine the shortest route in time windows while accounting for the probability of incidents. This fact makes researchers model the realities and propose varieties of models to solve the problems. The models cover different transport modes. Erkut et al. [26] have provided a classification of papers reviewing different problems. Table 21.3 presents an extended version of what they have done. Not all of the research shown in the table concentrates on energy transportation, but some of it discusses models of transporting hazmats such as energy.

Some research has also focused on designing a transportation network. The networks are used to transporting hazardous materials in general, but they may also be applicable for energy freight transportation. Some of them are as follows: Berman et al. [65]; Erkut and Alp [66]; Erkut and Gzara [67]; Erkut and Ingolfsson [39]; Kara and Verter [68]; and Verter and Kara [69].

Although the cost of a transportation network is a significant factor, other parameters also act on the network. A transportation network service problem which is in the operational level consists of deciding on some elements. The elements include the characteristics (frequency, number of intermediate stops, etc.) of the routes to be operated, the traffic assignment along these routes, and the operating rules and laws at each terminal [20].

Service network design models can be classified into frequency-based and dynamic models. Variables in frequency-based models express how often each transportation service is operated in a given time horizon, while in dynamic models a time-expanded network is used to provide a more detailed description of the network. Models of service network design in both categories are fixed-charge network design models, the linear fixed-charge network design model, the weak and strong continuous relaxation.

21.3 Case Studies

Some researchers have attempted to model the components of a real case and apply the models in order to achieve proper solutions. It would be difficult to account for all the parameters dealing with a problem, but the researchers have done their best to approximate reality while making models. The more realistic the model, the more it can achieve.

21.3.1 Case: A Pricing Mechanism for Determining the Transportation Rates

Farahani et al. [70] developed a systematic method for calculating the transportation rates for tanker trucks of the National Iran Oil Product Distribution Co. (NIOPDC). The objective of the research was to design a computer-based system for calculating transportation rates and estimating the required budget. Determining appropriate transportation rates is critical, because of the cost of transporting oil products.

The researchers first reviewed and classified studies that determined transportation rates. Afterward, the current supply chain of oil products was described, and

Table 21.3 A Classification of Energy/Hazmats Transportation Model

	Road	Railway	Marine	Deterministic Models	Stochastic Models	Single Objective	Multiple Objective	Local Routing Models	Global Routing Models
Akgün et al. [30]	*								*
Batta and Chiu [31]				*					
Bowler and Mahmassani [32]								*	
Chang et al. [33]								*	
Corea and Kulkarni [34]					*				
Carotenuto et al. [35]									*
Darzentas and Spyrou [36]			*						
Dell'Olmo et al. [37]									*
Erkut and Ingolfsson [38]					*				
Erkut and Ingolfsson [39]						*			
Erkut and Vecter [27]						*			
Fagerholt and Rygh [40]			*						
Frank [41]					*				
Fu and Rilett [42]					*			*	
Glickman [43]		*							
Glickman [44]	*	*							
Gopalan and Kolluri [45]									*
Haas and Kichner [46]			*						
Hall [47]					*			*	
Iakovou et al. [48]			*						*
Iakovou [49]			*						
Kara et al. [50]	*								
Kulkarni [51]					*				

Lindner-Dutton et al. [52]								*
Marianov and ReVelle [53]	*					*	*	*
Miller-Hooks [54]					*		*	
Miller-Hooks and Mahmassani [55]					*		*	
Mirchandani [56]					*			
ReVelle et al. [57]				*				
Richetta and Larson [58]			*					
Sherali et al. [59]						*		
Turnquist [60]					*			
Verma and Verter [61]		*						
Weigkricht and Fedra [62]	*	*	*					
Wijeratne et al. [63]					*	*		
Zografos and Androutsopoulos [64]							*	

then the process used for calculating the transportation rates for the company was explained. In the remainder come the purpose, input and output of the case, which is accompanied by the introduction of the new developed system and the techniques used to calculate the transportation rates. The work contains an example to approve the model, the necessary data, and the final results.

There have been two generic approaches for transportation rates determination; one of them is based on learning from previous patterns and behaviors, and the other is based upon total transportation cost. The researchers have used the second approach because it estimates the total transportation cost while taking into account time value of money. The designed technique to determine transportation rates is a combination of two methods; time value of money methods and engineering economics methods.

To estimate the next year's transportation budget, information from previous years is used because the monthly depot-to-depot transportation is estimated only for the next month. The budget is divided into two parts that are calculated separately. It consists of the procurement transportation (depot-to-depot transportation) budget and the depot-to-retailer transportation budget.

The system that supports the models of rate calculation and estimation of transportation budget includes three parts. A central database collects and saves the information related to rate calculation and budget estimation. The main part is the processing system based on the models of determining liquid gas transportation rates, oil product transportation rates, and estimating transportation budget.

To compare the designed models with the previous ones, calculated rates are compared with current rates in the company. Also, a sensitivity analysis is carried out of main input parameters. In addition to that, the paper assesses the results of chosen routes and their rates. The sample routes are a combination of intercity and inner-city routes, including depot-to-depot and depot-to-retailer cases.

The results indicate the estimation of costs for freight transportation, loading, and unloading. The software developed in the paper estimates transportation rates and the final required budget based on the database.

21.4 Conclusions and Directions for Further Research

To summarize, we have explained the importance of energy throughout the world. We clarified that energy plays a vital role in today's human lives, so planning and scheduling the transportation of energy in an almost optimal situation is inevitable. Therefore, modeling the energy freight-transportation network requires modeling real problems and then solving them with methods such as operation research or fuzzy logic.

However, there are many components involved in a transportation network that affect network modeling. Modes of transportation alter the planning and modeling of a network. For instance, maritime transportation differs greatly from other modes of transporting energy—that is, trucks and trains. It includes specific characteristics and requires decision support models which would be appropriate to solve its problems. More research in maritime transportation problems has been done recently than ever before, but the field still needs much attention compared to other modes. To model and solve more realistic problems in maritime transportation, there has to be development of optimization algorithms and computing power.

While trying to design a model that is able to minimize the cost, travel destination, and time, it is necessary that the risk of such models when becomes applicable should be at the lowest possible level. Some presented models in previous sections were specific to energy transportation, and others were general in transporting hazardous materials of which energy is a part. Some models were explained, whereas for other problems we confined ourselves to a review of research about the transportation of energy that has been categorized in different fields from modes of transportation to single- or multiple-objective models. Readers were referred to sources that deal more extensively with the problems.

Erkut et al. [26] suggest that researchers emphasize global routing problems on stochastic time-varying networks because it has received almost no attention. The problem is so close to reality and most of maritime transportations are global and goes through international waters.

Furthermore, risk models and the probabilities of risk during freight transportation still need to be studied. The field would be rather difficult to survey, because there is no agreement on general accident probabilities and conflicting numbers are reported by different researchers. Lack of essential data limits improvements in such fields, and perhaps more attention should be paid to quantifying and modeling perceived risks. In general, risk and its relevant topic in energy freight transportation is of high importance, but unfortunately it has attracted little attention.

During the previous decade, attacks on energy freights increased as the price of energy rose. Energy freight can be a significant target to terrorists around the world. This fact raises the interest in the security of such freight. The US federal government, for instance, now requires hazmat truckers to submit to fingerprinting and criminal background checks [71]. However, security as an important factor in freight transportation has not yet received much attention from operations researchers. Obviously, the problem is complex, and many parameters should be considered while modeling. Erkut et al. [26] propose three dimensions for operation researchers to focus on security issue: rerouting around major cities, changes in the modeling of incidence risks, and route-planning methodology.

In addition, as fleets become larger, network planning problems become harder. So the need arises for a new generation of researchers and planners who have less practical but more academic backgrounds. As computer technology advances, new software and optimization-based decision-support systems are introduced for the varieties of applications in energy freight transportation. These advances make it easier to model all of the important problem components. This new generation of planners is more adapted to computers and software and therefore is capable of modeling realistic issues and finding good solutions to hard problems in a reasonable amount of time.

References

[1] J. Munoz, N. Jimenez-Redondo, J. Perez-Ruiz, J. Barquin, Natural gas network modeling for power systems reliability studies, in: IEEE Bologna PowerTech Conference, Bologna, Italy, 2003.
[2] C. Unsihuay-Vila, J.W. Marangon-Lima, A.C.Z. de Souza, I.J. Perez-Arriaga, A model to long-term, multiarea, multistage, and integrated expansion planning of electricity and natural gas systems, IEEE Trans. Power Syst. 25(2) (2009) 1154–1168 (May 2010).
[3] IEA, South American Gas—Daring to Tap the Bounty, IEA Press, France, 2003.
[4] L.A. Barroso, H. Rudnick, S. Mocarquer, R. Kelman, B. Bazerra, LNG in South America: the markets, the prices and the security of supply, in: IEEE Power and Energy Society General Meeting—Conversion and Delivery of Electrical Energy in the 21st Century, Pittsburgh, PA, 2008.
[5] G. Shi, Y. Jing, X. Zhang, G. Zheng, Prospects of Natural Gas Storage and Transportation Using Hydrate Technology in China, IEEE, Baoding, China, 2009.
[6] J.R.G. Whiteford, G.P. Harrison, J.W. Bialek, Electricity and Gas Interaction: A UK Perspective and Risk Assessment, IEEE, Edinburgh, UK, 2009.
[7] T.G. Crainic, G. Laporte, Planning models for freight transportation, Eur. J. Oper. Res. 97(3) (1997) 409–438.
[8] P. Nijkamp, Globalization, international transport and the global environment: a research and policy challenge, Transport. Plan. Technol. 26(1) (2003) 1–8.
[9] N. Stchedroff, R.C.H. Cheng, Modeling a continuous process with discrete simulation techniques and its application to LNG supply chains, in: Winter Simulation Conference, 2003.
[10] C.-L. Lee, M.-H. Shu, Fault-tree analysis of intuitionistic fuzzy sets for liquefied natural gas terminal emergency shut-down system, in: Third International Conference on International Information Hiding and Multimedia Signal Processing, IEEE Computer Society, 2007.
[11] UNCTAD, Review of Maritime Transport, United Nations, New York and Geneva, 2003.
[12] B.E. Okogu, Issue in global gahlral gas: a primer and analysis, 2002.
[13] Council Directive 2004/67/EC, Council of the European Union, 2004, 1–5.
[14] R.W. Hall, Handbook of Transportation Science, second ed., Kluwer Academic Publishers, New York, 2003.
[15] E. Cascetta, second ed., Transportation Systems Analysis—Models and Applications, vol. 29, Springer Science + Business Media, New York, 2009.
[16] D. Lowe, Intermodal Freight Transport., Elsevier Butterworth-Heinemann, Oxford, Great Britain, 2005.
[17] H.S. Marcus, Marine Transportation Management, Croom Helm and Auburn House Publishing Company, London and Dover Massachusetts, 1987.
[18] M. Christiansen, K. Fagerholt, B. Nygreen, D. Ronen, Maritime Transportation, in: C. Barnhart, G. Laporte (Eds.), Handbook in Operation Research and Management Science, vol. 14, Elsevier B.V, 2007.
[19] USDOT, List of Hazardous Materials, The Office of Hazardous Materials Safety, US Department of Transportation, Washington, 2004.
[20] G. Ghiani, G. Laporte, R. Musmanno, Introduction to Logistics Systems Planning and Control, John Wiley and Sons, West Sussex, England, 2004.
[21] A. Rushton, P. Croucher, P. Baker, The Handbook of Logistics and Distribution Management, third ed., Kogan Page, London, 2006.

[22] M.D. Abkowitz, J.P. DeLorenzo, R. Duych, A. Greenberg, T. McSweeney, Assessing the economic effect of incidents involving truck transport of hazardous materials, Transport. Res. Rec. 1763 (2001) 125–129.
[23] FMCSA, Comparative Risks of Hazardous Materials and Non-Hazardous Materials Truck Shipment Accidents/Incidents, Federal Motor Carrier Safety Administration, Washington, DC, 2001.
[24] G. Brønmo, M. Christiansen, B. Nygreen, Ship routing and scheduling with flexible cargo sizes, J. Oper. Res. Soc. 58 (2007) 1167–1177.
[25] K. Fagerholt, Ship scheduling with soft time windows—an optimization based approach, Eur. J. Oper. Res. 131 (2001) 559–571.
[26] E. Erkut, S.A. Tijandra, V. Verter., Hazardous materials transportation, in: C. Barnhart, G. Laporte (Eds.), Handbook in Operation Research and Management Science, vol. 14, Elsevier B.V., 2007.
[27] E. Erkut, V. Verter, Modeling of transport risk for hazardous materials, Oper. Res. 46 (5) (1998) 625–642.
[28] H.H. Jin, R.J. Batta, M.H. Karwan, On the analysis of two new models for transporting hazardous materials, Oper. Res. 44(5) (1996) 710–723.
[29] H.H. Jin, R. Batta, Objectives derived from viewing hazmat shipments as a sequence of independent Bernoulli trials, Transport. Sci. 31(3) (1997) 252–261.
[30] V. Akgün, E. Erkut, R. Batta, On finding dissimilar paths, Eur. J. Oper. Res. 121(2) (2000) 232–246.
[31] R. Batta, S.S. Chiu, Optimal obnoxious paths on a network: transportation of hazardous materials, Oper. Res. 36(1) (1988) 84–92.
[32] L.A. Bowler, H.S. Mahmassani, Routing of Radioactive Shipments in Networks with Time-Varying Costs and Curfews, Technical Report ANRCP-1998-11, Amarillo National Resource Center for Plutonium, TX, 1998.
[33] T.S. Chang, L.K. Nozick, M.A. Turnquist, Multi-objective path finding in stochastic dynamic networks, with application to routing hazardous materials shipments, Transport. Sci. 39(3) (2005) 383–399.
[34] G. Corea, V.G. Kulkarni, Shortest paths in stochastic networks with arc lengths having discrete distributions, Networks 23 (1993) 175–183.
[35] P. Carotenuto, S. Giordani, S. Ricciardelli, Finding minimum and equitable risk routes for hazmat shipments, Comput. Oper. Res. 34(5) (2007) 1304–1327.
[36] J. Darzentas, T. Spyrou, Ferry traffic in the Aegean Islands: a simulation study, J. Oper. Res. Soc. 47 (1996) 203–216.
[37] P. Dell'Olmo, M. Gentili, A. Scozzari, On finding dissimilar Pareto-optimal paths, Eur. J. Oper. Res. 162 (2005) 70–82.
[38] E. Erkut, A. Ingolfsson, Catastrophe avoidance models for hazardous materials route planning, Transport. Sci. 34(2) (2000) 165–179.
[39] E. Erkut, A. Ingolfsson, Transport risk models for hazardous materials: revisited, Oper. Res. Lett. 31(1) (2005) 81–89.
[40] K. Fagerholt, B. Rygh, Design of a sea-borne system for fresh water transport—a simulation study, Belgian J. Oper. Res. Stat. Comput. Sci. 40(3–4) (2002) 137–146.
[41] H. Frank, Shortest paths in probabilistic graphs, Oper. Res. 17 (1969) 583–599.
[42] L. Fu, L.R. Rilett, Expected shortest paths in dynamic and stochastic traffic networks, Transport. Res. Part B 32(7) (1998) 499–516.
[43] T.S. Glickman, Rerouting railroad shipments of hazardous materials to avoid populated areas, Accid. Anal. Prev. 15(5) (1983) 329–335.

[44] T.S. Glickman, Benchmark estimates of release accident rates in hazardous materials transportation of rail and truck, Transport. Res. Rec. 1193 (1988) 22–28.
[45] R. Gopalan, R. Batta, M.H. Karwan, K.S. Kolluri, Modeling equity of risk in the transportation of hazardous materials, Oper. Res. 38(6) (1990) 961–975.
[46] T.J. Haas, J.J. Kichner, Hazardous materials in marine transportation—a practical course, J. Chem. Educ. 64(1) (1987) 34–35.
[47] R.W. Hall, The fastest path through a network with random time-dependent travel times, Transport. Sci. 20(3) (1986) 182–188.
[48] E. Iakovou, C. Douligeris, H. Li, C. Ip, L. Yudhbir, A maritime global route planning model for hazardous materials transportation, Transport. Sci. 33(1) (1999) 34–48.
[49] E. Iakovou, An interactive multiobjective model for the strategic maritime transportation of petroleum products: risk analysis and routing, Safety Sci. 39(1–2) (2001) 19–29.
[50] B.Y. Kara, E. Erkut, V. Verter, Accurate calculation of hazardous materials transport risks, Oper. Res. Lett. 31(4) (2003) 285–292.
[51] V.G. Kulkarni, Shortest paths in networks with exponentially distributed arc lengths, Networks 16 (1986) 255–274.
[52] L. Lindner-Dutton, R. Batta, M. Karwan, Equitable sequencing of a given set of hazardous materials shipments, Transport. Sci. 25 (1991) 124–137.
[53] V. Marianov, C. ReVelle, Linear, non-approximated models for optimal routing in hazardous environments, J. Oper. Res. Soc. 49(2) (1998) 157–164.
[54] E.D. Miller-Hooks, Adaptive least-expected time paths in stochastic, time-varying transportation and data networks, Networks 37(1) (2001) 35–52.
[55] E.D. Miller-Hooks, H.S. Mahmassani, Optimal routing of hazardous materials in stochastic, time-varying transportation networks, Transport. Res. Rec. 1645 (1998) 143–151.
[56] P.B. Mirchandani, Shortest distance and reliability of probabilistic networks, Comput. Oper. Res. 3 (1976) 347–355.
[57] C. ReVelle, J. Cohon, D. Shobrys, Simultaneous siting and routing in the disposal of hazardous wastes, Transport. Sci. 25(2) (1991) 138–145.
[58] O. Richetta, R. Larson, Modeling the increased complexity of New York City's refuse marine transport system, Transport. Sci. 31(3) (1997) 272–293.
[59] H.D. Sherali, L.D. Brizendine, T.S. Glickman, S. Subramanian, Low-probability high-consequence considerations in routing hazardous materials shipments, Transport. Sci. 31 (1997) 237–251.
[60] M. Turnquist, Multiple objectives, uncertainty and routing decisions for hazardous materials shipments, in: L.F. Cohn (Ed.), Proceedings of the 5th International Conference on Computing in Civil and Building Engineering, ASCE, New York, 1993, pp. 357–364.
[61] M. Verma, V. Verter, Rail transportation of hazardous materials: population exposure to airborne toxins, Comput. Oper. Res. 34(5) (2007) 1287–1303.
[62] E. Weigkricht, K. Fedra, Decision-support systems for dangerous goods transportation, INFOR 33(2) (1995) 84–99.
[63] A.B. Wijeratne, M.A. Turnquist, P.B. Mirchandani, Multiobjective routing of hazardous materials in stochastic networks, Eur. J. Oper. Res. 65 (1993) 33–43.
[64] K.G. Zografos, K.N. Androutsopoulos, A heuristic algorithm for solving hazardous materials distribution problems, Eur. J. Oper. Res. 152(2) (2004) 507–519.

[65] O. Berman, V. Verter, B. Kara, Designing emergency response networks for hazardous materials transportation, Comput. Oper. Res. 34(5) (2007) 1374–1388.
[66] E. Erkut, O. Alp, Integrated routing and scheduling of hazmat trucks with stops en route, Transport. Sci. 41(1) (2007) 107–122.
[67] E. Erkut, F. Gzara, A Bi-level Programming Application to Hazardous Material Transportation Network Design, Research Report, Department of Finance and Management Science, University of Alberta School of Business, Edmonton, Alberta, Canada, 2005.
[68] B.Y. Kara, V. Verter, Designing a road network for hazardous materials transportation, Transport. Sci. 38(2) (2004) 188–196.
[69] V. Verter, B.Y. Kara, A Path-Based Approach for the Hazardous Network Design Problem, Faculty of Management, McGill University, Montreal, 2005.
[70] R.Z. Farahani, E. Miandoabchi, S.M. Mousavi, M.B. Baygi, A Pricing Mechanism for Determining of Transportation Rates: A Case Study for Oil Industry, Tehran, 2009.
[71] M. Glaze, New security requirements for hazmat transportation, Occup. Health Saf. 72 (9) (2003) 182–185.

Printed in the United States
By Bookmasters